Theaters of Time and Space

Theaters of Time and Space

American Planetaria, 1930–1970

Jordan D. Marché II

Rutgers University Press
New Brunswick, New Jersey, and London

Library of Congress Cataloging-in-Publication Data

Marché, Jordan D., 1955–
Theaters of time and space : American planetaria, 1930–1970 / Jordan D. Marché, II.
p. cm.
Includes bibliographical references and index.
ISBN 0–8135–3576–X (hardcover : alk. paper)
1. Planetariums—United States—History—20th century. I. Title.
QB82.U6M27 2005
520'.7473—dc22

2004016425

A British Cataloging-in-Publication record for this book is available from the British Library

Manufactured in the United States of America

To my parents

Contents

Preface

Audience members preparing to watch a planetarium program enter a lighted, round room and take their seats, waiting expectantly. A hemispherical screen looms above them, illuminated by soft lights around its base. At the center of the room stands an unusual-looking device, which resembles nothing they've ever seen before. Music emanates from every direction.

On cue, the lights fade and the stars magically appear overhead, just as they would if one were actually outside, away from lights of the city. The voice of a narrator directs viewers' attention to objects in the sky as the program gradually unfolds. Time flows effortlessly, and the viewers are transported in their imaginations to places far out in distant space. They might be taken far back in time to witness the birth of our universe, or be swept into the future to watch the death throes of a star like our Sun. The portrayal of events achieves a sense of realism that is nearly unsurpassed.

Out-of-this-world experiences like these have long earned planetaria their high reputation among visitors of all ages. The creation of superlative illusions explains much of their popularity and success. The versatility of planetaria allows them to manipulate time and space in a theatrical sense, but ensures that such flights of fancy remain educationally sound. It is little wonder, then, that planetaria helped to initiate a broad resurgence of popular interest in astronomy during the late 1920s and early 1930s.

As theaters of time and space, planetaria have enabled the average citizen to experience the majesty of the night sky. As long as planetaria continue to reveal the mysteries of the universe, they will inspire future generations to reflect upon their lives with a newfound sense of identity and purpose.

*M*y first experience of a planetarium program occurred in December 1968 when I was an eighth-grade student, almost a year after I became interested in astronomy. I visited the Lakeview Center for the Arts and Sciences in Peoria, Illinois, and attended its Christmas Star presentation. What has remained especially memorable from that performance, and which I have not seen duplicated elsewhere, was its well-crafted ending. After the standard scientific accounts were offered to explain the appearance of the Star of Bethlehem, an adaptation of the children's story *The Littlest Angel* proved a very fitting and satisfying conclusion to the narrative.

A more prolonged and influential exposure to planetaria came after my family relocated to southeastern Michigan. There, I attended programs offered by the Robinson Planetarium and Observatory of Adrian College, then directed by Joseph B. Noffsinger. His casual yet professional manner of delivering live programs allowed me to envision that I, too, might become a future planetarium director, an ambition that was eventually realized. For how many others has the adolescent exposure to an observatory or planetarium led toward a career in astronomy or space exploration?

While employed for roughly a decade in the American planetarium community, I did not reflect much upon the discipline's history. Instead, like most others, I simply came to know its broad outlines and the major initiatives. Of more immediate interest were current issues of astronomy education and show production techniques. For five of those years, I had the opportunity to edit the discipline's quarterly professional journal, *The Planetarian.* Along the way, I was privileged to meet a host of very creative, talented, and dedicated individuals from facilities located across the country and beyond.

My return to graduate studies at Indiana University in the 1990s provided me with the perspective needed to investigate the rise of American planetaria as a social phenomenon. That same period coincided with renovations conducted at almost every major U.S. planetarium, which seemed to underscore the need for a better understanding of the origins of this popular cultural attraction. My research produced a doctoral dissertation that became the foundation for this book, which addresses a number of questions concerning the social roles of planetaria and their directors.

What motivated a handful of private donors to construct the first American planetaria in the 1930s? A Zeiss projector then cost around $75,000, and many times that figure was expended on the building that housed it. Beyond the communication of astronomical facts, what other cultural values did these programs convey to audiences?

The projection planetarium was a new and wondrously complex device, yet one created expressly for public education rather than scientific research. How was it to be used most effectively in that capacity? Almost from the beginning, astronomers recognized the theatrical qualities of planetarium demonstrations. To this day, planetaria impart their information to audiences through visual and verbal means, regardless of the technology employed. What criteria were essential for individuals to become planetarium directors? Was each facility operated autonomously, or did a sense of community arise among prewar directors and their institutions?

How did the advent of a simplified, pinhole-style projection system, widely introduced after the Second World War, affect American planetaria? Was this innovation seen as a welcome sign of change, or as a potential threat to the integrity and identity of the discipline's major Zeiss installations? Did this innovation result in a lasting distinction between large- and small-domed facilities?

More importantly, how was the planetarium community influenced by the dawning space age? The launch of *Sputniks I* and *II* triggered a third phase of development that completely overshadowed the earlier periods. By the end of the 1960s, American planetaria had come of age by taking steps toward creating an international professional association and beginning publication of its supporting journal.

Finally, how have planetaria been incorporated into broader aspects of popular culture? Apart from their primary task of communicating astronomical information to the public, these institutions have captured the imaginations of artists, sculptors, novelists, and filmmakers. For the millions of visitors who attend their programs each year, planetaria constitute "the greatest teaching aid ever invented."

Acknowledgments

Many individuals and institutions have contributed toward this study. I take much pleasure in thanking them here and acknowledging the scholarly debts I have incurred in the course of my research and writing.

Of the members of my dissertation committee at Indiana University, I owe a special debt of gratitude to James H. Capshew, its chairman, for his unflagging support and continued guidance of this project. John Lankford, while nominally an outside member, played a considerably larger role as my primary content specialist. To him I am deeply indebted, not only for providing an exemplary model of how a community-level study should be conducted, but also for insisting upon the highest standards of scholarship and writing. Professors Frederick B. Churchill and Richard J. Sorrenson provided critical commentary and introduced me to some valuable sources during the research phase.

The Graduate School of Indiana University generously provided a Doctoral Student Grant-in-Aid of Research award that enabled me to conduct archival studies at the original Hayden Planetarium in New York City. I hereby acknowledge its kind support.

Thomas R. Williams, Rice University, furnished an advance copy of his dissertation on the role of amateurs in American astronomy, but more importantly, corrected my thinking in regards to the administrative shortcomings of Hayden Planetarium director G. Clyde Fisher. David H. DeVorkin, National Air and Space Museum, Smithsonian Institution, and Marc Rothenberg, Joseph Henry Papers Project, Smithsonian Institution Archives, offered critical commentaries and many valuable suggestions for improving my research and strengthening the writing in this book. Richard Berendzen, American University, provided further bibliographic references and recommendations.

George F. Reed, West Chester University and Spitz, Inc., read the entire dissertation and provided helpful contacts with former Spitz employees. Jack Spoehr, IMAX Corporation, freely offered historical insights on post-Sputnik developments within Spitz Laboratories. Grace Scholz Spitz shared valuable manuscripts and recollections of her late husband's accomplishments. Brent P. Abbatantuono furnished a copy of his master's thesis on Armand N. Spitz.

William H. Bolles, formerly of the Pennsylvania Department of Public Instruction, reassured me that no state-level mandate had ever been passed requiring new secondary schools in the Commonwealth to include planetaria in their architectural designs. Glenn A. Walsh, former Buhl Planetarium staff member, has kept me informed on historical preservation issues concerning the Buhl's Zeiss Model II planetarium instrument and siderostat telescope.

At planetaria across the country, my requests for historical information were greeted with a cooperative spirit. John Hare IV of the Bishop Planetarium, Bradenton, Florida, allowed me complete access to the International Planetarium Society (IPS) archives. D. David Batch offered similar privileges regarding the CAPE Conference files held at Abrams Planetarium, Michigan State University in East Lansing. Anthony Cook of the Griffith Observatory, Los Angeles, deserves special thanks, not only for directing me to archival sources but also for sharing a large body of his own research on that institution's early history. Lee T. Shapiro of the Morehead Planetarium in Chapel Hill, North Carolina, supplied abundant files on NASA's astronaut training program and offered key insights on the program's eventual demise. Steven B. Craig, Morrison Planetarium in San Francisco, graciously copied materials pertaining to that institution's early foray into planetarium light shows. Lance Roberts of the Rosicrucian Planetarium in San Jose, California, took a personal interest in my search for the first American-built planetarium projector.

From other observatories, museums, libraries, and archives, I received the assistance of individuals almost too numerous to mention. For contributions along many historical fronts, I thank Judith L. Bausch, Yerkes Observatory; Kate Desulis, Adler Planetarium and Astronomy Museum; Laura H. Graedel, Museum of Science and Industry; Julia Sniderman, Chicago Park District; Patrick M. Quinn, Northwestern University Library; Thomas Baione, American Museum of Natural History; Sandra Kitt, Richard S. Perkin Collection, Department of Astrophysics, American Museum of Natural History; Brenda Hearing, Columbia University Rare Book and Manuscript Library; John Alviti and Irene D. Coffey, Franklin Institute Science Museum; Gail Pietrzyk, University of Pennsylvania; the Historical Society of Pennsylvania; Carolyn Schumacher and Steve Doell, Historical Society of Western Pennsylvania; the Carnegie Institution of Washington and the California Institute of Technology; Ron Brashear, Huntington Library; Michele Mednick, UCLA Special Collections Library; Melinda Gilpin, Otterbein College; George Brightbill, Temple University Library; Annette Pringle, Einstein Papers Project; Lonny Baker, Morrison Planetarium; Gretchen Young-Weiner, Cranbrook Institute of Science; the Cleveland Museum of Natural History; and the Wisconsin State Historical Society Library.

From the outset, Audra J. Wolfe, acquisitions editor at Rutgers University Press, expressed enthusiasm for this project and provided much valuable criticism and advice toward its publication. Marvin P. Bolt, Adler Planetarium and

Astronomy Museum, critically reviewed the amended draft and suggested further stylistic revisions throughout the manuscript.

On a personal level, I have two wonderful parents to thank for their encouragement of my scholarly endeavors. Finally, my wife, Teri, has provided sound intellectual as well as financial support through the years required to complete and revise this study. Without those vital ingredients, it would have remained a work in progress.

Theaters of Time and Space

Introduction

Historians and sociologists of science have increasingly paid attention to the importance of collective biographies, or community-level studies, over the past generation. Scholars have examined the social structures and research practices within a broad range of American and European scientific disciplines.[1] Analyses have appeared on the emergence of nationalistic communities in nineteenth-century American and British science, while other studies have documented the barriers faced by women scientists of every discipline.[2] Collective biography looks beyond the knowledge-production activities of individual scientists and regards the disciplinary community to which these men and women belonged as "a fundamental unit of social/historical analysis."[3]

Throughout this study, I use the phrase "American planetarium community" to mean that body of workers whose professional activities were directed toward the fulfillment of a planetarium's mission, namely, the production and presentation of scientifically accurate programs and exhibits for audiences of all ages. Members of this community acquired such formal job titles as directors, associate and assistant directors, lecturers, technicians, photographers, artists, exhibit designers, and curators.

The scientific community that most nearly resembles (and indeed, partly overlaps) the one reported here is the body of professional astronomers active in America between 1859 and 1940. As was the case in historian John Lankford's study of that community, gender-equity issues played a significant role in restricting the employment opportunities available to women planetarium directors.[4] My examination of the planetarium community, however, departs from Lankford's approach in several respects that deserve further explanation.

One of the major differences between our two studies concerns the function and usage of chronological narratives. Lankford has coined the term "sociological narrative" to describe his hybrid style of presentation, which supports and extends the generalizations reached through quantitative analysis with samplings of individual biographies gathered from archival sources.[5] Because many key aspects of American planetarium directors (e.g., their highest educational attainments) remain unknown, my interpretations rely less heavily on such numerical data and are presented through a more traditional narrative approach. Nonetheless, over 700 permanent institutions and more than 900 individuals comprised the cohort I studied.

For reasons pertaining to the initial small size and relative youth of this population, an analysis of Lankford's triad of community traits—career trajectories, the reward system, and power—cannot realistically be applied to the cohort of American planetarium directors. Prior to 1970, almost no discipline-wide system of rewards existed that recognized levels of professional achievement.

Fundamental differences in the nature of the astronomy and planetarium communities demand separate standards of analysis. With minor exceptions, planetarium personnel were employed in the popularization of science. Rather than embracing those means by which scientific knowledge remained the exclusive province of scientists, the planetarium community has fostered a diametrically opposing set of values. The transmission of knowledge from research astronomers to the public remains the primary social function of the planetarium. This study examines the origin, development, and rise to significance of the American planetarium community within the larger spheres of astronomy education and popular culture through 1970.

By their design and use, planetaria are vehicles for the diffusion of scientific knowledge. Each year, millions of visitors attend programs at hundreds of independently operated planetaria that feature appropriate combinations of education and entertainment. Since their inception, planetaria have provided a significant component of the public's understanding of science for audiences of all ages. Nonetheless, it remains impossible to reconstruct historical measures of that understanding, even under local conditions. Attendance figures convey little more than the number of visitors exposed to particular scientific ideas. Few planetaria conducted routine assessments of their audience's knowledge of the heavens or surveyed their attitudes toward such issues as future space exploration.

By contrast, historians and sociologists of science have addressed markedly different questions regarding scientific knowledge (and institutions) across a host of social contexts. Their efforts have focused upon the broader meanings that science has attained among various interest groups (and publics) through time. Among those investigators, historian Steven Shapin has offered an essential framework for the study of "science and the public." Rejecting conventional wisdom behind the "self-evident" notions embodied in the terms "scientist" and "layperson," Shapin stresses the need to explain the historical construction of those categories.[6]

In turn, historians Roger Cooter and Stephen Pumfrey have proposed that broader consideration be given to the "plurality of . . . sites for the making and the reproduction of scientific knowledge." The hundreds of American planetaria established through 1970, together with the actions of their directors, offer a strong validation of those social spaces constructed by and for them. Cooter and Pumfrey urge that greater attention be directed toward those disciplinary "practice[s] and object[s]" that signify the status of science among popular cultures.[7] By these prescriptions, planetaria are well suited for such analysis, and a recognition of their cultural roles offers many potential contributions toward a fuller understanding of the popularization of science.

As instruments crafted for the demonstration of basic astronomical concepts, planetaria prompt us to revise our thinking regarding traditional notions of scientific instruments.[8] The Zeiss projector was manufactured to exacting tolerances, a point that was well understood by visitors to its performances (see chapter 1). Its representation of celestial positions and motions bore such a striking resemblance to the actual heavens that it drew near-universal acclaim and approval by professional astronomers. They readily accorded to planetaria the mantle of scientific legitimacy and authority that was usually reserved for instruments (and institutions) dedicated to research. These perceptions enable us to better understand the negative reactions expressed among Zeiss planetarium directors to the introduction of a more simplified projection system (see chapters 2 and 5).

This book illuminates a significant transformation that occurred within twentieth-century astronomy education. The enormous popularity of planetaria led to their widespread establishment as cultural attractions. This development fostered the creation of a professional community of planetarium workers whose goals became the dissemination of astronomical knowledge to young and old alike. Along with the mass media, planetaria act as one of the leading bridges of communication between professional astronomers and the public. Together with observatories, museums, and science centers, they continue to play an ever-widening role in the public's understanding of science and technology by conveying a deeper sense of humanity's place in the cosmos.

This book is organized into four principal parts. In the first part, the origins of planetaria are traced to the junction of two streams of conceptual and technological expression. For centuries, large hollow, perforated globes replicated the appearances of the stars, while mechanically driven models of the planets and their satellites displayed regularities of the Newtonian world system. When the two streams were united by the principle of image projection techniques, the Zeiss planetarium instrument was born, and the age-old quest of enclosing the universe inside a room was achieved.

Tremendous audience response led to the construction of planetaria in leading European cities and in five metropolitan centers in the United States. The book's second part explores the formative period (1930–46), when Zeiss-equipped facilities donated by private philanthropists dominated the American planetarium scene. Four of those installations reveal the influence of astronomer George E. Hale, who argued that planetaria should be institutions where active research is conducted. Two smaller planetaria, featuring homebuilt projectors, were constructed at San Jose, California, and Springfield, Massachusetts. Peer reactions toward the latter facility are examined according to "boundary-work" analysis.

Planetarium workers' training and careers are subjects of additional consideration. At Chicago and New York, the directorships of Maude V. Bennot and Marian Lockwood revealed that attainment of gender equity was but a temporary measure, later erased by the postwar ideology of male superiority. Thereafter, the typical career pathway followed by men, which led to the directorship of a major planetarium, remained virtually closed to women.

Audience members who regularly visited planetaria received an introductory course in descriptive astronomy, an approach dubbed the "American practice." Signs of innovation, however, soon became apparent. James Stokley, director of Philadelphia's Fels Planetarium, crafted Depression-era programs that featured imaginary trips to the Moon and scenarios depicting the end of the world. These productions earned Stokley the reputation of "greatest showman in planetariana." Stokley's influence was extended to New York's Hayden Planetarium, and served to inspire theme park entertainment spectacles shown at the 1939 New York World's Fair.

The book's third part covers the middle or postwar period (1947–57), when Philadelphia entrepreneur Armand N. Spitz produced and marketed inexpensive pinhole-style projectors. Spitz's innovation revolutionized the availability of artificial skies and brought the planetarium experience to much wider audiences. Through the establishment of many new planetaria, women gained readmission to the planetarium community. In turn, the first viable association of planetarium educators was founded under the auspices of the American Association of Museums. Plans to launch the first artificial satellites, as part of the International Geophysical Year, sparked tremendous public interest in astronomy.

The community's final developmental period (1958–70), described in the book's fourth part, came in response to the "crisis of confidence" triggered by launch of the Soviet *Sputniks*. This largest phase of planetarium building resulted from the impact of federal assistance, set in motion by the National Defense Education Act of 1958. Astronomy and space science were suddenly viewed as academic disciplines to be mastered in the nationwide rush to catch up with the Soviets. The necessity of offering a space science education to students of all ages justified the construction of hundreds of school planetaria through the 1960s. Planetaria likewise became training grounds for the country's newest breed of folk heroes, its Mercury, Gemini, and Apollo astronauts.

Rapid institutional growth fostered urgent needs of formalized training in planetarium instruction. The advent of "space science classrooms" encouraged educational research on the most effective teaching strategies to be employed among audiences of all ages. The wealth of new institutions and personnel redefined the social structures of the planetarium community as a whole. By decade's end, the first continent-wide association of planetarium educators was founded, whose thirtieth anniversary was observed in the year 2000.

From their inception, planetaria have occupied a significant, if neglected, component of the American astronomical community. This account brings forward their historical developments to the era of the manned lunar landings and beyond.

Part I

Origins of the Projection Planetarium

CHAPTER 1

Zeiss Planetaria in Europe, 1923–1929

> The [projection] planetarium . . . made possible the realization of a two thousand-year-old idea—a perfect representation of the starry sky *inside a room.*
>
> —Mark R. Chartrand III, 1973

*H*uman attempts to create models of the universe extend back to antiquity and beyond. Over the past two millennia or so, many of the proffered solutions have tried to represent the positions of the fixed stars upon the surface of a globe, while the apparent motions of the brighter celestial objects (the Sun, Moon, and planets) have been replicated through mechanical devices of various sophistication. Despite considerable ingenuity, none of these innovations offered more than an imperfect and incomplete rendering of the phenomena. Oddly enough, a full realization of this goal had to await the arrival of the twentieth century, even though much of the relevant technology predated the planetarium's invention by several decades or more. What was crucially lacking was the inventive solution for displaying both stellar and planetary movements at once from the same device. An entirely novel approach to the problem had to be adopted before the planetarium was born.

The projection planetarium was jointly conceived in 1914 by Walther Bauersfeld and Werner Straubel of the Carl Zeiss optical firm in Jena, Germany, and first demonstrated publicly on 21 October 1923 at the Deutsches Museum in Munich. The word "planetarium" usually refers to "an optical projection instrument that shows upon the inner surface of a hemispherical dome the stars, planets, sun, moon, and in most cases, additional astronomical effects." It may also signify the "room or building in which the projector is housed."[1]

Charles F. Hagar, former director of the Planetarium Institute, San Francisco State University, has argued that the planetarium originated from a remarkable synthesis of those twin streams of conceptual and technological expression.[2] Both streams can be traced backward, if sporadically, into the ancient world. Because the literature that describes the two streams is large, only a brief summary of the leading developments is recounted here.[3]

Models of the Heavens

The theory of a spherical heaven is often attributed to the pre-Socratic philosopher Parmenides of Elea, but was likely anticipated by his teacher Pythagoras of Samos.[4] Today, however, the "best preserved and scientifically most significant celestial globe of [Hellenistic] times," adorned with forty-two allegorical constellation figures, is that supported by the Farnese Atlas.[5] This sculpture dates from the first century B.C., although its globe depicts the heavens as they would have appeared in roughly 300 B.C. Such an artifact might be the last surviving example of an Aratus globe, designed to illustrate the poem "Phenomena" composed by the third century B.C. poet Aratus of Soli. Notable celestial globes were later fashioned by Alexandrian astronomer Claudius Ptolemy in about 150 A.D. and by Danish astronomer Tycho Brahe in about 1584. What all of these devices had in common was a representation of the heavens as seen from *outside.* Constellation figures were depicted in *reverse* from their actual appearances in the night sky.

To overcome this drawback, hollow globes large enough to admit small audiences were constructed. Around 1650 the twelve-foot-diameter Gottorp Globe was commissioned by Frederick III, Duke of Holstein. Its exterior was decorated with a world map, while its interior was painted with constellations. Although largely destroyed by fire in 1747, it was rebuilt completely and is displayed at the Lomonosov Museum in St. Petersburg. Similar efforts were expended by Erhard Weigel, professor of mathematics at the University of Jena, and by Roger Long, professor of astronomy at Cambridge University, in the eighteenth century. The most recent example of a hollow globe was unveiled in 1913 by Wallace W. Atwood, then museum director at the Chicago Academy of Sciences. During the 1990s, the Atwood Sphere was moved to Chicago's Adler Planetarium, where it has been completely refurbished for demonstrations.

Atwood's motivations in building this globe were as follows: "Many of the mathematical conceptions necessary for the study of descriptive astronomy, which often discourage the beginner, are made, with this sphere, perfectly simple." He added, "There is now no reason why anyone, including the younger school children, cannot become familiar with the chief constellations, their apparent movement, the brighter stars and the real and apparent movements of the sun, moon, and planets."[6] It might not be a coincidence that Atwood's device was completed shortly after the publication of Anna Botsford Comstock's *Handbook of Nature Study* (1911).[7] Fabrication of the Atwood Sphere was possibly linked to the popular nature study movement and was well suited for displaying the lessons on "Sky Study" featured in Comstock's text. Atwood's globe was the last of its kind to be fashioned before the projection planetarium's debut.[8]

An instrument demonstrating the motions of planets is usually termed an *orrery* or a mechanical planetarium, and such a device possesses a long heritage of ingenious craftsmanship equal to that of the celestial globe. In 1901 a team of sponge divers recovered the encrusted fragments of an ancient scientific instrument. From his study of its construction, Derek J. de Solla Price concluded that the instrument, called the Antikythera mechanism, was fashioned in about 80 B.C. and functioned as a sophisticated calendrical device.[9] Historian Henry C. King has written that it "may represent a technological development fairly widespread in the Aegean and Greek mainland," whose recovery "makes untenable the traditional notion that Greek science, especially astronomy, discouraged the use of mechanical devices."[10]

Comparable technical ingenuity did not reemerge until the fourteenth century A.D., in the form of Giovanni de Dondi's seven-faced astronomical clock (ca. 1348–64). Besides solar and lunar indicators, the clock's faces depicted each planet's epicyclic motion according to Ptolemaic theory. From two surviving manuscripts, several working replicas of de Dondi's clock have been fabricated.[11] One of these manuscripts contains "the first such use of the word [planetarium] . . . to describe an actual machine that demonstrated planets' movements."[12]

Following acceptance of the Copernican theory, new devices were constructed to portray heliocentric planetary phenomena. Dutch scientist Christiaan Huygens applied the technique of "continued fractions" to the replication of orbits. Huygens's orrey, or mechanical planetarium, fashioned in 1682 by Johannes van Ceulen, is preserved at the Boerhaave Museum in Leiden. Enlightenment conceptions of a "clockwork universe" inspired the creation of similar devices in eighteenth-century England, France, and the United States. Their popularity is attested by the painting of artist Joseph Wright of Derby, "A Philosopher Giving a Lecture on the Orrery" (ca. 1763–65). David Rittenhouse, one of America's foremost colonial astronomers, completed two orreries: one for the College of New Jersey (now Princeton University) and a second for the College of Philadelphia (now the University of Pennsylvania).

A working model depicting current planetary positions was brought to fruition by Dutch mechanic Eise Eisinga in 1781. Later preserved by the town of Franeker as a public museum, Eisinga's ceiling planetarium is the only one that operates in real time.[13] The last major ceiling orrery to be constructed before the projection planetarium's debut was the Copernican Planetarium erected at the Deutsches Museum in 1923. Executed by Franz Meyer, chief engineer of the Zeiss firm, this device was enclosed in a circular room twelve meters in diameter. It featured a mobile "observer's platform situated beneath the model of the earth." By means of a periscope device, observers could sight other planets and note their apparent retrograde movements from an earthly perspective.[14]

While both celestial globes and orreries were important forerunners to the projection planetarium, each model lacked an essential element. That missing ingredient was the technique of image projection. As both device and name, the "laterna magica" (or "magic lantern" projector) originated with Christiaan Huygens, who correctly described the instrument in private correspondence exchanged in 1659.[15] Huygens recognized little scientific value in the device and relegated its use to entertainment. No one living before the twentieth century conceived of projecting star or planet imagery onto a fixed hemispherical screen to simulate the natural heavens.

Oskar von Miller and the Deutsches Museum

The individual chiefly responsible for creation of the projection planetarium was Oskar von Miller, director of the Deutsches Museum. A graduate of Munich's Polytechnium in 1878, Miller resolved to fashion a career "in the new electrical engineering field and to investigate the application of . . . waterpower . . . to the production of electricity." Miller's early experiments with three-phase alternating current demonstrated that "power could be transmitted great distances safely and economically" and helped to establish the feasibility of such distribution grids.[16]

Earlier in his career, Miller inspected two of the world's leading science and technology museums, the Conservatoire National des Arts et Métiers in Paris and the Patent Museum at South Kensington in London. Those visits instilled a desire to establish a museum that would highlight the achievements of scientists and engineers from many nations. Miller's dream began to take shape in 1903 when he announced plans for the creation of an industrial museum in Munich. The Deutsches Museum was opened (in temporary quarters) on 13 November 1906.

The museum's mission, according to Miller, was to exhibit "the [historical] development of the various branches of natural science and technology by means of original apparatus and machines, as well as by . . . models and arrangements for demonstration, in a manner easily understood by all classes of people." Its purposes were to instruct visitors "as to the effects of the multifarious applications of science and technology to the problem of human existence," and "to keep alive in the whole people a respect for great investigators and inventors."[17] The museum's success was enormous. Miller's innovations included the employment of trained interpreters and the adoption of audience-participation techniques. Many full-scale models were operated by turning a crank or pressing a button. Miller attempted to teach scientific concepts and laws in the most direct ways possible. Museum exhibits were designed "to clarify the underlying *theories* and yet convey the variety and excitement of a world's fair."[18]

Opening of the museum's permanent home on the River Isar was delayed until 7 May 1925. Its collections and exhibits were tailored to attract a diversity of cultural and socioeconomic groups. Miller wished visitors to understand that "all classes of people take part in the achievements of modern life. Workmen will see what laborious mental work was necessary by the greatest investigators in order to lay the foundations of technology," while in turn, "[m]embers of the privileged classes will learn . . . how toilsome and difficult is the work of laborers." Miller sincerely hoped that "a more intimate sense of partnership" could be attained among "all working classes," and that enhanced social stability might prove an additional outcome.[19]

As early as 1905, Miller sought to procure two mechanical devices for the museum's astronomical department. The first was a Copernican (heliocentric) planetarium, while the second demonstrated apparent movements of sky objects from a Ptolemaic (geocentric) perspective. The Sendtner instrument company of Munich fabricated both table-top-size instruments.[20]

Despite their adequate performances, Miller remained unsatisfied and strove to have much larger versions constructed. In 1912 he applied to the Carl Zeiss firm for a room-size Copernican planetarium. Subsequently, Miller acted upon a suggestion from Max Wolf, former director of the Baden Observatory in Heidelberg, to try to replicate performances of the Gottorp and Weigel globes. Neither Wolf nor Miller was yet aware of the recently completed Atwood sphere in Chicago. Late in 1913 Miller again approached the Zeiss firm with plans for executing a hollow globe capable of demonstrating celestial rotation around a stationary audience.

Progress on the latter soon became stalled, however, because of the difficulties in simulating planetary movements and reducing noise and vibrations. The optimal solution demanded a complete reconceptualization of the problem.[21] At a meeting held in March 1914 between Miller and the Zeiss firm, Walther Bauersfeld proposed that the cumbersome mechanical design for producing planetary motions be abandoned and replaced "by optically projecting the pictures of the heavenly bodies on the interior surface of the sphere." Upon hearing this suggestion, Bauersfeld's colleague on the board of management, Werner Straubel, exclaimed, "Then, of course, also the fixed stars should be projected from the central apparatus."[22] From these two crucial insights, the projection planetarium was born, a concept that was first made public on 3 February 1917.[23]

The twin solutions reached by Bauersfeld and Straubel likely represent a psychological phenomenon known as a "gestalt-switch." This is the sudden transformation of an individual's perceptions from one mental framework or model to another. According to Gestalt psychology, an observer's recognition of ordinary figure-ground relationships may undergo a spontaneous reversal, leading to a highly altered perception of the visual field.[24] Philosopher of science Thomas S. Kuhn employed the analogy of a gestalt-switch in his explanation of paradigm shifts that accompany the creation of new scientific theories.[25] That a dramatic shift in thinking had occurred among the Zeiss engineers is supported by Bauersfeld's description of the projection planetarium.

> The great sphere shall be fixed; its inner white surface shall serve as the projection surface of a multiplicity of small projectors which will be placed in the centre of the sphere. . . . [T]he little images of the heavenly bodies thrown upon the fixed hemisphere [shall] represent the stars visible to the naked eye, . . . just as we are accustomed to see them in the natural clear sky.[26]

Starting in 1919, Bauersfeld spearheaded the design and construction of such a projector (the Zeiss Model I), whose realization demanded the skills of opticians, electricians, and machinists. A sixteen-meter-diameter concrete-shell dome was fabricated on the Zeiss factory roof, where in August 1923 the planetarium's stars and planets were first projected.[27] At the Deutsches Museum, a nine-meter dome was erected, wherein Bauersfeld gave the first public planetarium demonstrations. The instrument was returned to Jena for refinements before being installed at Munich in August 1924. During this period, Walter Villiger, manager of the astronomical instruments division, gave countless planetarium demonstrations on the factory roof. News of the "Wonder of Jena" spread throughout Europe and to the United States. For his design of the projection planetarium, Bauersfeld was awarded the Elliott Cresson Medal of the Franklin Institute" *in absentia* at the 1933 dedication ceremony of the Fels Planetarium in Philadelphia.[28]

One fundamental question about the planetarium's origin has remained unanswered. Why did Miller wish to have such an instrument constructed in the first place? It is clear that he did not set out to build a planetarium based on the Ptolemaic (geocentric) perspective solely as a means of demonstrating apparent celestial motions. The most significant fact is that Miller wanted audiences to recognize the complementary nature of both Copernican and Ptolemaic representations, starting with the small-scale Sendtner models. Substantial innovations were incorporated into the room-size Copernican mechanical planetarium, which allowed visitors to experience both "external" and "internal" views of solar system objects and motions. The most plausible explanation is that Miller pursued this approach in order to offer pedagogical confirmation of scientific theories.[29] Nowhere is this better illustrated than in a 1923 account of how museum visitors were to receive successive demonstrations of the Ptolemaic projection planetarium and the Copernican mechanical planetarium. Franz Fuchs, a division chairman at the Deutsches Museum, described the outcome as follows.

> After the visitor has taught himself the apparent motion of the stars above him in the Ptolemaic planetarium, he then proceeds into the Copernican Planetarium, in which the true motions of the planets around the sun will be presented to him. . . . [T]he visitor can position himself on a revolving platform situated directly underneath the earth, and by means of this, be carried around the sun. The observer then sees . . . how the sun, in reality stationary, apparently travels through the zodiac; how Mercury and Venus stand, sometimes as morning, sometimes as evening stars, against the sky; how Mars, Jupiter and Saturn execute their retrograde loops against the sky. Herewith is given, by means of direct observation, the otherwise difficult explanation of the connection between apparent and actual planetary motions.[30]

The enormous popularity of the projection planetarium, however, quickly overshadowed Miller's complementary demonstrations. There is no evidence that professional astronomers or American planetarium directors ever grasped the full significance of Miller's plan, nor that any of them attempted to emulate his pedagogical scheme. Costs alone prohibited the acquisition of large-scale Copernican mechanical planetaria by all but two major facilities in the United States, namely the Hayden Planetarium of New York and the Morehead Planetarium of Chapel Hill, North Carolina. Nonetheless, a third institution, Eastern Mennonite College at Harrisonburg, Virginia, constructed a mechanical ceiling planetarium to augment its purchase of the first Spitz pinhole-style projector (see chapter 5).

Rather than being operated in tandem with a large mechanical planetarium, the projection planetarium became a stand-alone educational and entertainment spectacle. Across Europe and America, its use as a versatile tool for teaching descriptive astronomy drew widespread acclaim. The planetarium's strikingly realistic depiction of the night sky offered a respite for city dwellers from the growing impact of light pollution associated with modern urbanization and industrialization. As reported in the American media, each of these factors contributed to the planetarium's enormous popularity and growth.

Planetaria and the American Press

The earliest notice regarding the planetarium to appear in an American periodical was a three-paragraph summary released in *Science—Supplement* on 5 September 1924. This description of the Model I instrument installed at Munich was framed in a host of militaristic, biological, and Biblical metaphors. Termed "unusual, even weird and startling" in appearance, the device was said to resemble "a small anti-aircraft cannon" with a battery of projectors arrayed on "a large sphere studded with high-power lenses, [like] a gigantic insect's eye." The planetarium's operator was endowed with "the power of a Joshua, for he [could] bid the sun and moon to stand still, and cause the stars to run backward in their courses." A related journalistic theme stressed the planetarium's extraordinary accuracy. Gearing was designed to such tolerances that, after an equivalent of five thousand yearly turns, planetary positions might deviate from the truth by "less than two degrees." These claims served to reinforce the high reputation for excellence associated with the Zeiss name. The projection planetarium was "expected to have great influence on the teaching of popular astronomy."[31]

Two significant accounts of the planetarium's construction and use appeared during the following year. Lt. Commander Haller Belt (U.S. Navy), who had witnessed a planetarium demonstration, secured an English translation of Bauersfeld's description of the Model I instrument. There, an unexpected theatrical quality of the planetarium's performance was first disclosed. As the room lights were gradually lowered and one's eyes became dark adapted, visual perceptions of the hemispherical dome finally vanished. Observers were left with an impression that projected stars were located infinitely far away, as if real stars were being examined. This illusion of infinite space was to remain one of the most widely cited attractions of the planetarium experience, although its realization was attainable only in the largest domed facilities.

Bauersfeld admitted that the costs of producing planetaria were "necessarily high," although "not so prohibitive that they could not be erected in large cities," especially if admission fees were collected.[32] This prediction was borne out across the German republic and much of Europe in coming years. Such capital expenditures, however, restricted the purchase of American planetaria to the liberality of private patrons.

In February 1925 a popular article describing the Munich planetarium, "Die Geheimnisse der Sterne" ("The Secrets of the Stars"), appeared in a Braunschweig periodical, *Westermanns Monatshefte.* This drew the attention of *Scientific American* editors, who promptly translated and reprinted the account, giving American readers their first literary and visual impressions of a planetarium demonstration.[33] A notable illustration featured a cutaway diagram of the dome erected on the Zeiss factory roof, into which people were crowding. In the dome's center stood the projector and, beside it, the operator. A darkened sky showing the Moon and stars was framed by a panorama of the Munich skyline, adding realism to the scene. The projector's appearance was likened to a "porcupine's back," wherein each "quill" contained a group of lenses. Author G. H. Morison noted how, under those darkened conditions, a "wonderful optical illusion" was produced: "[O]ne [could not] escape the feeling of looking at the infinitely distant real sky and at real stars."[34]

The first American astronomer to describe the planetarium's educational and theatrical potential was Amherst College Observatory director-emeritus David Todd. In his judgment, the planetarium enabled "the 'man on the street' to comprehend quite as fully as the learned professor, the seemingly intricate, though actually simple, workings of the celestial mechanism." More explicitly, Todd enthused that its performances were "sufficiently spectacular" to "arrest the attention of the indifferent and fascinate those dwelling on the threshold of enquiry." He argued that "an apparatus which compresses the cosmic happenings of many years into a few minutes, so that the eye may follow and the mind comprehend the motions of heavenly bodies," could not fail to "arouse popular interest in the majesty of the universe." While public institutions such as libraries, museums, and art galleries strove to "improve the mind and elevate the moral and esthetic values" of the nation's populace, so too could the planetarium boost its audiences' perspectives, "intellectually, ethically, and esthetically." Metaphorically, planetaria allowed visitors to experience the "influence of a vision" that enabled them to "hold Infinity in the palm of your hand and Eternity in a hour."[35]

An anonymous contributor to the periodical *The Outlook* wrote in 1926 that "[p]robably no other piece of pedagogical apparatus has ever attracted such world-wide attention." The planetarium "appeals to the imagination, quite apart from its instructive value, and deserves high rank among scientific 'illusions.'" *The Outlook* emphasized its unmatched versatility at "teaching the elements of astronomy and stimulating popular interest in the oldest of the sciences." Planetaria were judged to have set a "new pace" in the production of "scientific working models."[36]

In September 1925 G. Clyde Fisher, then in charge of astronomy education at the American Museum of Natural History in New York, traveled to Europe to inspect its leading astronomical institutions. His itinerary included the Deutsches Museum and its Zeiss planetarium—visits undertaken "with a view to its suitability for our proposed Hall of Astronomy." Fisher witnessed demonstrations of both the Copernican ceiling and Ptolemaic projection planetaria, but did not grasp (or was not instructed on) their significance in Miller's pedagogical scheme. While Fisher appraised the former as the "best [of its kind] ever constructed," he complained that "these old-fashioned . . . orreries were very crude and unsatisfactory at best," and offered performances "conspicuously inadequate" by comparison with the latter. Inside the projection planetarium, Fisher declared, planetary motions were "visualized much more satisfactorily than by the [Copernican] planetarium."[37]

Fisher noted that "[w]hether the audience was made up of children or of adults, when the fixed stars appeared, an involuntary 'Ah' swept over the assembly and they were spellbound." No one, himself included, was prepared "for such a realistic representation; . . . [t]he illusion of the immensity of space [was] perfect."[38] Further evidence of this spontaneous reaction came from Harvard astronomer W. J. Luyten, who attested that "the whole audience[,] whether trained astronomers who know what to expect, or school children eagerly awaiting the unexpected, cannot suppress an involuntary exclamation of delight." Albert G. Ingalls, associate editor of *Scientific American,* witnessed a performance at the Jena planetarium, reporting that one could "hear it ripple across the room and back, a genuine kind of an 'Ah.'" In perhaps a mild parody of Luyten's remarks, Ingalls added that "[e]ven a case-hardened professional astronomer will say it under his breath."[39] Testimony concerning the illusion of infinite space appears in several other popular accounts. Author O. D. Tolischus remarked how "[t]he walls seem[ed] to have been removed by magic hands, and the starry . . . canopy of the heavens . . . apparently stretch[ed] itself in infinite space above us." Ingalls noted how, when the "planetarium spectacle" was on, "the confining dome retreat[ed] to infinity. Thus perfect [was] the verisimilitude. The dome seem[ed] to vanish by magic."[40]

One of few design limitations possessed by the Model I instrument was its inability to show the effects of latitude variation on appearances of the heavens. Both the Deutsches Museum and Zeiss factory models operated according to fixed latitudes. To allow the entire sky to be witnessed from anywhere on Earth's surface, a vital modification was suggested by Walter Villiger and sketched by Bauersfeld.[41] The spherical star projector was divided into twin hemispheres. Planet projection apparatus was placed in "cages" between the hemispheres. These changes resulted in the Zeiss Model II instrument, which sported an entirely different appearance. Waldemar Kaempffert, a former *New York Times* science journalist turned director of Chicago's Museum of Science and Industry, portrayed the Model II as being "so unlike anything with which even engineers [were] familiar that it might be taken for the fantastic creation of some Martian inventor in H. G. Wells's 'War of the Worlds.'" Albert G. Ingalls likened the "giant 'dumb bell'" to a pair of "diver's helmet[s]."[42] Separation of the star hemispheres, however, demanded that larger domes be constructed to minimize parallax effects. These actions significantly escalated costs for all later Zeiss planetaria.

Several trends were apparent in American media coverage of European planetaria after 1926. Gone were the cutaway diagrams or photographs of crowds standing on the factory roof. One of the first freestanding planetaria (with the Model II instrument) was erected in Jena to alleviate audience pressures exerted on Zeiss personnel. Photographs depicted visitors seated, awaiting demonstrations inside the domes. Planetarium operators now assumed their command at a specially designed console, no longer placed beside the great instrument but instead located toward the chamber's edge, where a flashlight pointer was wielded to illustrate the lecture-demonstrations. But whether that individual should be considered a scientist, technician, educator, or entertainer was not yet clear; no concise role definition for the operator yet existed. Until the planetarium's own mission was clarified, as an institution devoted to the popularization of astronomy, the social role and career of a planetarium operator could not easily be defined. Still, a much more settled and professional decorum attended this newer media coverage, enhancing the desire to replicate these educational experiences on American soil.

As the first freestanding planetaria were constructed, their designs received special architectural treatment. The task of enclosing their large, concrete-shell domes within buildings that were both functional and attractive proffered a diversity of innovative solutions. The facilities erected at Jena, Düsseldorf, Berlin, Leipzig, Dresden, Barmen and other cities incorporated practical and aesthetic design criteria.[43] Many achieved a satisfying integration with their surroundings.[44] These factors, however, significantly raised the costs of such installations and discouraged the establishment of any temporary or makeshift structures within American cities.

Not all information conveyed by the press was entirely accurate. At least one misconception was introduced by comparing planetarium demonstrations with motion picture theaters. In 1927 *The World's Work* stated that "[t]his miracle of art and science is accomplished by the utilization of the moving picture principle. For the modern planetarium is really a moving picture of the sky." But apart from the optical projection of their images, there was little resemblance between a planetarium instrument and motion picture projector; no film of any kind was transported through the former. Yet that author's probable intention was to communicate by analogy; elsewhere, quotation marks were placed around the expression "movies of the stars." Albert G. Ingalls, meanwhile, drew a more explicit comparison between the planetarium's simulated compression of time and a time-lapse motion picture. The latter, he noted, allowed study of an otherwise "slow happening such as the gradual growth of a root or tendril." In the planetarium, motions of heavenly bodies could be "accelerated and multiplied many fold," allowing one to "visualize them clearly as a whole." But in speaking informally about the planetarium experience, Ingalls remarked, "It is the best 'movie' I have ever seen!"[45]

Reports suggested that a revival of popular interest in astronomy might be attributable to the planetarium. O. D. Tolischus argued that one of the drawbacks suffered by "modern city man" was his "gradual loss of understanding and appreciation" for that "most beautiful and overpowering spectacle," the starry sky. "[C]hildren of a jazz age" sought other attractions than "the slow procession of the stars. . . . But if the mechanical age took the heavens from us," he mused, "modern science [was] beginning to bring them back to us." This statement supports the notion that expanding urbanization and industrialization, together with artificial illumination, had rendered popular study of descriptive astronomy obsolete. But inside a planetarium, it "look[ed] as if in a jazz age even the heavens were moving in jazz time," allowing modern man to "chase all the heavenly bodies through the course of days, of years and even centuries, in as many minutes." Tolischus made it clear to readers, however, that planetaria were something more than just a novel form of entertainment. The spectacle presented "[was] not only realistic, it [was] as scientific and exact as mathematics."[46] So, while the planetarium might provide a theatrical experience of sorts for audience members, the astronomical basis of its performance was unassailable.

Professional astronomers were uniformly positive in their assessment of the planetarium's educational merits. John Jackson, vice-director of the Greenwich Observatory, asserted that "[t]he value of the [p]lanetarium for giving the public an insight into the motions of the heavens can hardly be exaggerated."[47] Still higher admiration was expressed by S. Elis Strömgren, director of the Copenhagen Observatory, who enthused that "never before [had] a means of entertainment been provided that [was] so instructive as this, never one so fascinating, never one with so wide an appeal. It [was] a school, a theatre, a cinema in one: a schoolroom under the vault of heaven, a drama with the celestial bodies as actors."[48] W. J. Luyten wrote allegorically of the planetarium's chief illusion: he affirmed that a spectator "[could] not help realizing that he ha[d] for a moment, at least, peered into the mysteries of the infinite."[49]

That planetarium demonstrations could offer an appropriate blend of entertainment and education was affirmed by several authors. Visitors of every age and background could gain something of value from the experience. O. D. Tolischus urged, "[t]o the merely curious, the show [was] as entertaining as a movie. To the student of astronomy, [however], it [was] a laboratory of the heavens where the stars and planets [could] be moved at will with scientific accuracy."[50] Albert G. Ingalls voiced the opinion that audiences "ha[d] not only been instructed but entertained. . . . Intrinsically the performance [was] esthetic. It provide[d] thrills while it educate[d]."[51] Waldemar Kaempffert, keenly aware of the Broadway-style theaters against which planetaria must compete, declared that the latter could hold their own. "A planetarium is nothing more nor less than a playhouse in which the majestic drama of the firmament is unfolded." At the start of each performance, the "lights are turned down gradually, just as in a theater before the curtain rises on a play." Once the stars have suddenly appeared, to the audience's delight, "[a] hush falls over the spectators. No play is ever more intently followed than this" in which the familiar celestial bodies "enact their parts."[52] By presenting astronomical facts in a visually compelling yet educationally sound manner, Kaempffert argued, the planetarium could flourish as a viable popular attraction for cultured American citizens. If one needed further proof, attendance figures from European institutions convinced most skeptics.

European Planetaria: Attendance, Expenditures, and Patronage

By the spring of 1930, a total of fifteen planetaria were established in Europe, twelve within the Weimar Republic and one each in Vienna, Rome, and Moscow. All but the Munich facility possessed domes of at least twenty to twenty-five meters in diameter and would be considered major planetaria today. The largest was erected at Düsseldorf, a thirty-meter structure that accommodated some 600 people.[53] As measured by public attendance, these facilities were outstanding successes.

Although attendance records from this period are not widely available, published accounts reveal a consistently high level of popular interest. At the Deutsches Museum, which operated the smallest planetarium, more than 80,000 visitors witnessed performances in less than two years. During G. Clyde Fisher's visit in September 1925, two lecturers gave nine demonstrations per day. Fisher also reported that after the factory roof planetarium had opened, as many as 600 persons were squeezed into each performance, with a maximum of twelve demonstrations given per day! Multiple sources confirm that more than 100,000 visitors attended programs at the Jena planetarium during its first year of operation. Comparable yearly attendance was reported for the Berlin planetarium, to which an entrance fee of one mark (25 cents) was charged. By the close of 1927, more than one-half million people had attended demonstrations at German planetaria; that figure doubled by the middle of 1928.[54]

Research and development costs associated with the Deutsches Museum's projector were underwritten by the Zeiss firm. Mass production of the Model I instrument was never undertaken, however, because of design changes introduced to accommodate latitude variations. The resulting Zeiss Model II projector was the first to be commercially distributed within the German republic and to surrounding countries. Published prices for the Model II varied widely from $80,000 in 1927 to only $40,000 one year later. Whether that 50 percent reduction, if accurate, can be explained by improved efficiency in production or large fluctuations in the domestic economy remains unknown. By 1929 the Model II was again priced near $75,000.[55]

Expenditures for the construction of planetarium buildings were sparsely reported. After Zeiss's concrete-shell techniques became proven, planetarium domes grew to extraordinary sizes. In 1925 Lt. Commander Belt projected costs of $125,000 for instrument and building at the planned Berlin facility. Belt's figures were based on an eighteen-meter dome and Model I instrument, whereas the actual size was increased to twenty-five meters (using the Model II projector) at a cost of $140,000. Estimates for similar buildings approached $160,000 by 1928.[56] How did the German cities finance so many planetaria in such a short time?

Most planetaria throughout the Weimar Republic were owned and operated by local municipalities. Government sponsorship permitted their operations to be conducted by city school boards, which delivered services to student and public audiences alike. No analogous situation existed within the United States, where long-standing resistance to federal support of education relegated all decisions and resource allocations to state and local authorities. Only during the third phase of American planetarium development (1958–70) would widespread government support (usually provided on a matching basis) become available to construct hundreds of new planetaria, at last exceeding the financial generosity bestowed toward this new educational medium during the Weimar era. Between the wars, however, patronage of American planetaria was undertaken almost exclusively by an elite body of private donors and foundations, whose actions are recounted in the next chapter.

Summary

The origins of projection planetaria are traceable to a junction of two streams of conceptual and technological expression. Previously, those twin currents spurred the development of hollow, perforated globes and mechanically driven models of the Sun, Moon, and planets. Due to Oskar von Miller's prominent influence, both types of teaching devices were synthesized and brought to fruition by Walther Bauersfeld of the Carl Zeiss firm. Optical projection techniques, devised centuries earlier as "magic lanterns," were the catalyst behind this remarkable innovation.

By some mental process akin to a gestalt-switch, Zeiss engineers Walther Bauersfeld and Werner Straubel reconceptualized the planetarium's hemispherical dome as a fixed projection screen for the depiction of planetary and stellar images. When operated in conjunction with a large ceiling orrery, the planetarium offered pedagogical confirmation of the heliocentric theory. But tremendous audience response to the planetarium's reproduction of the heavens overshadowed Miller's original objective. The Deutsches Museum's exhibit was thereby transformed into a celebrated pedagogical tool, around which arose a new professional discipline and career.

Since their conception, planetaria have excelled at two primary instructional tasks. They realistically portray the positions of celestial objects as seen from any location on Earth, and they permit astronomical events to be unfolded before audiences in as little as a few minutes, thereby simulating the acceleration of time. These features offer spectators an entertaining, yet scientifically accurate replica of the night sky as viewed from a geocentric perspective. Over the years, their programs have become tremendously popular with audiences of all ages. As cultural institutions, planetaria serve as bridges of communication between the research of professional astronomers and the public. Planetaria operate as educational theaters of time and space, presenting facts of descriptive astronomy through the creation of superlative illusions, including that of infinite space itself.

Under the auspices of government and municipal patronage, German cities erected a dozen planetaria by 1930. This trend sparked a wave of architectural designs for planetaria as buildings. Much of this success originated from the Zeiss firm's innovation of concrete-shell construction techniques. Professional astronomers embraced the planetarium concept because of its versatility and accuracy, though none grasped Miller's pedagogical aims. As judged by reported attendance figures, planetaria became important public attractions for citizens of their communities.

Word of these developments was relayed to Americans via popular media. Press coverage of European planetaria created high expectations that Zeiss projectors would be erected on American soils. These devices were praised for delivering many of the elements of high culture associated with a burgeoning entertainment industry. Through point-by-point comparisons with theatrical productions, planetaria were expected to deliver performances that would entertain as well as educate their audiences. Didactic arguments supported wider moral and religious dimensions associated with planetarium instruction.

Most important of all, planetaria helped to initiate a broad resurgence of popular interest in astronomy. By offering a temporary respite from viewing restrictions imposed by urban modernization, this teaching tool allowed city dwellers to reexperience the intricacy and majesty of the preindustrial heavens. Almost miraculously, human ingenuity and scientific technology had produced a stunning optical, electrical, and mechanical device whose man-made skies offered a near-perfect replica of the cosmos. Through the magic of this apparatus, humanity at last achieved its ancient quest of enclosing the universe inside a room.

Part II

Zeiss Planetaria in America, 1930–1946

CHAPTER 2

Planetaria, Patrons, and Cultural Values

. . . the heavens portrayed in great dignity and splendor, dynamic, inspiring, in a way that dispels the mystery but retains the majesty.
—Philip Fox, 1932

The extraordinary success of European planetaria raised hopes of duplicating these institutions on American soil. Even before the Model II instrument was mass produced, movements were underway to secure a Zeiss planetarium for the United States. A host of social and cultural factors underlay the acquisition of each new institution. Public expression of the planetarium's educational potential was intermixed with the personal motivations of donors in promoting popular study of the heavens. Astronomer George E. Hale envisioned planetaria as scientific institutions at which astronomical research, as well as public education, was to be conducted. Major differences were apparent in the financial support of European and American planetaria, strongly limiting comparisons between those two cultural groups. With the exception of New York City, no government or municipal funding was used to erect an American planetarium. Creation of these institutions depended strongly on the backing of private business and foundation resources.

Five major American Zeiss installations comprised an elite core that strongly shaped the emergent self-identity, authority, and influence of planetaria. Two smaller American-built projectors also made their debuts during this formative period. Table 2.1 gives the locations, dates, and types of star instruments installed or fabricated by these institutions.

Planetaria and Museums: New Contexts for Astronomy Education

The advent of planetaria in the United States helped to fuel a resurgence of popular interest in astronomy beginning in the late 1920s and continuing through the 1930s.[1] One year after the nation's fourth Zeiss planetarium was opened (1936), a census revealed that some four million persons had attended demonstrations of their artificial skies. Exposure to this novel form of astronomical

TABLE 2.1 *Prewar American Planetarium Institutions*

Name	City	Opening date	Star instrument
Adler Planetarium	Chicago	10 May 1930	Zeiss Model II
Fels Planetarium	Philadelphia	1 November 1933	Zeiss Model II
Griffith Observatory	Los Angeles	14 May 1935	Zeiss Model II
Hayden Planetarium	New York City	2 October 1935	Zeiss Model II
Rosicrucian Planetarium	San Jose, Calif.	13 July 1936	Lewis
Seymour Planetarium	Springfield, Mass.	2 November 1937	Korkosz
Buhl Planetarium	Pittsburgh	24 October 1939	Zeiss Model II

Source: Author.

teaching occurred within an informal context largely unrelated to school curricula or collegiate instruction, namely public museums. To better understand the setting in which this revitalization of U.S. astronomy education took place, it is worthwhile to review the changing exhibit philosophies of American museums in the era that preceded the planetarium's introduction.

Historian Joel J. Orosz has argued that conflicting demands between popular exhibition and professional conduct prevented the establishment of a consensus within American museum practices before the year 1870. But the emergence of major metropolitan museums in New York City and elsewhere signified that both scholarly research and popular education had become "fully accepted as equal goals of the American museum"—a viewpoint that Orosz has labeled the "American Compromise."[2] By pursuing these often-contradictory objectives, museums have attempted not only to bridge the chasm between scientific knowledge and the public (via their displays and exhibits), but also to devote significant resources and support to the advancement of natural knowledge. This dual approach, which served as a guiding principle in subsequent museum administration, affected the decisions of early planetarium directors (and their advisors) to combine astronomical research with popular education.

A trio of the most important of museum functions, namely, the collection, curation, and exhibition of specimens, is the subject of an important study by historian Steven Conn. In his analysis of late-Victorian museums, Conn argues than an "object-based epistemology" strongly influenced the practices of both curators and directors. When properly displayed, museum objects embodied particular types of knowledge and meanings, which were thought capable of relating, even to the untrained observer, a host of separate narratives, possibly including a "metanarrative of evolutionary progress." Artifacts were commonly arranged along trajectories leading from simple (primitive) to complex (civilized) in order to instill a "positivist, progressive and hierarchical view of the world," supported by the investigations of Western science. Natural history museums conveyed a host of didactic truths to visitors, among which was that nature's bewildering complexity, once it was understood through a proper classification scheme, "illuminat[ed] God's plans for the world and humans' place in it."[3]

At the Milwaukee Public Museum, and later at Chicago's Field Museum, preparator Carl E. Akeley reshaped museum exhibits with his perfection of large dioramas, or "habitat groups." There, animals were realistically posed in precisely reconstructed habitats (including native plants, rocks, and soils) that merged seamlessly with curved background paintings.[4] Scenes containing elements of high drama added significant visual impact to these lifelike portrayals. Visitors were transported to some faraway corner of the world and placed at the center of an unfolding spectacle. Every effort was taken to enhance the realism depicted in the vista, which simulated extension of the immediate surroundings to the observer's visual horizon. Dioramas formed a major departure from the densely arrayed collections of specimens that epitomized a Victorian penchant for order and ushered in the next major phase of museum exhibition techniques.

Historian Ronald Rainger has argued that construction of large-scale dioramas enabled realization of a "vicarious experience of the outdoors" and its potential to effect a "personal transformation" on the visitor. American Museum of Natural History president Henry Fairfield Osborn even claimed that the "whole theory and practice" of museum education lay in the ability "to restore to the human mind the direct vision and inspiration of nature" as it existed in all parts of the world.[5] How, then, did this exhibit philosophy relate to the planetarium's realistic depiction of the natural sky?

When viewed from the confines of an urbanized, industrial landscape, the innate starry sky had become another of those elements that vanished from the natural world. A Zeiss planetarium was capable of removing the self-imposed blinders and providing audiences with a "vicarious experience" of the heavens' panoramic splendor as it had been witnessed by our preindustrial ancestors. The planetarium's entrancing replica of the night sky offered a striking fulfillment of Osborn's pedagogical scheme and readily illustrates the desire of museum directors, curators, and educators to unite astronomy with other museum exhibits and programs.

In turn, most planetaria have adopted one or both of the following museum-style exhibit schemes to supplement their cadre of programs: (a) the mounting of large astronomical photographs or transparencies taken through telescopes at major observatories; (b) displays of historic or current astronomical instruments, such as sextants, telescopes, and spectrographs, intended to teach physical principles of astronomical observation. These images and artifacts provide visitors with something interesting to examine, either before or after the formal planetarium program. Finally, direct views of the Sun, Moon, planets, and stars might be offered through moderate-size telescopes, in the tradition of collegiate observatories.

Overview: Institutional Characteristics

When America's first planetaria were constructed, no one expected that these institutions (and their personnel) might one day comprise a unified community. The projection planetarium was still a relatively new and imperfectly understood medium of communication for the popularization of astronomy. How could it be used most effectively? Each installation arose from a host of local political and economic circumstances in which all facilities were operated autonomously. Significant geographic barriers separated West Coast and Midwestern planetaria from their East Coast counterparts, limiting the exchange of professional know-how. While the first American planetarium directors occasionally looked to European colleagues for guidance, they devised their own solutions to the administrative problems of operating a public planetarium.

Putting aside for the moment the two smaller planetaria that featured American-built projectors, what were the institutional characteristics of the nation's five Zeiss installations? How and by whom were they operated? Two of those five planetaria arose through association with established metropolitan institutions whose scientific activities and collections originated in the nineteenth century: Philadelphia's Franklin Institute and New York City's American Museum of Natural History. These planetaria were physically and administratively linked with their parent organizations. Philadelphia's planetarium was conceived through an important redefinition of purpose that accompanied the Franklin Institute's centennial celebration. There, it received the benefits of a short-term visionary origin as well as the strengths of a venerable antecedent. In New York City, however, initial efforts to secure a Zeiss planetarium proved unsuccessful. The planetarium concept fell victim to over-ambitious goals, inappropriate timing, and the onset of national economic depression—factors that delayed its ultimate realization for a decade.

Zeiss planetaria in Chicago, Los Angeles, and Pittsburgh were operated independently of any institutions devoted to research or the popularization of science. All were administered by municipal governments. In Chicago and Los Angeles, these facilities fell under the jurisdiction of the city's parks department. Each board of park commissioners exercised considerable influence over the planetarium's creation and its subsequent operations. By extension, they oversaw the appointment and (on occasion) the termination of a director.

Administrative characteristics are not the only way by which planetaria may be differentiated. Another category is the association with the nation's first *industrial* museums. At Chicago, Philadelphia, and Pittsburgh, these centers were linked with American replications of Oskar von Miller's Deutsches Museum. Chicago businessman Max Adler originally slated his gift of a Zeiss planetarium for that city's new Museum of Science and Industry. But when faced with lengthy delays and possible competition from other patrons, Adler enlarged his proposal and thus established a freestanding installation. At Philadelphia, the planetarium became a noted addition to the newly fashioned science museum of the Franklin Institute. In Pittsburgh, by contrast, creation of an adjoining scientific and technological museum was envisioned only secondarily to the planetarium itself. Nonetheless, the Buhl Institute of Popular Science was closely patterned after those examples set by the Franklin Institute and the Deutsches Museum.

Planetaria may also be categorized according to their original means of financial support. Zeiss projectors and buildings erected at Chicago and Philadelphia were the products of living donors, businessmen Max Adler and Samuel S. Fels, respectively. New York City acquired its planetarium instruments from investment broker Charles Hayden, while the building was secured through a federally administered Reconstruction Finance Corporation (RFC) loan.

American planetaria also arose from the provisions of two deceased benefactors. Los Angeles philanthropist Griffith J. Griffith stipulated in his will that an astronomical observatory be constructed for the public's benefit. A committee of scientists from the California Institute of Technology oversaw the observatory's posthumous completion through the execution of Griffith's estate. At Pittsburgh, a private foundation established by the will of merchant Henry Buhl Jr. left no specific instructions for the utilization of its assets. The Buhl Foundation's executive director, Charles F. Lewis, led the campaign to construct America's fifth (and final) prewar Zeiss planetarium.

In short, no overall pattern describes the creation, support, or administrative control of those five Zeiss planetaria established during the formative period. Key incentives were undertaken by various institutional, committee, or foundation leaders in conjunction with the stated goals of noted philanthropists, either living or deceased. A better understanding of these institutions requires a more thorough knowledge of their contextual origins. Brief accounts of the principal developments and larger issues surrounding their establishment are sketched below, arranged in chronological order of their completion.[6]

Zeiss Planetaria in Historical Context

Chicago

Julius Rosenwald, president of Sears, Roebuck and Company,[7] pledged three million dollars in 1926 to establish the nation's first industrial museum (the Museum of Science and Industry) in the renovated Palace of Fine Arts Building, a temporary structure erected for the 1893 World's Columbian Exposition in Chicago.[8] Rosenwald's brother-in-law, Max Adler (Sears vice president and general manager), quietly determined to endow the future museum with a Zeiss planetarium. Adler followed the suggestion of Suzanne Joachim, wife of the violin maestro with whom he studied in Berlin, that his gift should take the form of a projection planetarium.[9] But the Fine Arts Building proved a hindrance to Adler's objective. Renovations necessary to convert that structure into Rosenwald's industrial museum consumed almost a decade. Frustration over those delays stimulated Adler to look elsewhere.[10]

A more suitable location was procured on Northerly Island, adjacent to the Field Museum of Natural History and the Shedd Aquarium. The board of

South Park Commissioners accepted Adler's formal offer on 20 June 1928.[11] With a donation of $500,000, Adler left a substantial legacy to the citizens of Chicago as a reminder of his own personal success. Adler's cousin, architect Ernest A. Grunsfeld Jr., designed the planetarium in the shape of a regular dodecagon. Grunsfeld's conception was awarded the Gold Medal of the Chicago chapter of the American Institute of Architects.

Philadelphia

By performing the functions of a "national technical institution," the Franklin Institute became one of the most important centers for advancing American technology during the nineteenth century.[12] After the Institute celebrated its centennial in 1924, secretary Howard McClenahan announced a bold initiative that sought to transform its mission from that of applied research to public education. McClenahan, a former dean and alumnus of Princeton University, had earned bachelor's and master's degrees in electrical engineering. His plans envisioned a new "scientific, technical and industrial museum, similar in character to the world famous Deutsches Museum at Munich," although they did not yet include a planetarium.[13] Unknown to McClenahan, Philadelphia businessman Samuel S. Fels was taking independent steps to provide the city with just such a device. Fels's humanitarian spirit was influenced by his friendship with Julius Rosenwald.[14]

One of Fels's closest associates was physician David Riesman, an instructor on the medical faculty of the University of Pennsylvania. During a trip to Vienna in 1927, Riesman witnessed a performance at that city's planetarium and described his experience before the Rittenhouse Astronomical Society. Riesman noted how Fels had become "deeply interested in astronomy and . . . erected a small telescope in his summer home."[15] On account of Riesman's association with the University, Fels originally slated his donation for that institution. He described to Philadelphia journalist Steven M. Spencer how he had "heard about planetariums, read about them, thought it would be well for Philadelphia to have one, and so, I ordered one."[16] But University officials were reluctant to accept the unknown responsibilities of operating a Zeiss planetarium, regarding as "doubtful whether this would be self-supporting and would not be a burden" — in retrospect, perhaps a wise decision.[17] (Over two decades were to elapse before the University of North Carolina at Chapel Hill became the first American institution of higher education to operate a major planetarium. See chapter 5.) After McClenahan learned of Fels's gift, he convinced him that the Franklin Institute's new science museum offered a more suitable location for the Philadelphia planetarium.

Los Angeles

America's third planetarium was established by one of the most unusual patrons of public recreation and education, Griffith J. Griffith. A Welsh immigrant, Griffith amassed a considerable fortune as a mining consultant and real estate speculator. In 1882 Griffith purchased a large tract of land outside the Los Angeles city limits. This estate, Rancho Los Feliz, became Griffith's permanent home and the inspiration for his greatest philanthropic gesture.[18] In 1896 Griffith donated 3,015 acres from Rancho Los Feliz to the city of Los Angeles as a public park.[19] His plans did not yet envision an astronomical observatory on its highest point.[20] In 1912 Griffith submitted a proposal to city officials in which he offered "$100,000 for a public observatory, to be erected on . . . Mt. Hollywood, in Griffith Park, and to be fully equipped on or before the year 1915."[21] No action, however, was taken on Griffith's proposal. After Griffith's death in 1919, his will directed that a Greek theater be erected in Griffith Park, and that his estate's remainder be expended "towards the erection and completion of a Hall of Science and Observatory." The latter was to feature "a large motion picture theatre or hall," and to house "at least a twelve inch telescope."[22]

To fulfill Griffith's final wishes, Erwin W. Widney, trust counsel of the Security-First National Bank and executor of Griffith's estate, sought advice from officials at the California Institute of Technology.[23] An advisory committee, consisting of Mount Wilson Observatory director Walter S. Adams, astronomer George E. Hale, and Caltech physicist Robert A. Millikan, decided to make "public interest in the enterprise center around a Zeiss planetarium" as a substitute for Griffith's ill-defined motion picture theater.[24] Southern California thus became the third metropolitan center in the United States to embrace this novel device for purposes of astronomical instruction.

New York City

The American Museum of Natural History, chartered in 1869, arose from ambitions of local businessmen to create an institution whose collections and prestige matched those of museums located elsewhere.[25] Starting in 1891, paleontologist Henry Fairfield Osborn became in succession department head, trustee, and museum president.[26] Cosmic evolutionary principles supported Osborn's deterministic model of the history of life on Earth, leading him to broaden the Museum's scope and encompass the heavens themselves. A portion of his master plan, issued before the planetarium's invention, was to showcase "the most recent discoveries from the great astronomic observatories of America, especially the work of Hale at Mount Wilson."[27] Osborn's exhibit philosophy reflected a deliberate attempt to recapitulate the known evolutionary process, starting from its stellar and planetary origins.[28] To meet this objective, Osborn repeatedly solicited funding from the newly chartered Carnegie Corporation of New York.[29] He also employed freelance artist and architect Howard Russell Butler to design the proposed five-story Astronomic Hall, the plans of which by 1925 included a projection planetarium and circular ambulatory beneath its dome.[30] But no amount of pleading convinced Carnegie executives of the worth of Osborn's scheme.[31]

Instead, Museum officials began to consider the possibility of creating a less expensive, freestanding planetarium. F. Trubee Davison, formerly assistant secretary of war in charge of aeronautics, was chosen as Osborn's successor in 1933. Previously, Davison had been elected to the New York State Assembly (1921) and served as a museum trustee since 1923.[32] Under Davison's initiative, a federal loan was secured from the newly chartered Reconstruction Finance Corporation (RFC).[33] But purchase of the planetarium instruments from abroad was later found to be forbidden under the auspices of the loan, and unless the Museum secured additional funds, the project would be jeopardized. After Davison personally explained the Museum's precarious situation to him, investment broker Charles Hayden agreed to contribute $150,000 to purchase both Zeiss and Copernican (ceiling) planetaria.[34] Acquisition of the Copernican planetarium furnished the only prewar American replication of Oskar von Miller's tandem exhibit concept. While the latter was said to "splendidly and adequately complement . . . the Zeiss projection planetarium," no evidence has been found that the device was used for the teaching of theory confirmation in the style envisioned by von Miller (see chapter 1).[35]

Pittsburgh

Pittsburgh's planetarium was established without any prior directives supporting astronomy or science education from the late Henry Buhl Jr. Buhl graduated in 1869 from Duff's Business College and founded a dry goods store in Allegheny (as Pittsburgh's North Side was then called) with Russell H. Boggs, a childhood friend. Boggs and Buhl's department store grew into that city's largest. After Boggs's death in 1922, Buhl became the company's sole president.[36] Buhl likewise entrusted the bulk of his estate towards establishment of "the BUHL FOUNDATION, to be managed and controlled by a Board of four Managers." He specified that this board "shall not be restricted in any manner or extent in their selections of the uses, objects and purposes to which the said trust fund and its income shall be applied."[37] Buhl's only concern was that his foundation's activities be directed toward the greater Pittsburgh area in which he had prospered.[38] The post of foundation director was filled by Charles F. Lewis, a graduate of Allegheny College and former chief editorial writer for the *Pittsburgh Sun.* Before being tapped by the Buhl Foundation, Lewis had edited the University of Pittsburgh periodical *The Pittsburgh Record.*[39]

The earliest movement to establish a planetarium in Pittsburgh was headed by two amateur astronomers, Chester B. Roe and Leo J. Scanlon. Roe was president of the astronomical section of Pittsburgh's Academy of Science and Art, Scanlon its secretary-treasurer. Both men had traveled to Chicago in 1930 to see the new Adler Planetarium and conversed with director Philip Fox.[40] Initially unsuccessful, their efforts were reinvigorated by city councilman George E. Evans, who supported redevelopment of a site on the city's North Side, augmenting the "renaissance of Old Allegheny."[41] Evans's suggestion found ready support in the Buhl Foundation.[42] However, Lewis's report to the Board of Managers omitted an important consideration, namely, whether the planetarium should become part of a larger "center of general interest, such as a museum," as proposed by Andrew W. Robertson, chairman of the Westinghouse Electric and Manufacturing Company.[43] Thereupon, C. V. Starrett, the Foundation's associate director, assumed much of the responsibility for designing the Buhl Planetarium and the Institute of Popular Science,[44] which comprised the largest single gift awarded by the Buhl Foundation.[45] Total expenditures for the building, equipment, and exhibits reached nearly $1,100,000.[46]

Motivations of Planetarium Donors

We have seen how five Zeiss planetaria were erected in major American cities through the actions of donors and foundation leaders. But what motives guided these philanthropists to endow their cities with the relatively unknown form of public education represented by projection planetaria? Prior interest in the subject of astronomy is traceable only to Griffith J. Griffith and Samuel S. Fels. But their patronage must be carefully distinguished from that of scientific research itself. On the other hand, no prior interests in astronomical matters are found within the public or private lives of Max Adler, Charles Hayden, or Henry Buhl Jr. Thus, we must look beyond the strictly pedagogical purposes of the planetarium in order to find the underlying motivations of these men.

Neither Griffith nor Fels elected to fund basic research in astronomy; instead, each supported the diffusion of astronomical knowledge among the broader public. By their estimates, the backing of institutions for the acquisition of scientific knowledge stood to benefit only a handful of like-minded investigators. Using a similar argument, American Museum of Natural History president Henry Fairfield Osborn had lobbied the Carnegie Corporation of New York. In Osborn's opinion, its Mount Wilson Observatory had yielded few, if any, tangible benefits for the average person. By contrast, a Zeiss planetarium held the potential of spreading an understanding of the workings of the heavens, along with newer astronomical discoveries, to much wider audiences.

Griffith was one of many potential donors solicited by George E. Hale to provide support for construction of the Mount Wilson Observatory.[47] Griffith declined Hale's proposal but later sought to popularize astronomical observation among the public. It was likely Walter S. Adams, a "friend of the benefactor for many years before his death," who invited Griffith to look through the completed Mount Wilson reflector,[48] whereupon Griffith reportedly exclaimed, "If all mankind could look through that telescope it would revolutionize the world."[49]

A generation later, Fels noted (with apparent reference to Hale's Palomar Mountain reflector) how "telescopes of greater and greater power" were being constructed in the search for "more light on the old question of the cause and scope of creation." Provision of such funds did not greatly interest Fels because, in his judgment, those "great tools [remained] the tools of the experts." His patronage was directed towards a middle ground of recipients who might accrue the largest share of benefits. Fels was attracted to the planetarium by its ability to "translate the systems and the movements of the heavenly bodies" so that ordinary citizens might obtain a better grasp of their place within the universe.[50] Griffith and Fels supported the construction of public observatories and planetaria as institutions that could bridge the gap between research astronomers and lay persons. Each understood that telescopes or planetaria offered a more captivating experience for visitors than static museum exhibits of photographs or artifacts.

Broader social and religious issues are likewise discernable behind the creation of America's Zeiss planetaria. Two living donors expressed hopes that the experience of their programs might serve to mitigate human differences and unify strained relations between social groups divided by race, class, and gender. Both the Chicago and Philadelphia planetaria were established by noted Jewish businessmen. Formal announcement of Max Adler's gift conveyed the "deeper purposes than that of teaching astronomy" that he intended: "I wish to emphasize through this instrument that all mankind, rich and poor, here and abroad, constitute part of one universe, and that, under the vast firmament, there is no division or cleavage but rather interdependence and unity."[51] Samuel Fels urged that "[o]ur job . . . is so to organize our world that we can live together without fears and without hatred of each other. All men and women are fundamentally alike, regardless of where they live or what color they are."[52] In an era of deepening anti-Semitism, Adler and Fels shared an optimism that planetaria might allow human differences to seem inconsequential by comparison to the immensity of the cosmos in which all Earth's inhabitants are enjoined.

Both Fels and Charles Hayden were strongly attracted by the planetarium's purported affirmation of spiritual values. Planetaria became linked, in the minds of donors as well as audiences, with an intellectual tradition that gained ascendancy in the seventeenth to nineteenth centuries, the so-called "argument from design." This Protestant natural theology movement espoused that God's power, wisdom, and benevolence could be discerned through observed regularities of the heavens.[53] During the Age of Enlightenment, widespread notions of a clockwork universe characterized by Newtonian mechanics encouraged the popular use of orreries for demonstrating the law-like behaviors of planets and their satellites. Although discredited by twentieth-century science, natural theology and the argument from design did not entirely disappear from popular discourse. As purveyors of the modern understanding of space and time, Zeiss planetaria were deemed capable of imparting to audience members a profound sense of the magnificent structure and divine purpose of our universe.

Charles Hayden graduated from MIT in 1890, where he studied economics and mining. In 1892, with Galen L. Stone, he formed the partnership of Hayden, Stone & Company, a brokerage firm that specialized in copper stocks. After Stone's death in 1926, he assumed complete charge of the company. Like Henry Buhl Jr., Hayden survived the loss of a lifetime business partner in a venture that brought him considerable financial success.

Hayden's willingness to provide New York City's planetarium instruments carried a ready explanation. On a business trip to Chicago, Hayden attended an Adler Planetarium program during the 1933–34 Century of Progress Exposition (see chapter 4). That experience reportedly left the sixty-three-year-old financier deeply moved and prompted a reexamination of his views concerning humanity's role and place in the universe.[54] Strongly countering the claim that "science has a tendency to make one less religious," Hayden affirmed his belief that the planetarium "should give [a] more lively and sincere appreciation of the magnitude of the universe and of the belief that there must be a very much greater power than man which is responsible for the wonderful things which are daily occurring in the universe."[55] John D. Rockefeller Jr. privately applauded the seeming purity behind Hayden's motives: "Surely, people must realize more and more that spiritual values are the only ones that offer a solid foundation for the development of civilization if the world is to go on and mankind to become in any sense worthy of the Creator."[56]

Samuel Fels's characteristic optimism toward human nature (published at age 73!) was expressed in his book *This Changing World: As I See Its Trend and Purpose* (1933). His volume offered a critique of the nation's ideological and economic woes and presented an almost rudimentary philosophy of technological determinism. Fels was no Marxist, yet believed that the modern world had been caught up in the process of negotiating "one of the greatest industrial changes in history," while still equipped with "social habits, legal forms or economic apparatus" originally devised for an agrarian civilization.[57]

In the book's final chapter, Fels explained that his literary endeavor was undertaken to show "the possibility of, the necessity for, a Universal Life Purpose," which he identified as the "practicable, purposeful, never-ending endeavor towards the progress and betterment of the race."[58] Fels asserted that this conviction arose not from belief in a personal god, but from faith in the existence of an impersonal deity who nonetheless created our universe, and to whom Fels referred as the "Great Innovator." Reliance upon the logic of first causes prompted Fels to argue, "[I]t is difficult to comprehend how human beings can think of an effect without a cause, a wondrous world and world system without a Power behind it." In addition, Fels disclosed the most explicit answer of any planetarium donor to an important metaphysical issue arising from the investigations of modern science. He argued that a pervading sense of purpose offered a satisfying response to the recognition of "man's minuteness," or the dilemma of human insignificance, within the cosmos—an awareness derived from the "new discoveries of physics and astronomy."[59]

To a depression-ridden world laced with discord, animosity, and anxiety, Hayden and Fels offered an alluring prescription: a reassurance of supernatural purpose and design in the universe, to be fostered by attendance at the mechanized demonstrations of projection planetaria.

Hale's Influence on American Planetaria

George E. Hale (1868–1938) was a dominant figure in the establishment of American scientific institutions and one of the most important astronomers of his generation. Founding director of the Yerkes and Mount Wilson Observatories, he helped to create and organize much of the infrastructure of American (and world) astrophysics after 1890. Little recognized today, however, is the extension of Hale's influence, through direct and indirect means, upon the design, operation, and even personnel selection at several of America's earliest Zeiss planetaria. Still, Hale was both a remote and reluctant planetarium consultant, jealously guarding his time for the Palomar telescope project, delegating authority to others, and proffering advice by correspondence. Despite worldwide travels, Hale did not attend a planetarium demonstration until 1931 at Chicago. G. Clyde Fisher, who was present when Hale and his wife witnessed their first performance, reported to Osborn: "Dr. Hale made many enthusiastic statements of approval" and termed the optical reproduction "superb—far beyond his expectations."[60] Later in that decade, Hale's declining health prevented further engagements, while his limited expertise was superseded by Fels planetarium director James Stokley.

Hale's fundamental stance towards planetaria echoed precisely the tenets of Orosz's "American Compromise" as reached by museum directors after 1870. He believed that planetaria should be institutions devoted to astronomical research and not merely tools for the popularization of science. Under Hale's prescription, planetarium directors must be well-trained researchers in astronomy, physics, or a closely related discipline. In his judgment, a planetarium's mission of presenting astronomical information lacked scientific credibility and authority without a recognized scientist at its helm. In addition, Hale argued that the public would not be satisfied with viewing only the planetarium's reproduction of the night sky. He campaigned for the use of astronomical, and especially solar, telescopes in providing direct observations of celestial objects. Hale's plan of encouraging solar research among amateur astronomers (by using his invention, the spectrohelioscope) was popularized through Albert G. Ingalls's *Amateur Telescope Making* series.[61]

One of many institutions that sought Hale's advice regarding astronomical exhibits was Philadelphia's Franklin Institute. Even before a Zeiss planetarium was conceived as part of its new science museum, secretary Howard McClenahan strove to ensure that astronomy would be adequately represented. The Franklin Institute, in turn, was no stranger to Hale. In 1917, Hale lectured in Philadelphia on the subject of "Whirlwinds and Sunspots," explaining his theories of magnetic polarities and solar vortices.[62] Through McClenahan's influence, Hale was awarded the Elliott Cresson Medal in 1926 for "researches and discoveries relating to the sun and the solar atmosphere" and the Franklin Medal in the following year.[63] When McClenahan approached Hale about the selection of instruments and displays, the latter could hardly refuse to offer help, despite a host of competing obligations. Hale indicated a willingness to comply, but argued,

"I am not sure that I can contribute any useful ideas."[64] He nonetheless issued recommendations comparable to those given to Adler Planetarium director Philip Fox (see chapter 3), and asked whether the Franklin Institute would take part in cooperative solar observations.[65]

Hale assumed a more prominent role in the founding of the Griffith Observatory than with any other American planetarium. This situation arose naturally from its proximity to the Mount Wilson Observatory. As a member of the committee appointed to advise the executors of Griffith's estate, Hale expressed hopes to Walter S. Adams that "the outcome will be not merely a museum, but also an observatory in which research can be conducted."[66] The institution's director, in Hale's opinion, should be "an astronomer of high standing as an original investigator and [someone] deeply interested in evolution and its value for public education."[67] Hale possessed an abiding interest in evolutionary theory as it applied to the formation of stars. His book, *The Study of Stellar Evolution* (1908), was published the same year in which the sixty-inch telescope at Mount Wilson went into service. For a choice of observatory director, the committee put its weight behind Caltech physicist Edward H. Kurth, a candidate deemed "nearly ideal for this purpose."[68] When Kurth was killed in an automobile accident in 1934, he was replaced by Caltech physicist Rudolph M. Langer. Hale also proposed increasing the telescope's aperture to thirty-six inches, on grounds that it "would make a much stronger appeal to the public than a smaller instrument," and the observatory's reputation could be "greatly increased if new discoveries were made there."[69] The advisory committee, however, rejected Hale's plans for a larger telescope because the night sky was already "so illuminated by the surrounding lights of Los Angeles" that astronomical research from the summit of Mount Hollywood remained an impossibility.[70]

After the two-hundred-inch Palomar telescope's contract had been awarded by the Rockefeller Foundation, Hale recruited a team of experts in Pasadena who would oversee every aspect of its completion. One of the most unusual men hired by Hale was Russell W. Porter, an optical consultant to the machine tool industry in Springfield, Vermont. Porter had been responsible for launching the amateur telescope-making movement in collaboration with *Scientific American* editor Albert G. Ingalls, but was equally noted for his artistic and architectural skills.[71] Working as a freelancer, Porter executed a series of conceptual drawings of the Griffith Observatory. Anthony Cook, Griffith's astronomical observer, has described a host of Porter sketches, including one that bears a strong resemblance to the actual construction.[72] Porter also designed Griffith's triple-beam coelostat, which permits visitors to study solar images in white or monochromatic light and projects a high-dispersion spectrum.

Hale also directed advice toward officials at New York City's Hayden Planetarium. Even before the RFC loan was approved, he voiced an opinion to American Museum of Natural History president F. Trubee Davison that "a Planetarium will accomplish only a part of your probable purpose." Hale suggested the addition of solar telescopes and related exhibits, "to bring the public into direct contact with natural phenomena in action." When Hale congratulated the staff on receipt of Hayden's gift, he urged that they obtain a twelve-inch telescope, arguing that "the public will not be contented with a Planetarium alone, but will wish for an opportunity to observe the sun, stars and planets." Hale's solar instruments, including a "coelostat, a spectroscope, and a spectrohelioscope," were said to be "included in the plans," although for practical reasons no large astronomical telescope was ever erected at the Manhattan facility.[73]

Pittsburgh's Buhl Planetarium was the only Zeiss installation that did not receive the benefit of Hale's advice. By the time the Buhl Foundation's plans were announced, he was in the final stages of illness that claimed his life in 1938. Instead, Fels Planetarium director James Stokley acted as technical consultant to the Buhl project. Several innovations, including use of an elevator to raise and lower the star instrument, along with design of a siderostat, or fixed-eyepiece telescope, are attributable to Stokley's influence. Stokley's adoption of the siderostat was derived from a similar telescope constructed by Wynnewood, Pennsylvania, amateur astronomer Gustavus Wynne Cook.[74]

Non-Zeiss Planetaria

Besides the nation's five Zeiss planetaria, two smaller facilities, sporting projection instruments designed by local craftsmen, became part of the American planetarium scene during the formative period. From the standpoint of patronage, these institutions differed only in scale from their Zeiss-equipped predecessors. One installation drew support from the subscribers to a fraternal organization; the other resulted from an unspecified bequest of a deceased benefactor. Little if any notice was taken of the first American-built projection device, while the second was denigrated for its shortcomings and regarded as an unauthorized replica of a Zeiss planetarium. The presence of American-made instruments, however, signaled the end of the Zeiss monopoly on projector technology and presaged a new era that developed rapidly after the Second World War.

San Jose, California

The first American-built projection planetarium was a little-known device constructed by Harvey Spencer Lewis, Imperator of the Ancient and Mystical Order of the Rosae Crucis (AMORC), in San Jose, California. A native of Frenchtown, New Jersey, Lewis founded the New York Institute for Psychical Research in 1904. Thereafter, Rosicrucian principles dominated his life. Lewis's efforts were devoted "to forming a worldwide fraternal organization teaching philosophical and mystical practices to develop the latent faculties of man, and selling literature by mail order."[75] The origins of his interest in acquiring a planetarium for the Rosicrucian order cannot be traced, but was possibly triggered by publicity surrounding the advent of European or American planetaria. When announcing plans to establish his own institution, however, Lewis spoke as if unaware of the Zeiss instrument's true capabilities and referred only to mechanical planetaria (orreries). His projection device, "designed and worked out in model form in the [Rosicrucian] scientific laboratories," reportedly exhibited both the "ancient principles of the geocentric as well as heliocentric theories of astronomy." With appropriate modification, even the "strange theory of cellular cosmogony" was demonstrable by Lewis's machine.[76]

While Lewis's projector no longer exists, a partial understanding of its operation can be inferred from a remaining sketch of the device. Upon its hemispherical surface were approximately two dozen lenses that optically projected stars upon the planetarium's forty-foot-diameter dome. Lewis's device was more sophisticated than the pinhole-style projector later devised by Armand N. Spitz (see chapter 5). The "starball" presumably rotated on an axis corresponding to the celestial poles; a horizon cutoff surrounded the projection assembly. Five rotary knobs were attached to the support system. The apparatus rested on a trapezoidal framework containing eight additional "switches" and other unidentified pieces of hardware.[77]

The Rosicrucian Planetarium was dedicated 13 July 1936 at the annual convention of the AMORC, wherein Lewis declared the mosque-shaped building given "to Rosicrucians and the world." Over $75,000 had been expended on the facility. Though press accounts did not mention it, construction of the planetarium as a "Memorial Educational Facility" was made possible "through the Generous Bequest of [Rosicrucians] Paul P. and Helen H. Merritt."[78] Programs for school children and adults began the following month. Ralph Maxwell Lewis succeeded his father as Imperator (and planetarium director) after the latter's death in 1939.

As this institution's governing body lay well outside of the mainstream community of museum professionals, American astronomers and planetarium directors either remained unaware of, or else deliberately ignored, the San Jose planetarium until after the Second World War. Mention of its existence was largely confined to a three-volume study of American museums prepared in 1939 by Laurence Vail Coleman.[79] Consequently, the Rosicrucian Planetarium exerted no measurable effect on other Zeiss planetaria during the formative period.

Springfield, Massachusetts

The notion of boundary-work in the field of science studies provides a useful conceptual tool for understanding Zeiss planetarium directors' reactions toward the American-built projector located at Springfield, Massachusetts. Sociologist Thomas F. Gieryn defines this concept as follows: "Boundary-work occurs as people contend for, legitimate, or challenge the cognitive authority of science—and the credibility, prestige, power, and material resources that attend such a privileged position." Gieryn describes what he terms "monopolization," or the contest waged by competing social groups for recognition of cultural authority. Here, "contending parties carve up the intellectual landscape . . . each attaching authority and authenticity to claims and practices of the space in which they also locate themselves, while denying it to those placed outside." Gieryn styles "the erection of walls to protect the resources and privileges of those inside" as an allied form of "protection."[80]

Boundary-work enables us to understand why two eastern Zeiss planetarium directors charged that the Springfield, Massachusetts, star instrument was not a real planetarium because it lacked planetary projectors to accompany its starfield. While this distinction might appear to reflect a semantic debate over a planetarium's definition, the argument illustrated a defensive posture against the spread of non-Zeiss projection instruments throughout the United States.

The Springfield instrument became operational on 2 November 1937. It was housed at the Museum of Natural History in Springfield, Massachusetts, and was America's sixth planetarium. Several years earlier, Frank D. Korkosz (1902–1987), the museum's technician, had proposed turning an unoccupied gallery space into a projection planetarium. Funds amounting to $12,000 were obtained from the bequest of Stephen Seymour, for whom the planetarium was named.[81] Korkosz's homebuilt projector signaled a return to the "starball" principle of Bauersfeld's Zeiss Model I instrument. In the Springfield projector, the heavens' south polar region was necessarily obscured. Korkosz enlarged the projection sphere to five feet in diameter and attached 41 lens assemblies to produce its "high fidelity" starfield. Three axes of motion, controlling diurnal, latitude, and precession motions, enabled the projector to simulate travel in space and time. No other celestial objects were shown by the device, although Korkosz hoped to add auxiliary projectors for depicting planetary motions.

Museum officials pointed out the "far-reaching effect" that such equipment could have "for educational purposes in museums, colleges, public schools, and even parks in smaller cities." The "low-cost installation" was viewed as a "precursor of many such in smaller institutions serving population areas of less than 250,000 inhabitants."[82] The *New York Times* argued that "[t]here is no reason now why any college or high school with an adequate domed hall may not teach astronomy at low cost in the most vivid and dramatic way ever devised."[83] Statements of this kind worried two Zeiss planetarium directors, who feared that their preeminence would be undermined by the spread of smaller, competing institutions.

Fels planetarium director James Stokley was the first to offer a critical appraisal of the Korkosz projector. He visited the Seymour Planetarium and received a thorough demonstration of its features. In a letter to the editor of *Popular Astronomy,* Stokley expressed "great respect" for the museum's effort but detailed the projector's various shortcomings. Stokley acknowledged that some essentials might be added by its designer, and "[u]ntil this is done . . . it seems that the name 'planetarium,' applied to this device, is a misnomer. . . . A name such as 'stellarium' might be more appropriate." Stokley, however, went on to raise a more serious issue concerning the possible infringement of patents held by the Zeiss firm. He recommended that "officials of any institutions [who] are considering building or purchasing a similar projector" secure "legal advice to insure that they will not be liable to prosecution for infringement of these patents." Stokley's letter implied that either Korkosz or the Springfield Museum of Natural History might be held accountable for "[u]nauthorized *construction, sale* or *use*" of a patented device.[84]

Although Stokley's letter might have put to rest a number of fears among Zeiss planetarium directors, it did not settle the issue. A letter to *The Sky* was submitted by William H. Barton Jr., newly appointed director of New York's Hayden Planetarium. Barton and his technician, Ernest Deike, visited the "so-called Planetarium" at Springfield and witnessed two demonstrations. Barton admitted that Korkosz had done a "clever job," but declared that "it is not a planetarium at all, since it displays no planets, no sun, no moon, no celestial circles." He decried the "glowing terms" by which the instrument was described in the press and found such statements "at [the] least misleading, and unfortunate." To Barton, Korkosz's projector lacked the "flexibility and range" of a genuine Zeiss instrument. Barring additional projection devices, the "finished shows of the foreign-built instruments" could not, in his opinion, be presented with it.[85]

Stokley directed a final insult toward the Springfield projector after he was appointed director of the Buhl Planetarium in Pittsburgh. He consistently described the Buhl's Zeiss-equipped facility as "America's Fifth Planetarium."[86] By omitting mention of the Korkosz instrument from his tally, Stokley reinforced the notion that only planetaria manufactured by Zeiss deserved such an appellation. Stokley and Barton's statements offer evidence supporting attempted monopolization and protection of the cultural spaces represented by Zeiss-equipped planetaria. Their admonitions, however, were never answered by Korkosz; no public debate over the demerits of the Springfield projector was ever staged. Neither Stokley nor Barton could deter visitors from attending programs at the Seymour Planetarium. The nature of public education overrode any rhetorical measures by which the Zeiss directors sought to discredit the stature of Springfield's projector.

No additional Springfield-type planetaria were ever manufactured. This outcome, however, does not establish the success of "protective" boundary-work. Nonetheless, from possible suspicions planted about patent infringements, Stokley's rhetoric might have produced the intended effect, even if the possibility of a lawsuit was greatly exaggerated. Whether the non-duplication of Springfield's projector was due to Stokley's warning, the coming of war, or Korkosz's own reluctance to become a planetarium entrepreneur cannot be determined.

Korkosz spent the remainder of his career at the Springfield museum, whose presidency he assumed in 1958. He was awarded an honorary doctorate by Western New England College in Springfield in 1964. Korkosz also designed and built the twin-hemisphere planetarium projector for Boston's Museum of Science, but this, too, remained a unique venture that was never duplicated elsewhere.[87]

Advent of the Springfield (and San Jose) projectors signaled the end of a monopoly by Zeiss planetaria in the United States. Neither instrument, however, was ever mass produced. Instead, a far simpler projection device, bearing an equally significant cost reduction, would transform the postwar American planetarium community. Arguments over the monopolization by Zeiss equipment resurfaced with the marketing of Armand N. Spitz's pinhole-style projector (see chapter 5).

Planetaria and Cultural Values

From the dedication speeches and publicity that attended the openings of America's Zeiss planetaria, we can obtain a clear picture of the cultural and symbolic expectations that were held of planetaria and their associated science museums. Along with self-congratulatory sentiments expressing benefactor munificence, these statements reveal a host of complex beliefs and attitudes, ranging from intellectual fulfillment to affirmations of contemporary moral, spiritual, and sociopolitical ideals.

Intellectual Enrichment

The planetarium's ability to illustrate concepts of descriptive astronomy was firmly endorsed by professional astronomers. Lick Observatory director Robert G. Aitken described the Zeiss projector as the "most remarkable instrument that has ever been devised to exhibit impressively, and with the illusion of reality, the motions of the heavenly bodies and the phenomena which result from these motions." Mount Wilson Observatory director Walter S. Adams seconded Aitken's opinion, stating that the instrument's "realistic and rather dramatic [portrayal] of the celestial objects would prove of great educational value, fixing in the minds as no description could do the simple astronomical principles which everyone should know." Otto Struve, director of the Yerkes Observatory, declared the planetarium to be of "very great importance in stimulating public interest in Astronomy. It is also valuable to schools and colleges."[88]

In pitching his proposal to the Board of Managers, Buhl Foundation director Charles F. Lewis argued that "within the next generation," every American city whose population exceeded some 200,000 people would begin to consider a planetarium "as essential to its civic resources as a free public library was first considered a generation ago."[89] While the cost of procuring such a device was admittedly high, Waldemar Kaempffert assured readers that "any one who has ever spent an hour in a European planetarium cannot but be convinced that Chicago will get full value for Max Adler's money."[90] Los Angeles Board of Park Commissioners president Mabel V. Socha attested that the Griffith Observatory's mission was not to conduct "highly technical or scientific" research but was designed "to appeal to the masses." She expressed hope that Griffith's dream of providing "a better conception and clearer understanding of the universe" would be fulfilled for all who visited it.[91]

Adams bespoke of astronomy's "extraordinary appeal to the mind and imagination of men." No one, in his judgment, could observe the planetarium's reproduction of the heavens "without feeling a strong desire to know more about them." He encouraged greater cooperation between professional astronomers and science educators because "[t]he discovery of the laws of nature and the teaching of these laws to others are parts of the same true type of education, and neither process is complete without the other."[92] Maintenance of this symbiotic relationship was crucial to fostering the growth and development of the American scientific community. Two years earlier, at the Griffith Observatory's groundbreaking ceremony, George E. Hale had expressed anticipation that through the stimuli provided by public planetaria and observatories, "many new recruits will be added to the group of original investigators which has contributed so largely to the intellectual, social, and industrial development of the world."[93]

A major goal of the Griffith Observatory displays was to convey the relationship of scientific progress to American life—a stance reminiscent of Oskar von Miller's exhibit philosophy at the Deutsches Museum. California Institute of Technology president Robert A. Millikan explained that its exhibits were designed to teach citizens how the "progress . . . of our industrial world, our industrial life, too, has been very largely dependent on very specific physical techniques." In his judgment, the discipline of astronomy had acted as the "mother" of all the sciences that had nurtured many offspring, including modern-day physics. One could gain an impression, Millikan argued, "that the children of the mother are not wholly neglected." And in referring to astronomy's dependence upon physical laws, he attested that "[s]ome of them are actually the main supports of the mother today."[94]

Charles F. Lewis's address at the Buhl Planetarium's dedication ceremony reiterated critical facets of an ethos of science popularization. He affirmed that a "major objective" of the Institute of Popular Science was "to interpret to the people the amazing, almost miraculously swift onward advance of scientific progress." In venturing to explain why existing social relations were unable to keep pace with scientific and technological feats, Lewis advocated a bold hypothesis. "For too long new scientific discoveries were the prized and secret possessions of scientists who regarded popularization as vulgarization." One of the Institute's primary goals, he asserted, was to counteract that tendency. Not until a proper understanding had been gained of "what these [scientific] advances are, how they have been achieved, and where they may be expected to lead us," Lewis argued, could the average citizen be expected to make intelligent decisions regarding social actions.[95]

Moral Instruction

With its reproduction of the canopy of night, the planetarium's dramatic venue appealed to the emotional side of the human psyche, enabling a host of subtle messages to be conveyed to receptive audiences. The constancy of the stars provided an unmatched symbol for the renewal of inner strength and confidence. Samuel S. Fels urged that quiet contemplation of the heavens offered a valuable perspective on the materialistic aspects of life, illustrating that "much of importance is to be found outside the narrow confines of the daily routine." To Fels, the stars harbored "special meaning" for those beset with the uncertainties of a depression-ridden world. "The heavens constantly change," he mused, "but there is order in their movements." Fels urged men and women to draw inspiration from the celestial pageantry as they repeatedly addressed the "problems and promises of our changing world."[96]

Heber D. Curtis, director of the University of Michigan observatory, delivered the keynote address at the Fels Planetarium's 1933 dedication. In 1920 Curtis and Harvard astronomer Harlow Shapley had debated the nature of "spiral nebulae" and the Milky Way Galaxy before the National Academy of Sciences.[97] Curtis's keynote address was titled "The Importance of the Planetarium to Science, to Education, and to Recreation." He argued that "a widespread, genuine, and rapidly growing interest in popular astronomical knowledge" had taken hold, leaving professional astronomers with a "great deal of satisfaction." But why, Curtis asked, "are the majority of people apparently more interested in astronomy than in history, or medicine, or biology?" His conclusion pointed toward the spiritual nature of astronomy itself.

To Curtis, the public acquired closer contact with "the great 'unanswerable questions'" that came to everyone's mind, questions such as "why are we . . . here," "how did things start," and "[w]hat is the meaning and purpose of mankind?" Whether the universe and the origin of life were "just an accident, an aimless happening," or resulted from "a higher guiding purpose, some nonmaterial spiritual control," might never be answered, Curtis admitted. Astronomy's fundamental attraction arose from the fact that in "more than any other science, [it] gives us a glimpse of the infinite." By communicating new insights into those "unanswerable questions," planetaria displayed a "far higher value" than as simple pedagogical tools. Audiences, Curtis believed, emerged from such demonstrations as "better citizens of the universe."[98]

Modern astronomy had revealed the enormous extent of the physical universe and confirmed the recognition of human insignificance in the cosmos. With improved understanding of celestial motions, astronomical predictions bore accuracies of almost mythical proportion. This notion was borrowed by Charles F. Lewis to convey another reason why the "popular study of astronomy" had drawn support from the Buhl Foundation. Planetaria demonstrated that "everything in the universe takes place in compliance with eternal and unchanging laws." In contrast to the variable norms enacted by human societies, celestial events allowed "no referendum, no amendment, no repeal. There [was] only certainty." Once that concept had been grasped, Lewis argued, it was "difficult to see how ever again [human beings] can be other than humble."[99]

Speaking at the Hayden Planetarium's dedication ceremony, John H. Finley, New York State's former commissioner of education, remarked that its programs might impart a kind of "geographical planetary consciousness," by which audience members grasped "the common fate of the human race in one spherical boat out upon the boundless ethereal sea."[100] Finley's concept seemingly presaged an environmental ethic not widely recognized until the late 1960s, when the distant Earth was first glimpsed from space by the Apollo astronauts. Stewardship over "spaceship earth" became a hallmark of the fledgling environmental movement and underlay proclamation of the first Earth Day ceremony in 1970. Finley's notion was perhaps one of the more profound social messages to be communicated by American planetaria.

Argument from Design

In the minds of several observers, planetarium lessons conveyed almost sacred experiences that placed them nearly on a level with churches or cathedrals. Horace J. Bridges of the Chicago Ethical Society styled the Adler Planetarium a "shrine of the intellect" and extolled how its "majestic spectacle of the heavens" aptly befitted Francis Bacon's declaration of purposes in acquiring scientific knowledge: ". . . for the glory of the Creator and the relief of man's estate."[101] In reference to the planetarium's reproduction of the heavens, Griffith Observatory acting director Philip Fox declared that a "discerning eye" could glimpse "Creation's plan." Spectators might "wander out through the depths of space and measureless time and touch the divinity that broods there."[102] American Museum president F. Trubee Davison echoed donor Charles Hayden's sentiments by noting that, beyond the visible manifestation of celestial bodies, there "lay a pervading spiritual force which cannot fail to impress every human being."[103] Pittsburgh's Buhl Foundation explicitly characterized this stance through their inscription on the Institute's east wall of the Biblical verse: "The heavens declare the glory of God; and the firmament sheweth his handywork. Day unto day uttereth speech, and night unto night sheweth knowledge." Attendance figures compiled from the Buhl's first thirty-two months of operation disclosed that over one million people visited the Institute, while over four hundred thousand attended its planetarium programs. In a 1942 report, Charles F. Lewis noted with evident satisfaction, "If for those who come into this building there may be shown a bit of this knowledge, and for them there may be caught a bit of this glory, the donors will be content."[104]

American Democracy

Dedication of the Buhl Planetarium occurred shortly after Hitler's armies launched the Second World War. Those circumstances offered Lewis an unmatched opportunity to affirm the purposes of a popular science institute. To his mind, such facilities deserved a ranking alongside other cultural attractions that included libraries, museums, orchestras, and art galleries. All represented "the finest flower of the economic, social, and political systems" upon which American democracy was founded. Lewis defended the privately endowed educational institution as a "bulwark of American liberty." These facilities, he argued, "are dedicated to truth. By their very nature they can serve no selfish purpose. In them is no facility for the wiles of a propagandist." To Lewis, "so long as free institutions endure . . . we may know that America is safe from threat of internal usurpation."[105] In his acceptance speech, Pittsburgh mayor Cornelius D. Scully noted ironically that the "skilled hands and brains" which constructed the Zeiss projector had been redirected to "forging weapons of destruction for a war of conquest and subjugation" whose mission was to spread "the divine right of dictators."[106]

At a November 1941 ceremony to dedicate the Buhl's siderostat telescope, an exhibit in the Octagon Galley asked the rhetorical question, "Can America Be Bombed?" A planetarium program, titled "Bombers by Starlight," explored the navigational uses of astronomy in conducting aerial warfare.[107] Few visitors to that exhibition or program realized how serious such a threat was to the United States until the attack on Pearl Harbor less than three weeks later. The Buhl's first years of existence were surrounded by the rhetoric of war. Along with other Zeiss planetaria, one of its principal tasks became the training of servicemen in methods of celestial navigation.[108]

Summary

During the formative period, five major Zeiss planetaria were erected in Chicago, Philadelphia, Los Angeles, New York, and Pittsburgh. Several of these installations were linked with efforts to replicate Oskar von Miller's Deutsches Museum on American soil. The planetarium's realistic depiction of the night sky gave audience members a dramatic, if vicarious, experience of the heavens in a manner analogous to the large dioramas of museum exhibits. Planetaria allowed visitors to escape from the light-polluted confines of an urbanized landscape and to reexperience the splendors of the stars as they had been known in preindustrial times.

The styles of patronage responsible for creating American planetaria differed markedly from those that supported the production of astronomical knowledge. Philanthropy was directed toward a larger middle ground of recipients by whom the largest share of benefits might be derived. In one case, a foundation's outreach was enlarged from that of a charitable organization to embrace the civic and cultural improvements enriching a cross section of the city's population. Planetaria offered venues wherein the findings of research astronomers were communicated directly to the public. As such, these institutions established new, informal contexts for U.S. astronomy education.

Professional astronomers greeted the planetarium's debut with enthusiasm, praising its ability to popularize the oldest of the sciences. Nonetheless, George E. Hale strove to impart a research agenda into the operation of American planetaria by urging that these facilities be used for more than pedagogical purposes. Reflecting the dual-purpose stance of Orosz's "American Compromise," Hale believed that a planetarium should actively engage in research as well as teaching. In his opinion, unless a planetarium was administered by a professional scientist, the institution's scientific credibility and authority were open to question.

One of Hale's legacies, which achieved its fullest realization at Los Angeles's Griffith Observatory, was the association of moderate-size astronomical, and especially solar, telescopes with major planetaria. He argued that visitors would want to experience direct views of celestial objects and not merely the planetarium's reproduction of the starry skies.

Two smaller institutions, featuring American-built projectors, were established in San Jose, California, and Springfield, Massachusetts. Efforts by two prominent Zeiss planetarium directors to denigrate the Springfield projector reflected the protection of special interests according to "boundary-work" analysis. Asserting that Zeiss projectors were the only authentic planetarium technology, they sought to prevent the spread of other, less-versatile projection instruments. The Springfield projector nonetheless signaled the end of a monopoly on Zeiss planetaria and alerted other institutions to the possibilities of duplicating its results at substantially reduced costs. Still, no other instruments of this kind were ever manufactured.

Donors expressed optimism that the planetarium experience might serve to mitigate human differences and ease strained relations between groups divided by race, class, and gender. By conveying notions of the immensity of the cosmos and the immutability of natural laws, planetaria were thought capable of imparting a humbling yet inspirational experience to visitors. Many of the arguments of natural theology were emphasized, namely that the regularities of celestial bodies' motions reflected an underlying supernatural purpose and design of the universe. These pronouncements offered reassurances of humankind's place in the natural world and encouraged audiences to look beyond the anxieties and realities of economic depression. Such moral and spiritual uplifts were among the intangible benefits to be acquired through attendance at the mechanized demonstrations of American planetaria.

As cultural attractions, planetaria were likened to civic institutions such as public libraries that provided intellectual enrichment for citizens of all ages. Among scientists, they were viewed as potential tools for the recruitment of new workers into the science and engineering fields, and thus assisted the production of new knowledge. As institutions devoted to the popular exposition of science, they were expected to communicate a broader understanding of recent scientific and technical advancements. These goals, in turn, were believed to help an educated public make better-informed judgments regarding future developments. As the 1930s drifted toward another global conflict, planetaria and their associated science museums were looked upon as emblems of American democracy, upholding and strengthening the values of free institutions and the freedom of inquiry that they embodied.

Chapter 3

Personnel, Training, and Careers

*B*y the middle to late 1930s the notion of a community of American planetaria, their directors, and support staff had begun to take hold. One of the leading factors behind this emergence was the accumulation and interchange of employment opportunities associated with each institution. Experienced lecturers and assistants came to fill temporary or permanent slots created within the newer facilities. The similarities in Zeiss projection hardware and public programming philosophies encouraged the transfer of knowledge and skills between institutions. More importantly, directors of the Fels (Philadelphia) and Hayden (New York) planetaria organized the first pair of meetings to which their colleagues were invited. While no long-term success may be credited to this venture, the attempt nonetheless signified an awareness of and respect for facilities beyond one's own and foresaw the value of professional interactions arising from shared goals and objectives. In turn, planetarium directors participated in the wider arena of professional venues within the astronomical community as the nation endured continuing economic hardships and prepared for war.

Planetarium Directorships

The fundamental social unit on which the planetarium community rests is the planetarium director.[1] Between 1930 and 1945 fifteen individuals attained directorships, thirteen men and two women. Only two directors oversaw more than one installation on a permanent basis. Six directors held doctorates, three master's degrees, and three bachelor's degrees, while the remainder were high school graduates (two acquired postsecondary coursework). Directorships extended from less than one year to eight years. In a cumulative total of just over seventy years of employment, the directors held their positions for an average of nearly five years.[2]

Projection planetaria were unprecedented teaching tools whose full potential was only beginning to be understood. As a consequence, the social role and career trajectory of a planetarium director had to be constructed accordingly. Apart from having a strong familiarity with astronomy, requirements for becoming a director remained largely undefined. The first official job description (1928) declared a planetarium director to be "a person learned in astronomy, able properly to operate the planetarium instrument in said Planetarium, and qualified to explain to visitors the celestial display and operation of said instrument."[3]

American planetaria were operated in the hierarchical style of public museums or science centers. Accordingly, the nearest models from which the roles of a director and supporting staff emerged were to be found in the professional museum community. While more than a dozen European planetaria had been established before 1930, they offered limited guidelines on the administration of American planetaria. Nonetheless, before becoming directors, Philip Fox, James Stokley, and G. Clyde Fisher examined European planetaria to acquaint themselves with their operations.

The planetarium director's duties consisted of a mixture of highly specialized tasks and a host of ordinary administrative functions. Directors presented educational yet entertaining lectures to an assortment of audiences ranging from school children to adults. Selection of program topics and the design of audiovisual techniques required practical as well as theoretical knowledge.[4] Directors were responsible for training and supervising assistants, lecturers, and other support staff.[5] They coordinated publicity regarding future programs and astronomical events through appropriate media channels. Administrative skills were devoted to budgetary matters and reports detailing expenditures, attendance records, and ticket sales. Directors fielded questions from the public, press, and colleagues. Planetarium support for telescope-making classes was commonly sought by amateur astronomical societies. Writing articles for scientific and popular journals demanded further attention. Services within professional associations kept directors immersed in broader scientific communities beyond the planetarium specialty.[6]

The choice of a professional astronomer to head an American planetarium facility was hardly surprising, given the specialized nature of the assignment. Some candidates had backgrounds as university professors or observatory directors and possessed skills in theoretical or observational astronomy along with teaching experience. Four planetarium directors were qualified by these criteria, although two were relatively fresh out of graduate school and sought scarce astronomical jobs during the Great Depression. The latter became directors only after serving in an apprentice's capacity for several years. Planetaria thus provided additional (if very limited) employment opportunities for astronomers who were starting their careers. A closely related preparation lay in the physical sciences. The astronomical knowledge imparted to audiences was considered within the training and expertise of a working physicist.

The planetarium career of Philip Fox (1878–1944) exemplified those traits that were most readily sought in a professional astronomer, and served as a role model for American planetarium directors. Fox earned both bachelor's and master's degrees from Kansas State College and a second bachelor's degree in physics from Dartmouth College in 1902. From 1903–1905 he was appointed a Carnegie Assistant at Yerkes Observatory and worked on its Rumford spectroheliograph. After a year of study abroad, Fox returned to Yerkes and taught astrophysics until 1909, when he replaced George W. Hough as director of Northwestern University's Dearborn Observatory, monitoring binary stars with its long-focus instrument.

In 1911 Fox replaced the telescope's tube and mounting with superior equipment that allowed him to undertake a more sophisticated research program on stellar parallax. As recounted by Oliver Justin Lee, Fox's successor at Northwestern, Dearborn was one of seven institutions (coordinated by Allegheny Observatory director Frank Schlesinger) devoted to the photographic determination of stellar parallaxes. University of Wisconsin astronomer Joel Stebbins reported that Fox's activity grew to involve the help of "no fewer than twenty-four assistants and students" who had been trained for the task. A number of these assistants were women astronomers, and as former Lick Observatory director Robert G. Aitken noted, Fox gave "scrupulous credit . . . to the part every one of the considerable number had taken."[7]

Fox might have continued this line of research except that his situation at Northwestern grew increasingly unstable, and when the opportunity arose to secure alternative employment, he resigned. This incident revolved around one of Fox's unmarried graduate students, Maude V. Bennot (1892–1982). She had graduated as valedictorian of her class at age sixteen from Thornton Township High School in Harvey, Illinois. Bennot was originally admitted to Northwestern University in 1912, but receipt of her bachelor's degree was delayed by wartime employment until 1919. Bennot returned to pursue graduate work in astronomy in 1924.[8] By 1927 she had completed requirements for a master's degree through a determination of the proper motions of forty stars. Her results were published in the *Astronomical Journal*.[9]

Fox, however, was unsuccessful in having Bennot supported beyond her terminal year and began a campaign of insubordination when she was denied access to the Dearborn Observatory.[10] Max Adler was to have appointed Fox as the planetarium's full-time director at an annual salary of $7,500, starting 1 January 1930. But Fox convinced Adler to employ him as "your special agent in the capacity of Director of the Planetarium on half-time and half-salary basis," retroactive to 1 January 1929.[11] Fox also secured an understanding with Adler that his staff would include "a research assistant to the director." This move allowed Fox to appoint Bennot as the planetarium's assistant director, offering her a salary more than commensurate with what she would have received at Northwestern University.[12]

Other professional astronomers who became planetarium directors during this era included Dinsmore Alter (1888–1968), chosen by Fox as Griffith

Observatory's first permanent director in 1935. Alter had spent eighteen years as a professor of astronomy at the University of Kansas, following receipt of his doctorate from the University of California at Berkeley.[13] Clarence H. Cleminshaw (1902–1985), a Harvard-trained lawyer who earned a doctorate in astronomy from the University of Michigan, became Alter's assistant and later acting director when Alter took a military leave of absence in 1942.[14] Roy K. Marshall (1907–1972), once described by Harvard astronomer Harlow Shapley as the "best-read young astronomer in the country," became an assistant to F. Wagner Schlesinger (b. 1901) at Philadelphia's Fels Planetarium in 1939.[15] Marshall likewise received a doctorate from the University of Michigan and assumed Schlesinger's post when the latter departed for Chicago's Adler Planetarium in 1945.

The field of museum education proved another fertile ground from which future planetarium directors sprang. The foremost exemplar in this category is G. Clyde Fisher (1878–1949), first director of New York's Hayden Planetarium. Fisher received a doctorate in botany from Johns Hopkins University in 1913 and joined the American Museum of Natural History, where he became chairman of its education department. Museum naturalist (and later director) Roy Chapman Andrews pronounced Fisher's views on child education to be "the cornerstone of modern visual education" that served as a model "for museum education throughout the country."[16] This attribution readily explains Fisher's enthusiasm for the planetarium.

Science journalist James Stokley (1900–1989) was chosen as the Fels Planetarium's first director. From his lack of credentials as a research astronomer, it might seem surprising that Stokley was offered this prestigious assignment. Stokley had received a bachelor's degree in education and a master's in psychology from the University of Pennsylvania. He taught science at Philadelphia's Central High School and wrote articles for local newspapers. An opportunity to cover the Centenary Dinner of the Franklin Institute proved a turning point in his life. On this occasion, Stokley made contacts that led to an invitation from Science Service in Washington, D.C., which he joined in 1925. By the time of his Fels appointment, Stokley had risen to the position of astronomical editor at Science Service.[17] On assignment in 1927, he visited the planetaria at Berlin and Jena, returning with the conviction that he, too, would become a planetarium director. Through personal interest and initiative (rather than formal academic training), Stokley acquired the practical astronomical knowledge and public speaking skills needed for operating a planetarium instrument.[18]

Apprentice-Style Training

Given the absence of any formalized training program in planetarium education or administration, an apprentice-style approach was developed out of necessity during the formative period. The normative practice of promoting employees from within an institution's ranks indicates how successful this practice became. Possession of a terminal degree was almost always waived in favor of on-the-job experience. Within the museum profession, a hierarchy of appointments offered candidates entrée and advancement. A typical career pathway consisted of assistant, associate, and (full) curator positions, analogous to the rankings of university professors. This career ladder was widely adopted by larger planetaria in the selection and promotion of employees. Apprentice-style training reflected the complex process of mentoring, by which promising junior planetarians of both sexes were taken into the confidence of senior directors and taught the skills of planetarium showmanship. Attainment of a directorship was a process usually requiring years to be accomplished, and often consummated only by a supervisor's retirement, lateral move to another institution, departure from the planetarium community, or death.

Perhaps the clearest example of an apprentice-style approach was that established by curator G. Clyde Fisher at New York's Hayden Planetarium. Fisher appointed William H. Barton Jr. (1893–1944), recipient of bachelor's and master's degrees in civil engineering from the University of Pennsylvania and a lecturer at Philadelphia's Fels Planetarium, as associate curator. The remainder of Fisher's full-time staff consisted of three assistant curators: Arthur L. Draper (1905–1973), a graduate of Cornell University's astronomy department; Dorothy A. Bennett (b. 1909), a University of Minnesota graduate and member of the museum's education department; and Marian Lockwood (b. 1899), formerly a student at Wellesley College and secretary of the Amateur Astronomers Association. Each of Fisher's assistant curators received identical salaries, regardless of gender or level of educational attainment.[19] The Hayden staff was the largest of any American planetarium and employed the greatest number of women. Apart from their duties as lecturers, these men and women authored popular articles and textbooks on astronomy and provided editorial assistance on *The Sky* (see chapter 4).

Two years after the Hayden Planetarium opened, significant restructuring affected its staff. These changes were prompted by Fisher's inability to provide sound management practices overseeing the planetarium and its coordination of the burgeoning Amateur Astronomers Association. Under Fisher's guidance, fiscal responsibilities and legal liabilities escalated beyond the parental institution's capacities to offer continued support. In response, American Museum of Natural History vice-director Wayne M. Faunce initiated a reorganization, whereby Barton was appointed curator of the Hayden Planetarium and Fisher elevated to curator-in-chief within the Department of Astronomy.[20] There, Fisher assumed editorial responsibility for *The Sky* and charge of the Museum's meteorite collection. In coming years, all but one of Fisher's original staff received promotions or else attained the directorship of another Zeiss-equipped facility. Draper left to manage Pittsburgh's Buhl Planetarium in 1940, while Lockwood succeeded Barton following the latter's wartime death. Only Bennett, the last of Fisher's assistant curators, who resigned in 1939, did not follow this career pattern.

Mentor-apprentice relationships likewise existed between Philip Fox and

Maude Bennot at Chicago's Adler Planetarium; Dinsmore Alter and Clarence Cleminshaw at Los Angeles's Griffith Observatory; and James Stokley and F. Wagner Schlesinger at Philadelphia's Fels Planetarium. Schlesinger, in turn, acted as a mentor to Roy K. Marshall. All four apprentices were equally fortunate and succeeded their mentors within those respective institutions. Apprentice-style training, however, was not limited to the formative stages of American planetarium development. Through the years, one of the most successful apprentice-style programs has been conducted by the Griffith Observatory; such institutions have been styled "planetarium colleges."[21]

However, powerful gender biases restricted women in the years after 1945. Apprentice-style training was virtually closed to women at major American planetaria. Only a single postwar record is known of a woman who attained a temporary leadership position beyond that of a lecturer's role. Catherine E. Barry, hired in 1945 by Marian Lockwood at the Hayden Planetarium, lasted through the directorships of Gordon A. Atwater (1945–51) and Robert R. Coles (1951–53). After Coles's departure, Barry was briefly appointed co-chair of the Planetarium, as part of a triumvirate with Joseph M. Chamberlain and publicity director Frank Forrester. After two months, however, this arrangement was dissolved and Barry was reassigned to another department in the Museum.[22]

Career Management Strategies

Becoming the director of a major planetarium has been viewed within the discipline itself as earning a mark of distinction. Much of this status was acquired during the formative period, when America's five Zeiss planetaria constituted almost the entire community. As such, the career analysis sketched below is limited to a consideration of the personnel within those five institutions.

After one's formal education was completed, an apprentice-style approach became the most common pathway toward a directorship. Once that position was secured, further career trajectories were restricted to lateral moves within the community, defined as a transfer to the directorship of another major planetarium, whether a new or existing facility, or exiting from the community altogether, for either voluntary or involuntary reasons. No reverse career moves, or transfers to smaller planetaria, occurred in this period.

Only two planetarium directors undertook lateral moves. These transitions, however, carried no guarantee of success; opportunities could be fraught with uncertainty or disaster. When James Stokley arrived at the Buhl Planetarium, he brought with him Roy K. Marshall as his assistant. Despite Stokley's prior experience as a consultant to the Buhl project, an undisclosed source of friction arose over C. V. Starrett's appointment as "science coordinator." That situation escalated rapidly and Stokley abruptly resigned in 1940, and Marshall returned to his earlier post at the Fels Planetarium. Stokley then accepted an associate lectureship at the Hayden Planetarium, along with writing assignments from Science Service. Disillusioned by the Pittsburgh affair and with no managerial prospects in sight, Stokley permanently withdrew from the planetarium community and became chief publicist for General Electric in Schenectady, New York. Following this shakeup, Arthur L. Draper succeeded Stokley and reported to Starrett. This arrangement was so congenial to Draper that he remained as director for thirty years.

Apart from planned retirement or death, directors sought more lucrative appointments outside of the community. Some of these career advancements, however, proved to be ill fated. On 1 May 1937 Philip Fox became the new executive director of Chicago's Museum of Science and Industry, at an annual salary of $15,000. Writing to Walter S. Adams at Mount Wilson, Fox admitted that he was facing "a task of very considerable magnitude," as the Museum had only partially opened in 1933–34. Ever the researcher at heart, Fox enthused, "I may be over optimistic but still hope I shall have time and energy to do some astronomical work."[23]

Although devoting five or six days a week to the museum, Fox evidently followed his instincts, because after only three years, he and several department heads were summarily dismissed after a change of the Museum's governing board.[24] Probable causes for Fox's dismissal can be traced to a lack of professional demeanor as well as the political machinations of his employer. An appraisal of Fox's inability to take administrative work completely seriously comes from Museum of Science and Industry historian Herman Kogan, who notes that "[Fox] seemed less concerned with attracting larger crowds than with converting the Museum into an institution for scholars and educators." He reportedly gave the impression of being "more interested in his specialized field of binary stars than in most of the . . . Museum exhibits."[25] While pursuit of a research agenda had been acceptable to Fox and his employers at the Adler Planetarium, this outlook was not shared by the Museum's governing board, which evidently saw its primary mission as one of exhibition for the general public. In 1941 Fox was recalled to active military duty and made a commanding officer of the U.S. Army Signal Corps school at Harvard University. He died from a cerebral thrombosis on 21 July 1944.

By the time America's fourth Zeiss planetarium opened in New York City, the notion of a planetarium community had received mention in a popular artistic venue. An anonymous commentator in *The Literary Digest* counted "[a]bout twenty persons . . . [who] are now professional planetarium-lecturers," and implied the existence of another dozen nonprofessionals. Neither theatrical nor theological qualities of planetarium demonstrations were overlooked. Lecturers were said to perform an "uncanny task" of explaining the behaviors of celestial objects while "standing in a kind of darkened pulpit." As such messages contained overtones of natural theology, "pulpit" denoted a more literal than metaphorical interpretation, with lecturers akin to priests. The control panel's "bewildering intricacy" was compared to the keyboard of a pipe organ, and the skills needed to manipulate the Zeiss instrument were termed a "new art." In more picturesque language, Hayden Planetarium assistant curator Arthur L. Draper styled the console a "dashboard of our space-and-time ship." Inside this "make-believe universe," a lecturer or "pilot" assumed "complete and perfect mastery" over its movements by remote control. Draper admitted that possession of such artificial powers was "calculated to foster a superiority complex!" The planetarium, in his estimate, deserved the appellation of "the eighth wonder of the modern world."[26]

Planetaria and Gender

An egalitarian attitude toward women's roles in planetaria, embraced by G. Clyde Fisher and Philip Fox, was not widely shared by male colleagues at other Zeiss installations. James Stokley refused to consider two successive endorsements from Fisher to hire a woman as lecturer at the Fels Planetarium. Elizabeth Thurber, founder of the Junior Astronomy Club (a youth-oriented division of the Amateur Astronomers Association of New York), had relocated to Philadelphia. Fisher praised her speaking and organizational skills to Stokley and expressed his confidence in her abilities to "master the Planetarium and present it most acceptably to your audiences." Stokley, however, replied that he would utilize "local men" as lecturers. Fisher persevered, and even risked sounding "officious," reiterating his opinion that Thurber possessed all the "qualities that would insure success as a lecturer"—traits that he had had "abundant opportunity" to witness. Although a meeting with Thurber was arranged, Stokley's attitude remained unchanged.[27] He refused to hire a woman planetarium lecturer, regardless of her qualifications or the recommendations she could muster.

Historian Margaret W. Rossiter has argued that an institutional "logic of containment" long enforced segregated employment and under-recognition for women scholars. Behind this scenario lay "the basic social desire to restrict women to the lower levels of the academic hierarchy." Such actions contradicted the supposed meritocratic practices of hiring and advancement, which in reality applied only to men. Rossiter's analysis of gender biases found within the scientific reward system details how "extrascientific assets" were needed for women to achieve parity with men. "Chief among these additional factors was the enthusiastic backing of powerful and politically astute male colleagues, without whose support even the most meritorious work would go unrewarded."[28] The planetarium careers of Maude Bennot and Marian Lockwood support these assertions and signify an extension of Rossiter's concepts to the popularization of science. But reliance upon male allies constituted a precarious arrangement for women. If that support system collapsed, guarantees of sustained employment were lost as well.

Maude Bennot and the Adler Planetarium

When Philip Fox left the Adler Planetarium in 1937 to direct Chicago's Museum of Science and Industry, his assistant, Maude Bennot, was appointed acting director. She thus became the first woman to head a planetarium facility in the United States (and probably the world).[29] Given the small pool of (male) directors from whom Fox's successor might be chosen, Bennot's lengthy appointment as acting director indicates a temporary condition that became "permanent" by default.[30] That arrangement, however, allowed the Chicago Park District to release Bennot without recourse when such a move was deemed politically expedient. Her responsibilities were *doubled* to include both director's and assistant director's duties, although her salary remained fixed at the latter's $400 per month level. Bennot's "promotion" carried little or no advancement, but was adopted as a cost cutting measure and constituted an obvious form of gender discrimination.

Both before and after her appointment as acting director, Bennot was subjected to the powerful effects of cultural stereotypes, which threatened to deny her the administrative duties. *Chicago Daily News* columnist Sydney J. Harris recounted these facts after conducting an interview with Bennot. "Park District officials," he noted, "were skeptical of this slim, fragile woman. Masculine astronomers shook their heads dolefully, said she was more in place in a tearoom than in an observatory." He added, however, that "[t]oday, they have chewed those words into very tiny bits." Harris described how Bennot had survived "[a] few abortive attempts . . . made to replace her with a man, but these plotters hastily withdrew under her withering fire." "The only limitations to a woman's ability," Bennot retorted, "exist in the minds of men!"[31]

Bennot operated the planetarium much as Fox had done, with monthly rotations of show topics virtually unchanged from their original formats.[32] She avoided controversy by refusing to adopt high-entertainment programs developed elsewhere by Stokley and Fisher (see chapter 4). But her policy of retaining lectures she and Fox devised years earlier harbored negative repercussions. Continued economic depression and the coming of war brought cuts in budget, personnel, and attendance, leaving Bennot as the one-woman staff by 1944. She encouraged the planetarium's use in teaching celestial navigation to U.S. Naval Academy midshipmen. But in spite of thrifty management policies, popularity with the public, and fifteen years of devoted service, Bennot was suddenly removed from her position in 1945 following the death of her mentor Philip Fox. This action was far from unique.

In her sequel volume on the career hurdles faced by women scientists, Margaret Rossiter chronicled the "detrimental impact" brought upon women's lives by the postwar period. "[W]omen's wartime accomplishments, rather than justifying an increased role for women in the postwar world, were quickly forgotten" or deliberately obscured. Many were displaced and demoted without adequate explanation or justification by men who possessed inferior credentials and experiences.[33] Like other women who had risen to high-ranking positions during the war, Bennot was slated for replacement by masculine authorities even before that conflict had ended.

The decision to have Bennot replaced with a man was engineered by Robert J. Dunham, Chicago Park District board president, and undertaken with the full knowledge and approval of Max Adler.[34] Announcement of this change came at a board meeting held 26 December 1944, after F. Wagner Schlesinger had been secretly appointed Bennot's successor. In Dunham's plan, Bennot would receive only three months salary in 1945. Afterward, the assistant director's position would be eliminated, preventing Bennot from reacquiring even her original means of employment. In response, Bennot charged, "this action constitutes a subterfuge, an evasion of the civil service laws." Dunham refuted this claim by arguing that the civil service rating on Bennot's position had been abolished several years earlier by the state legislature. Dunham's argument was challenged by Marvin J. Bas, an attorney for the civil service employee's association, who termed the board's failure to offer Bennot a full year's salary "a willful circumvention of the merit system." Bas, however, was unable to reverse the board's predetermined objective.

When pressed for an explanation, Dunham reported that Adler and the board felt that the planetarium "has not fulfilled all its possibilities; has not attained the position in the scientific world it deserves." In their judgment, astronomy education was an insufficient raison d'être for the planetarium's operation. In this respect, Bennot suffered from Fox's conception of American planetaria as institutions for teaching *and* research. But Dunham added cryptically that the facility "has not fulfilled its function to popularize astronomy." These words, although far from explicit, implied a second criticism, namely that Bennot had failed to originate any new programs during her tenure as director. Both charges, however, ignored the success with which she single-handedly administered the planetarium through prolonged national hardship and emergency. Nor was Dunham's pitch of Bennot's successor, a man called "outstanding as a scientist and administrator," any closer to the truth.[35] Schlesinger's hiring was predicated on the fact that he was a man.[36] One of his first actions as director, however, was to institute a number of new programs, a move that apparently found ready support.

To Maude Bennot, this forced removal must have seemed like a nightmarish flashback to the time, roughly two decades earlier, when she had been denied the privileges of Dearborn Observatory. What was different this time, however, was that her greatest supporter, Philip Fox, could no longer defend her. Embittered by the sudden dismissal, Bennot left the field of astronomy education forever. The Adler Planetarium once again became a male bastion.

Marian Lockwood and the Hayden Planetarium

William H. Barton Jr. succeeded G. Clyde Fisher in 1937 as curator of the Hayden Planetarium. With the outbreak of European hostilities, he devoted extensive efforts to teaching celestial navigation and reportedly trained some 30,000 midshipmen from the U.S. Naval Academy. Barton's desire to give all towards the war effort contributed to the breakdown of his health and premature death from heart failure on 7 July 1944. His place was filled by the last original member of Fisher's staff when Marian Lockwood was appointed acting curator, a move reflecting Bennot's tenuous position at Chicago's Adler Planetarium.

Lockwood had attended but did not graduate from Wellesley College, where she developed her interest in astronomical matters.[37] As one of Fisher's assistant curators, she distinguished herself as a lecturer and writer, publishing two books, *The Earth Among the Stars* (1935) and *The Story of Astronomy* (1939), in collaboration with Draper, and a third, *Astronomy* (1940), with Fisher. Lockwood also served as associate editor of *The Sky* from November 1936 to February 1938. Despite numerous accomplishments, she was not promoted to associate curator until May 1943. Following Dorothy A. Bennett's departure in 1939, Lockwood was the sole female presence on the planetarium staff.

Little evidence remains of Lockwood's activities during her single year as acting curator of the Hayden Planetarium. What is apparent, however, is that she suffered almost the identical fate that deprived Maude Bennot of the Adler Planetarium's directorship. On 1 September 1945 Lt. Commander Gordon A. Atwater, a naval officer who had taught navigation with Barton, replaced Lockwood as the planetarium's new chairman and curator. Before enrolling in the Navy, Atwater was a lumberman and avocational sailor who earned an engineering degree from Purdue University.[38] More concerned with matters of protocol and authority than past loyalty or competence, Atwater eliminated both associate and assistant curatorial positions, reducing Lockwood to nothing more than a lecturer by 1946. Likely demoralized and financially strapped by this demotion, she resigned several months later and obtained a more lucrative position as managing editor with the Grolier Society, a New York publishing house.[39] Like her counterpart Bennot, Lockwood never returned to planetaria or astronomical teaching.

Atwater's removal of Lockwood was neither as malicious nor sudden as the treatment Bennot received from the Chicago Park District board, although it was motivated by the same postwar cultural attitudes chronicled by Margaret Rossiter. Among the men who assumed control over America's Zeiss planetaria, women's roles were divested of the authority and autonomy they had exercised during years of national hardship and emergency. In the period that followed, American women only gained readmission to the planetarium community through the widespread adoption of Armand N. Spitz's pinhole-style projector (see chapter 5).

Teaching *and* Research?

Philip Fox recognized that the Adler Planetarium was to be situated close to the Field Museum of Natural History and the Shedd Aquarium on Chicago's lakefront. Together, these institutions comprised an unprecedented triad devoted to public understanding of the heavens, earth, and oceans. That each of its neighbors was devoted to scientific research as well as popularization aided Fox's own predilections toward the former. Fox envisioned that the planetarium itself should be operated as a research institution and not simply as a pedagogical device.

From the start, Fox made it clear to Max Adler that "part of my service in the planetarium may be devoted to a research program." Adler sought clarification on this point, and Fox explained that he wished to "have the privilege of maintaining a university affiliation where I may continue astronomical research work." Fox assured Adler, however, that this "will not in any way interfere with my duties toward the planetarium." Instead, he hoped that "the major part of the research work . . . can be done in my office at the planetarium and with such accessory instruments as may be installed." Fox declared his intention that "the Adler Planetarium shall be worthy in both respects to stand beside its neighbors."[40]

What Fox had in mind was a return to the solar research he had abandoned years before at Yerkes Observatory.[41] Fox wrote to Hale and Adams with the purposes of securing information about "a coelostat and vertical telescope with a spectrohelioscope" for the planetarium and arranging for a visit by Adler to Mount Wilson. Congratulating Fox on his appointment, Hale waxed enthusiastically about the renewed opportunities for research and exhibition of solar phenomena: "You will be able to do a valuable and much-needed public service in your new position, which has very wide possibilities." He argued that solar telescopes "would appeal to all visitors, help stir into action many amateur astronomers, and give you splendid opportunities for research."[42] Fox got his wish for the solar telescope, although it remains unclear whether any significant research was accomplished with the small instrument. Built by William Gaertner, its optics contained a six-inch objective lens of twenty feet focal length that fed a spectrohelioscope or projected solar images and a high-dispersion spectrum.[43] During his tenure at the planetarium, Fox published volume III of the *Annals of the Dearborn Observatory* (1935) and maintained professional ties with the scientific and astronomical communities, serving as secretary (1925–33) and vice president (1937) of Section D of the American Association for the Advancement of Science and vice president (1938–40) of the American Astronomical Society.

A very different type of research came to be associated with the Adler Planetarium. On 25 July 1929 Fox sailed for Europe to familiarize himself with its principal museums and planetaria. There, he became impressed with "many beautiful antiques—astrolabes, sundials, globes, armillary spheres, and so on." Writing to Hale after his return, Fox lamented, "It seems almost impossible to assemble at this time a representative collection." He then asked "where one might hope to find such things?" Whether from Hale's advice or elsewhere, Fox learned about the sale of an important collection of astronomical instruments by the Amsterdam antiques dealer W. M. Mensing. But Adler's purchase remained "contingent upon [Fox's] inspection and approval," which necessitated a return trip in January 1930.[44] The Mensing Collection proved to be a valuable resource for historical research and formed the nucleus of the Astronomical Museum, to which the planetarium's name would thereafter be associated.[45]

Fox was not the only planetarium director who voluntarily pursued scientific research. At New York's Hayden Planetarium, G. Clyde Fisher was an ardent student of meteorite craters and an early proponent of their extraterrestrial origin. Fisher studied the Coon Butte impact structure near Winslow, Arizona (known today as Meteor Crater), and secured photographs of it from an airplane.[46] On the 1936 expedition to observe a total solar eclipse from Kazakhstan, Fisher examined craters on an island of Estonia and published his conclusions regarding their occurrence. He assisted American meteoriticist Harvey H. Nininger in confirming the identification of a small impact crater near Haviland, Kansas, where samples of meteoric iron were collected.[47]

These studies, however, took Fisher away from the planetarium for weeks at a time and required delegation of authority to other staff members. In 1938 Fisher was assigned curatorial responsibility for the American Museum of Natural History's collection of meteorites, after being relieved of the planetarium's administrative role. Thus, if the proper balance was not observed among a director's professional obligations, the pursuit of a research agenda might prove detrimental to the successful operation of a major planetarium.

Planetaria, Attendance, and Population

On two occasions Philip Fox expressed the sentiment that the number of American planetaria should not exceed a half-dozen or so institutions. Soon after the Adler Planetarium was opened, Fox disclosed to Heber D. Curtis his wish that "we will not over-indulge in planetarium building in the United States." Apart from constructing new facilities in six other eastern cities, he allowed that "one should be added on the west coast, and perhaps one in the south." Fox did not give reasons for his assertion, although he harbored a "distinct feeling that too many planetaria have been erected in Germany, as far as demands of popular education are concerned."[48] Two years later, Fox reiterated these thoughts, while admitting that his own planetarium could no longer remain unmatched. "As other Planetaria are built . . . naturally our institution will lose something of its . . . unique standing. Unless they become too common I judge that the installation of others should enhance the interest."[49] While neither jealous nor mean spirited, Fox's attitude nonetheless reflected a growing protection of the professional interests of those who operated planetaria. During the 1930s, his wish remained fulfilled because of the paucity of donors necessary to match Chicago's lead. Fox's opinion was soon echoed within a larger public hearing and gained credence from reported attendance figures.

The 1936 conference of the American Association of Museums presented James Stokley with a forum for analysis of the economic conditions under which

America's four planetaria were administered. It marked the first occasion in which any comparative accounts of the incomes, expenditures, and attendance figures kept by those facilities were openly discussed. The symposium highlighted Stokley's prominence within the American planetarium community, along with a growing interest among museum professionals towards planetaria as public educational institutions.

Stokley reviewed costs of operating Zeiss planetaria and estimated $25,000 as the average annual expense. Because no planetarium was supported entirely by an endowment, Stokley conceded that "[w]e are all more or less dependent upon admissions." By adopting a standard twenty-five-cent admission fee, planetaria required an annual visitation of 100,000 people to break even. Attendance at all facilities, he noted, reflected an initial surge of popularity, which then declined over succeeding years. Stokley's Philadelphia planetarium attracted the least number of visitors (280,000) during its first year, while New York's topped the list with 850,000.

Stokley next compared those figures with the populations of each metropolitan area, and inferred that between 5.3 percent (Chicago) and 11 percent (Los Angeles) of residents had attended their respective planetaria. He attributed the abnormally high percentage found in California to a significant transient population. Finally, Stokley adopted a representative figure of $6^{2/3}$ percent to argue that no urban center of less than 1,500,000 inhabitants was likely to support a Zeiss planetarium. Voicing Fox's private sentiments, Stokley indicated that perhaps six other major eastern cities might profitably construct planetaria, while a unit serving the Baltimore-Washington area was not ruled out. From Stokley's figures, it is determined that over four million people visited American planetaria through the end of 1936. Roughly one quarter had attended the Adler Planetarium during the Century of Progress Exposition (see chapter 4).[50]

Two very different interpretations could be drawn from Stokley's cautious assessments. The first and most obvious conclusion argued for a tight restriction of planetarium numbers and diversity to the support of Zeiss instruments and facilities alone (recall Stokley's prominent role and the practice of "boundary-work" in chapter 2). According to these self-proclaimed standards, only a handful of comparable planetaria could be established and operated within the most populous urban centers. On the other hand, Stokley's report might have suggested to an imaginative few that a much larger potential market existed for the design of lower-cost planetaria, capable of serving university, school, and museum audiences throughout the nation's less-populous cities. Although glimpsed by Frank Korkosz at Springfield, Massachusetts, that realization belonged almost exclusively to Armand N. Spitz, who faced considerable opposition to this notion (see chapter 5). Spitz was not to be discouraged, however; his competing vision of a pluralistic yet educationally sound community of smaller planetaria completely overshadowed the limited numbers and elitist natures of Zeiss installations defended by Fox and Stokley.

Institutional Cooperation

Numerous interactions indicate that a close professional relationship developed between James Stokley and G. Clyde Fisher. This stemmed from their acquaintance that extended back to 1926. Stokley secured Fisher's nomination as a Fellow of the Royal Astronomical Society, a service he later performed for William H. Barton Jr. Stokley was invited on multiple occasions to lecture before the Amateur Astronomers Association of New York, while Fisher performed complementary favors before Philadelphia's Rittenhouse Astronomical Society. These exchanges extended to planetarium lectures given before public and school audiences at each other's institutions. Fisher supplied attendance figures to Stokley, and their institutions engaged in a friendly rivalry over which would achieve the first one-half million visitors. Stokley and Fisher sought each other's critical appraisal of forthcoming popular textbooks, and published reviews in their own institution's periodicals.[51] Thus, the Philadelphia and New York planetaria and their directors functioned as a dyad, contradicting stated assertions of competition or jealousy between such institutions.[52] Fisher repeatedly acknowledged Stokley's measures of support.[53] Strong personal bonds formed between the pair, whose social interactions included their wives. By contrast, larger personal as well as spatial distances separated directors Fox and Alter from their eastern colleagues.

The most far-reaching step toward institutional cooperation was an attempt to establish a series of regular, if informal, meetings between the Fels and Hayden personnel. That venture, however, proved short lived, as only two such meetings are known to have occurred. An insufficient level of interest and diversity existed to continue the practice; little that was new or different remained to be shared or discovered.[54] No efforts to involve the Adler Planetarium or Griffith Observatory directors are known. Whether such interactions might have been practical or beneficial cannot be answered.

The notion of creating this series of planetarium meetings was presented to Stokley by Frank Schlesinger, director of the Yale University Observatory and founder of the exclusively male association termed "Astronomical Neighbors." (Frank Schlesinger was the father of F. Wagner Schlesinger of the Fels and Adler planetaria.) After noting the period's "revival of interest in the teaching of Astronomy in colleges," Schlesinger argued, "would it not be a good idea for the [planetarium] leaders on the Atlantic coast to get together very much the same as the Neighbors do?" Schlesinger urged that an "interchange of ideas with regard to the presentation and general teaching of the subject" could offer mutual benefits, while such an "informal association" need not lack its "pleasant social side" as well.[55]

Stokley then distributed a form letter announcing the preliminary gathering. Therein, he clarified the "Astronomical Neighbors" as a group of astronomers, "primarily research men," from Middle Atlantic and New England states who met "very informally, four times a year for discussion of matters of interest." Stokley noted that for planetarium gatherings, there would "ordinarily be no formal meeting," but instead "more of a round table discussion." On this first occasion in Philadelphia, however, visiting University of Chicago astronomer Walter Bartkey projected new sound films developed for the teaching of astronomy, an opportunity Stokley expected to provide "great interest" to planetarium workers.[56]

Three months later, a second, but disappointing, meeting was held in New York, to which Schlesinger was invited but failed to attend. Lacking a focus similar to the previous meeting, the gathering's "small size" reflected little incentive for planetarium workers to attend. Barton expressed doubts to Stokley as to whether "the whole scheme is not worth pushing too hard." Initially, Stokley was unwilling to abandon the venture, arguing that "there has been enough expression of interest to warrant going ahead with the scheme."[57] But no further meetings of the planetarium neighbors were apparently held. Attention was soon diverted toward plans for observing the upcoming Peruvian solar eclipse from land and sea. Ironically, the opening of a non-Zeiss planetarium facility in Springfield, Massachusetts, that same year might have provided a vigorous forum of discussion, yet drew only the critical interests of Stokley and Barton. Nonetheless, Stokley's attempt to unite his eastern colleagues foreshadowed the postwar development of "planetarium executives" (see chapter 5), wherein a greater number of institutions and directors sought mutual involvement on an annual basis.

Planetaria and the Astronomical Community

Within the broader astronomical community, directors of major planetaria undertook a variety of tasks that paralleled cooperative strategies forged between their own institutions. While a majority of those duties performed were voluntary and routine, they nonetheless provide a nuanced picture of (a) the growing responsibility and acceptance of planetarium teaching methods; (b) participation in the observation of astronomical events; (c) traditional privileges associated with production and critical review of astronomical knowledge; and (d) the ongoing search for employment opportunities amidst unrelenting fiscal hardships. A necessity of helping astronomical colleagues abroad was triggered by unfolding sociopolitical events, as totalitarian governments drew nearer to the declaration of war. Answers sought were seldom easy, while outcomes reached were rarely fair, as promising or established careers were often placed in jeopardy.

American observatories furnished planetaria with astronomical photographs used in making large black-and-white transparencies for exhibition purposes. Photographs of all manner of celestial objects, along with those of major telescopes and observatories themselves, were requested. Leading suppliers were the Mount Wilson and Yerkes observatories, supplemented by others. Additional requests arose in preparation for the astronomical exhibits installed at Chicago's

Century of Progress Exposition (1933–34) and the New York World's Fair (1939). Photographs were also used in popular textbooks authored by planetarium directors and the monthly periodicals issued by their institutions (see chapter 4). Dramatic visual imagery brought further attention to the work of professional astronomers and aided the revival of public interest in astronomy. In turn, G. Clyde Fisher offered portraits of American astronomers to American Astronomical Society publications, including the *Astrophysical Journal.*[58]

Total solar eclipses drew widespread attention from both professional astronomers and planetarium directors. Some, like Philip Fox, possessed research credentials and assisted in the operation of apparatus designed to study the Sun's corona or flash spectrum. Others with fewer technical qualifications attended eclipses more for entertainment purposes and to obtain a photographic record of the natural spectacle. Fisher accompanied the Harvard-Massachusetts Institute of Technology eclipse expedition to Kazakhstan in 1936. New York's Hayden Planetarium mounted its own expedition in the following year to observe one of the century's longest total eclipses from Peru. William H. Barton Jr. headed the team that observed from a 14,000 foot elevation post in the Andes Mountains. James Stokley, by contrast, recorded that same eclipse from the *S. S. Steelmaker* in the Pacific Ocean.[59] Fisher was a pioneer in the use of aircraft to photograph eclipses, logging two successful flights before 1935. Eclipse expeditions yielded extensive publicity for American planetaria, with results highlighted in public displays, popular articles, and illustrated talks before groups of amateur astronomers.

Technological advancements, visitations from celebrity scientists, and notable commemorations offered sporadic opportunities for observation or participation within the astronomical community. Stokley and Fisher were invited to witness the first pouring of the two-hundred-inch mirror blank for the Palomar telescope, although the latter could not attend the event. In May 1935 Stokley received the honor of presenting the first planetarium demonstration to physicist Albert Einstein. Maude Bennot was appointed by the Midwest Committee of the Polish Institute of Arts and Sciences to serve as commentator at the 1943 observance of the four hundredth anniversary of the publication of *De Revolutionibus.* Yerkes Observatory director Otto Struve was invited to address the assembly.[60]

Apart from attendance and delivery of papers at meetings of the American Astronomical Society, planetarium directors furnished book reviews and other scholarly contributions. A partial list spans the range of biographical, technical, popular, and historical works produced on western astronomy. Fisher reviewed the autobiography of former Yerkes Observatory director Edwin B. Frost in *Natural History.* Dinsmore Alter agreed to review a book on statistical analysis for the *Astrophysical Journal.* Fox reviewed astronomer Forest R. Moulton's popular book *Consider the Heavens,* and after reenlisting for military duty completed a review of Northwestern University physicist Henry Crew's translation of Maurolycus's *Photismi de Lumine.* In preparing his own textbook, Stokley solicited technical knowledge from Otto Struve.[61] Future planetarium directors Roy K. Marshall, Israel M. Levitt, and Clarence H. Cleminshaw spent time on research at Yerkes Observatory or else submitted papers that were published in the *Astrophysical Journal.*

Employment opportunities for young astronomers remained fairly bleak, especially for those launching careers or struggling to complete their education. Planetarium lecturing offered a temporary position for qualified (male) applicants, provided valuable teaching experiences for novice instructors, and delivered a moderate compensation. As the economic depression wore on, however, both observatory and planetarium directors attempted to steer candidates toward financial resources available at each other's institutions.

For women students, the replies were frequently negative. Frances Sherman, a University of Chicago graduate student, was described by Otto Struve as "unusually quick in understanding difficult problems" and someone for whom it "would be a pity to terminate her training in astronomy because of lack of funds." While Sherman's tuition was paid by scholarship, her projected living expenses remained unmet. Struve appealed to Fox and Bennot, but neither was able to draw support on her behalf from the Chicago Astronomical Society. Fox himself was unsuccessful in attempting to find a placement for Winifred Sawtell. She had earned a bachelor's degree in education and was described as a frequent visitor to the Adler Planetarium.[62] Sawtell later married and, as Winifred S. Cameron, became a professional astronomer.

While domestic employment opportunities remained scarce for astronomers of either gender, colleagues abroad faced more difficult and frightening prospects. In 1938 Struve attempted to secure an American post for University of Prague physicist Erwin Finlay Freundlich, who had organized the Einstein Institute at Potsdam. According to Struve, Freundlich's position was "becoming unendurable because of the daily infiltration of Nazism" into the university. Struve appealed to Fox, asking whether Freundlich "might be of use in a planetarium organization or even in a scientific museum." Fox's institution could offer no help; Freundlich later found refuge at St. Andrews University in Scotland. Not long afterwards, Dinsmore Alter solicited aid on behalf of Prague Observatory astronomer George Alter (no relation). Struve could only recommend that Alter's European namesake contact Harvard astronomer Harlow Shapley, whom he intimated "takes a great interest in the fate of refugees from the totalitarian countries of Europe."[63]

Summary

As with any new discipline, the social roles and career trajectories of the planetarium director had to be constructed. Appropriate models were adopted from the American museum profession and the astronomical community. Out of necessity, an apprentice-style training approach was initiated, whose internal promotion of employees reflected the success of mentor-apprentice relationships.

The apprentice-style approach was not limited, however, to the formative period of planetarium development.

Directors were initially chosen from the ranks of professional astronomers, science journalists, and the field of museum education. Lack of a doctorate was not an insurmountable barrier for either men or women to attain directorships of Zeiss planetaria. Once a directorship had been attained, subsequent career opportunities were restricted to lateral moves among the handful of comparable facilities, or else exiting from the community altogether.

The later career failures of G. Clyde Fisher, Philip Fox, and James Stokley reveal that appearances of job security could prove illusory, either within or beyond the planetarium specialty. Inability to adopt sound fiscal management practices, a preference for elevating research over public education or administration, and personality conflicts led to the termination or resignation of these men from their highest professional attainments.

Gender bias strongly influenced the planetarium careers of Maude V. Bennot and Marian Lockwood. Before 1945 these two women attained the rank of acting director, a situation that was abruptly reversed at the war's end. Reflecting national norms experienced across disciplines, gender equity was but a temporary measure erased by the ideology of male superiority in the postwar period. Roughly a generation would pass before comparable career opportunities became available to women in America's major planetaria.

Notions of an emergent community, and the cultivation of professional interests, began to appear after the fourth Zeiss planetarium became operational. Institutional cooperation was limited to exchanges between Philadelphia and New York City planetaria (and their directors). Delivery of school and public lectures, along with addresses to societies of amateur astronomers, helped to cement the relationship between James Stokley and G. Clyde Fisher. However, the far-reaching attempt to establish an informal association of planetarium personnel, after the precedent of the "Astronomical Neighbors," did not succeed. Without a greater diversity of institutions, members, and interests, these gatherings remained too small to justify the practice. Colleagues at Chicago and Los Angeles planetaria remained separated by larger personal as well as geographical distances.

Planetarium directors undertook a variety of tasks that displayed their connections to the broader astronomical community, such as conducting research and hosting professional meetings. Those occasions strengthened support for the instructional capabilities of these devices. American observatories furnished astronomical photographs used for exhibition and publication purposes. Expeditions were mounted to observe and photograph total solar eclipses from land and sea. These efforts generated publicity for planetaria in the form of public displays, popular articles, and illustrated lectures. Traditional privileges associated with the production, review, and commemoration of astronomical knowledge further enriched professional lives.

By the end of the 1930s, America's five Zeiss planetaria offered modest opportunities for temporary or sustained employment. Planetarium lecturing provided an eclectic experience for astronomers just beginning their careers, or a supplementary form of income. Continued economic depression, however, forced both observatory and planetarium directors to steer promising candidates toward limited financial resources available at each other's institutions. Both groups attempted to secure exile for astronomical refugees fleeing the rise of European fascism.

CHAPTER 4

Planetaria and Popular Audiences

*A*t Chicago's Adler Planetarium in the 1930s and 1940s, Philip Fox and Maude Bennot devised a regularly changing schedule of monthly programs. They recognized that "[f]or the audience there are no intellectual prerequisites; there is a mixture of age and interest through wide range." Twelve lecture topics were developed in order "to show the various possibilities of the [star] instrument." The planetarium's original schedule consisted of the following program rotation:

January: Winter Constellations of the Home Sky
February: Time and Place
March: The Calendar
April: The Moon and Its Motions
May: The Way of the Planets
June: The Midnight Sun and The Heavens at the North Pole
July: Summer Constellations of the Home Sky
August: The Southern Sky and The Southern Cross
September: The Seasons and The Annual Journey of the Sun
October: The Great Precessional Cycle
November: Objects of Special Interest in the Sky
December: Architecture of the Heavens

These performances were occasionally supplemented by an "auxiliary projector for showing lantern slides." Celestial objects, viewed as if seen through large telescopes, added "great value" to the lectures, in Fox's judgment. However, "no strictly formulated progression" of topics was established. Audience members who attended the yearlong series, regardless of when it was begun, received a complete introductory course in descriptive astronomy.[1]

This programming style was dubbed the "American practice" by James Stokley. It stood in marked contrast to the approach of many German planetaria of giving "the same demonstration for a long period of time and to change it only after the attendance had dropped." Chicago's regularly changing fare offered visitors the necessary variety so that "they will come again and again."[2] Such a philosophy was emulated by every major American planetarium and explained the popularity of repeat visits by audience members. Fox wished all visitors to see "a stirring spectacle, . . . the heavens portrayed in great dignity and splendor, dynamic, inspiring, in a way that dispels the mystery but retains the majesty." It is noteworthy, however, that Fox and Bennot did not originate the Christmas Star program tradition, a popular topic among planetaria to this day. Their December program instead examined the Milky Way Galaxy.[3]

Fox and his staff found public response to the planetarium "very gratifying." First-year attendance in 1930–31 reached 730,000, and the one-millionth visitor was received during the sixteenth month of operation. Average daily attendance for the first nineteen months was almost 2,000 people.[4] Three days out of each week, and on holidays, admission was free; on other days, twenty-five cents was charged. Schoolchildren were admitted free every morning. "All of our museums—the Art Institute, the Field Museum, and the Shedd Aquarium," Fox noted, had adopted this uniform practice.[5] By November 1935 total attendance reached over three million people, although that figure was somewhat misleading because it included visitors to the 1933–34 Century of Progress Exposition. Despite allowances for this attendance booster, Fox declared that the Chicago facility "has maintained its attendance far better than any in Germany."[6]

Audience Impressions

Audience impressions of planetarium demonstrations are difficult to measure and evaluate. This situation was well understood by Fox, who recounted two years of experience in that endeavor. Fox related that "[t]he effect on visitors can best be judged by comments one hears among them as they are leaving the lecture room. For the most part these are highly enthusiastic." Praise, however, could be tempered by "occasional expression of disappointment." Some guests had come expecting to witness a different topic than was presented; others attempted to "criticise us as lecturers." Fox noted, however, that this sentiment came "mostly from people of little knowledge." Many spectators expressed opinions that the planetarium was the "most worthwhile thing in Chicago."[7]

Prior to May 1933 the Adler Planetarium conducted a survey of its visitors. While the actual number of questionnaires distributed and collected remains unknown, three typewritten pages of forty-six responses were prepared from the survey's final question: "Will you please write below your impressions and those of guests regarding the Planetarium and what caused you to visit it?" These remarks support Fox's contention of chiefly positive responses. Reactions varied from unequivocal enthusiasm ("I went to pass the afternoon in some way. I enjoyed every minute of it. I learned more in a half hour than I ever knew about heavenly bodies before.") to a more guarded appreciation ("I was curious to know what a planetarium was and . . . how . . . [it] could possibly do all that was claimed. The lecture was extremely interesting although too deep for me."). Many respondents praised the planetarium's uniqueness and stressed its significance as a "truly great landmark for Chicago."[8]

Two of the most valuable accounts of a visitor's planetarium impressions came from members of the Buhl Foundation staff. At the urging of director Lewis, each was encouraged to attend a planetarium demonstration while traveling on business. Because of their association with the foundation and its plans, however, neither individual can be regarded as a completely disinterested observer. Possibly reflecting an unconscious bias, their laudatory reports were utilized in the decision to fund Pittsburgh's Buhl Planetarium and Institute of Popular Science.

Mildred M. Lutz, secretary at the Buhl Foundation's office, witnessed two demonstrations of the Fels Planetarium's 1936 Christmas Star program. It was, she declared, her "first visit to a planetarium." She had "not studied astronomy" and "had no preconceived idea of what this demonstration might be." Witnessing the sudden lighting of stars, after acclimation to total darkness, was "truly an awesome moment. I had the feeling of being completely detached from people and things." This particular moment, she believed, "must awaken a greater amount of reverence than it would ever be possible to measure." Although she wondered whether the remainder of the program "might be anti-climactic or even tiresome," her reaction was "definitely not that." Instead, she found that "the spirituality attained was more than enough to carry one through forty-five minutes of changing skies." Concerned that the effect might not be repeated, Lutz attended the same program two days later and found "that I enjoyed my second visit equally as much as I had my first one."[9]

Detailed impressions were also reported by C. V. Starrett, then the Buhl Foundation's associate director. Starrett had seen one planetarium demonstration in Chicago and "was curious to see whether the second experience would impress me as strongly as did the first." He visited the Hayden Planetarium in 1937 and was "agreeably surprised; the magic worked again—and I am inclined to think that it might be repeated any number of times." Starrett remarked that, like good music, the planetarium experience "might even heighten with increasing familiarity."

The similarity of Lutz's and Starrett's reactions eliminates any question of gender difference, or even the deliberately religious nature of the Christmas Star program, upon a visitor's impressions. Starrett claimed to have attended the presentation "simply as John Citizen with a few hours to spend" and while in no particular "investigating mood." He experienced a "sense of the vastness and wonder of the universe . . . an emotional 'lift' . . . akin to that produced by the surging music of a fine symphony," not once, but several times during the lecture. Such a "spiritual uplift," in Starrett's judgment, could benefit anyone, along with providing an "added comprehension" of the "mechanics of our universe." He believed that a youngster attending such programs could not "help knowing more about the world in which he lives" and could not "help being the better for it."[10]

Another, somewhat selective, type of audience response to the planetarium experience comes in the form of a handful of poems composed during this period. Most verses were authored by women, although what significance this has on the conclusions drawn cannot be determined. Cultural attitudes might have encouraged more women than men to try their hand at poetic expression.

Several recurrent themes are found throughout this eclectic compilation. First, wonder and awe are expressed at human ingenuity in creating an exact, miniature replica of the heavens. The visible universe, Grace Maddock Miller enthused, had been "cupped within the circle of a room!" Second, the serenity of sitting beneath the stars enabled one to temporarily escape the distractions of a fast-paced world and be mentally transported into a different realm. Visitors emerged refreshed from this novel experience, having regained their mental equilibrium to help them face familiar tasks anew. Most importantly, a strong reverence was evoked toward the mechanical display of God's handiwork. This sentiment comprised a form of natural theology, stretching back centuries to the "argument from design." From these poetic effusions it is evident that the motivations that had underlain certain planetarium donors' actions (see chapter 2) were being registered among popular audiences.[11] The earliest of these poems, by London dramatic critic Horace Shipp (1891–1961), was reproduced by the Adler Planetarium:

In the Planetarium
The light dims: we take time and space to be
Playthings, and for a toy infinity.
A wheel turns: at our call the once proud stars
Flock like birds hungry—Venus, haughty Mars,
Sword-girt Orion, Mercury, the Plough,
Far-flung Arcturus of the flaming brow,
Neptune, Uranus, Jupiter; and now,
Swarming athwart the darkness like bright bees,
The glory of the choiring Pleiades.
It is like dwelling in God's mind and seeing
His bright thoughts poised in their eternal being;
It is like watching Him that great fourth day
And copying His vast labours for our play.
The lights go up:
We put our shilling universe away.
Horace Shipp[12]

James Stokley and the Fels Planetarium

The Christmas Star Tradition Is Born

Under James Stokley's guidance, monthly programs at the Fels Planetarium first emulated the cycle of topics devised by Fox and Bennot.[13] Signs of innovation, however, soon became apparent. Stokley's background as a science journalist gave him a sharper appreciation of audience needs than research astronomers possessed. Without direct influence from that peer group, he was less inhibited to try new and unconventional topics otherwise greeted with skepticism by scientific colleagues. In modern parlance, Stokley pushed the envelope of this new medium to greater heights and more fully exploited its theatrical and entertainment possibilities. Within a few years Stokley's provocative programs ranked him as the "[m]ost audacious of astronomical showmen."[14]

The planetarium's versatility makes it ideally suited for exploring theories relating to the Star of Bethlehem.[15] A near-perfect analog of the historical sky lies at the operator's disposal, around which a dramatic, yet scientifically defensible, performance is staged. Stokley was among the first to put theory into practice.[16] Movements of the planets that occur naturally over months can be unfolded before audiences in a few minutes. The Christmas Star program demonstrates another of the projector's grandest functions—the precession of the equinoxes.

In examining celestial events that occurred between 8 and 4 B.C., astronomers had adopted a theory of planetary conjunctions as the most likely explanation for the Christmas Star.[17] A triple conjunction of the two outermost planets known in antiquity, Jupiter and Saturn, was visible during 7–6 B.C. First recalculated by Johannes Kepler, this rare grouping acquired further astrological significance by the addition of Mars. Kepler was the first early modern astronomer to propose an association between these conjunctions and the appearance of a nova-like star, although he never claimed that the conjunctions were the Magi's Star.[18]

Stokley's one-hour planetarium program was titled "Skies of the First Christmas."[19] His presentation was divided into two parts. The first was a demonstration of the planetary conjunctions using the Zeiss instrument. Audience members were informed of the historical and scientific problems associated with the dating of Christ's birth. From skies of the present place and time, viewers were transported to the location of Palestine in the year 8 B.C. There, a succession of conjunctions involving Jupiter, Saturn, and Mars was depicted, and events proceeded toward an arbitrarily selected date for the Nativity. Several competing theories, regarding novae, meteors, and comets, were quickly dismissed. The Star's possibly miraculous occurrence was touched upon, with the qualification that such an event could not receive a scientific explanation. In the program's second half, a combination of scriptural readings, recorded music, and lighting effects, including a créche scene, were employed to create "the reverential atmosphere characteristic of the holy season."

Stokley's Christmas Star program was remarkably successful, not only from the standpoint of audiences who witnessed his performances, but for later generations exposed to planetarium demonstrations across the United States and elsewhere. The Christmas Star became the most widely presented astronomical topic in planetaria, large or small. A symbolic measure of this popularity appeared in the December 1964 issue of *Museum News.* Therein, a script prepared by the Alexander F. Morrison Planetarium in San Francisco elaborated the identical planetary conjunction theory explicated by Stokley. It was described by the editor as "one of the most logical and beautifully written explanations for man's age-old question: What was the Christmas star?"[20]

But why has the Christmas Star become such a popular program among American planetaria? What other influences are associated with its quest to decipher and link the Star's appearance to a genuine astronomical symbol or event? Stokley's (and related) efforts that propose a natural explanation for the Christmas Star reflect a pattern of investigation that originated at the dawn of early modern science and that strove to demonstrate purported harmonies between scriptural accounts and natural phenomena. Such Anglo-American approaches, under the label of "natural theology," have tended to reject contrasting arguments reached by nineteenth century "higher criticism" and its attempts to apply rigorous contextual analysis to the composition of sacred texts.

Scholars have characterized Renaissance learning as the eschewing of medieval thought and its concern for otherworldliness, combined with a rejection of the authority of the Church during the Reformation. Along with those factors came an increasing conviction of the intelligibility of natural phenomena. This notion, however, was derived from medieval scholasticism (and earlier Greek rationalism), wherein the orderliness and regularity of the heavens might be understood through a deductive approach.[21]

Another impetus to early modern studies of the natural world was the "argument from design." Glorification of God and the quest for natural knowledge were regarded as fully compatible endeavors. Such an approach toward science was grounded on Protestant beliefs that the twin "books" of revelation, those of nature (revealing God's works) and Scripture (His inspired words), could not conflict with one another. The attested harmony between natural and revealed truths became a persistent staple of Anglo-American natural theology traditions.[22]

By contrast, so-called "higher criticism," which reached its pinnacle in the writings of German historian David F. Strauss, postulated that "one could give a plausible account of the gospels without admitting the historicity of the miracles to which they referred." Drawn from increasingly scientific precepts, Strauss's approach employed philological and historical tools to evaluate the authorships, dates of composition, and contents of texts assembled from oral traditions. Under the tenets of this critical assessment, the Christmas Star's reported appearance is interpreted as a probable textual embellishment, having no physical basis in reality. Convergence of scientific and historical criticism opened a permanent rift between academic and popular receptions of theology, precipitating a crisis of faith among late nineteenth century intellectuals.[23]

Faced with the two alternatives of regarding the Star as a miraculous occurrence beyond the realm of scientific understanding or as nothing more than a scriptural embellishment, a majority of Anglo-American authors (and planetarium narratives) have opted for the middle ground of a naturalistic explanation. Neither alternative presents any serious hope of further study or rational explanation, but merely leads toward final acceptance or rejection of its occurrence. But among those who have chosen to accept the Star as an authentic (if misunderstood) astronomical phenomenon, such an avenue offers the greatest chance for reaching a plausible, and more intellectually satisfying, understanding of its lingering mystery.

By following this approach, however, planetaria themselves have fostered two related outcomes: first, support is apparently given toward the veracity of biblical statements; and second, the attested harmony between science and religion appears to be strengthened. Both results tend to affirm, rather than deny, an individual's faith concerning these matters. So while a literal account of the Star's appearance has not gained acceptance, the less-literal Christian public has accommodated a more moderate reading, with suitable allowances given for inaccurate translations and poetic license. After presenting the leading astronomical theories, most planetaria have tended to leave final interpretation of the Star of Bethlehem to their audiences. In this way, the widest range of possible explanations is accommodated without forcing unwanted dogmas on nonbelievers. That the host of natural explanations has not been exhausted is apparent from two additional astronomical theories presented at the turn of the millennium.[24]

Other Fels Planetarium Programs

Hoping to capitalize on the success of his "Skies of the First Christmas" lecture, Stokley inaugurated "The Easter Story" in March 1937, which explored astronomy's utility for timekeeping purposes. Stokley demonstrated how modern knowledge of lunar and solar calendars could be used to estimate the most probable dates of the Crucifixion and Resurrection. Origins of traditional symbols associated with springtime were likewise explored.[25] Without a recognized astronomical symbol, however, the Easter story never acquired the compelling attraction of the Christmas Star, and fewer planetarium directors emulated Stokley's initiative.

In April 1936 Stokley presented a program called "How Will the World End?" that earned him a reputation as the "greatest showman in planetariana."[26] Next to the Christmas Star, no other subject captured such media attention, was so widely copied by other planetaria, or reportedly drew such criticism from contemporary astronomers. In defending this controversial topic, Stokley argued that "no departure from scientific accuracy [was allowed], because we carefully pointed out that they were all merely theoretical possibilities and that other factors might prevent any one of them from coming to pass."[27]

More than one source of inspiration lay behind Stokley's "End of the World" show. For years, astronomical hypotheses had extrapolated the Earth's fate when the Sun's energy became exhausted. Freezing temperatures or intense heat from a sudden flare-up were regarded as possible consequences of the endpoint of stellar evolution. Astronomers had likewise predicted the eventual breakup of our Moon when tidal friction drew this satellite closer to the Earth. Finally, the remote chance of a collision between an asteroid or comet and the Earth had long prompted foreboding speculations. In Stokley's program, each of these theories was explained and dramatically illustrated.

What might have catalyzed Stokley's actions was the discovery of an earth-approaching asteroid, later named Adonis, in February 1936. Once astronomers determined the orbital properties of this object, they realized that humans had come close to experiencing the last scenario described above. The following year, a second earth-approaching asteroid, to be named Hermes, was detected, reviving apocalyptic discussions. A string of popular articles echoed how closely Earth had come to its brushes with disaster.[28] Through the use of artwork, recordings, and elaborate special effects, Stokley presented these theories "with a realism which [was] expected to surpass any previous Planetarium performance."[29] Based on nominal scientific foundations, Stokley's lectures reflected many of the broader social anxieties that swirled through a depression-ridden world poised on the brink of its second global conflict.

Stokley next adapted a topic originated in 1938 by New York's Hayden Planetarium. This was an imaginary "Trip to the Moon," slated to occur during the planetarium's centennial in the year 2033. Passenger-astronauts would arrive at the Franklin crater on a date that permitted them to witness an eclipse of the Sun as seen from the lunar surface. Stokley prepared visual effects that transformed the planetarium chamber into a space ship. For this, he sought the aid of Dick Calkins, creator of the Buck Rogers comic strip, who it was reported "personally designed the control panel which [was] seen on the navigator's bridge." Stokley admitted that the rocket trip was "pure fantasy," but defended its educational value on "the basis of absolute scientific knowledge."[30]

Stokley's assistant, F. Wagner Schlesinger, succeeded him as director of the Fels Planetarium in 1939.[31] Schlesinger continued Stokley's practice of presenting original, entertaining programs and developing new types of media. "What is Astrology?" explored the houses and signs employed by astrologers in practicing their "pernicious superstition." One month earlier, he had presented "Stars of Spring," using constellation figures created by Walt Disney. Some were reportedly adapted from Disney's *Fantasia,* while all were "drawn so that the figures would actually fit the stars."[32]

By moving well beyond the presentation of standard astronomical facts, Stokley (and his colleagues) developed the most provocative and controversial planetarium programs of the prewar period. He was seemingly the first to exploit the full entertainment capabilities of the new medium and thus became a true pioneer of the community. Stokley's audacious programming style, crafted during the middle to late 1930s, was very much imitated among major American planetaria during the 1950s and beyond (see chapter 5). Through his close association with Fisher and Barton, Stokley's influence rapidly spread to the staff of New York's Hayden Planetarium, with which a rivalry of sorts developed as to whose institution could stage the most dramatic program of that decade's closing years.

G. Clyde Fisher and the Hayden Planetarium

Beginning with "The Stars at Christmas" in December 1935, the Hayden Planetarium's monthly programs under G. Clyde Fisher closely resembled those in Chicago and Philadelphia.[33] After two years, however, the influence of Stokley's dramatic showmanship spread to the Hayden staff. In October 1937 Fisher premiered an adaptation of Stokley's "End of the World" lecture. This program garnered publicity in *Life* magazine; it was repeated the following autumn and numerous times thereafter. Accompanied by elaborate special effects, it portrayed four hypothetical scenarios: lunar disintegration, solar fade-out, solar flare-up, and "degravitation" (triggered by the passage of a massive celestial object near the Earth). Despite the speculative nature of these theories, Fisher likewise defended their formulation by reputable astronomers and application to popular education.[34] Though an unproven conjecture, partial inspiration for Orson Welles's famous radio broadcast on 30 October 1938, in which another fictionalized destruction of Earth's inhabitants was realistically portrayed and unwittingly believed by startled listeners, might have arisen from the Hayden Planetarium's "End of the World" program.[35]

Under William H. Barton Jr. the Hayden staff bested Stokley's imagination with their space travel program "The Mysterious Moon" in 1938. This futuristic voyage received stimuli from popular articles by Dorothy A. Bennett and G. Edward Pendray, secretary of the American Rocket Society.[36] Apart from its entertainment value, the program conveyed accurate scientific information about our satellite's airless, waterless surface and explored contrasting theories of its craters' origins. A fictitious game of baseball was played on the lunar surface under the reduced pull of gravity. Columnist Emile C. Schnurmacher described how, through "scenic, sound and lighting effects," visitors were treated to "a cross-country flight over craters, mountains, rifts and 'seas.'" The visual experience culminated in a landing and emergence onto the surface, "surrounded by tall white mountains glaring in blazing sunlight."[37] American Museum of Natural History vice-director Wayne M. Faunce enthused that planetarium visitors "were almost completely transported in their imaginations and universally thrilled and highly entertained." He added, however, that they nonetheless "acquired a good deal of worthwhile astronomical knowledge in a subtle, unconscious manner."[38] Both Hayden and Fels planetarium programs informed audiences about the possibilities of future space flight at the time when engineers and scientists were taking the first steps toward making that adventure a reality.

To a civilization that increasingly sought a means of escape from the realities of economic hardships, racial and ethnic tensions, and the massing of continental armies, great hopes were pinned on the future, where present difficulties vanished and spectacular, if routine, achievements beckoned. In the depths of the Great Depression, three East Coast planetarium directors propelled their medium beyond the confines of descriptive astronomy to embrace a handful of futuristic topics. Their well-publicized demonstrations thrilled audiences but annoyed or offended more conservative scientific colleagues. Through highly creative uses of audiovisual resources, Stokley, Fisher, and Barton demonstrated that American planetaria could significantly aid popular understanding of astronomy, space travel, and many other scientific wonders.

Popular Astronomical Journals

Another legacy from the formative period of American planetaria was the creation of two popular astronomical journals, both of which are still published today after an interval of nearly seventy years! Apart from supporting the public's understanding of astronomical matters, these journals significantly enhanced the reputations of their respective institutions. Since their creation, however, each publication has followed a significantly different pathway. A comparison of their means of production yields insights into the contrasting approaches taken toward the popularization of astronomy by their editors.

The Griffith Observer

At Los Angeles, Dinsmore Alter's lasting contribution was founding the monthly popular journal *The Griffith Observer,* which has increased substantially in size and circulation, and steadily gained respect. Alter's four-page mimeographed newsletter debuted in February 1937, wherein he explained, "In this bulletin . . . there is included an entirely non-technical statement of the astronomical facts . . . illustrated by the demonstration." Expanding well beyond Alter's narrow pedagogical purpose, the *Observer* has grown proportionately with the demands (and audiences) of its host institution, marking its fiftieth anniversary in 1987.[39] Feature articles and essays routinely fill its pages, making it a respected forum of astronomical communication throughout the world. Its editorial production has always been retained within the parent institution. By staying near to the vision of its founder, subsequent editors have never sought to transform its role into an outlet for mainstream astronomical journalism.

Sky and Telescope

From the outset, the Hayden Planetarium staff envisioned a monthly bulletin serving planetarium visitors and New York City amateur astronomers alike. More instructive than the newsletter of the Amateur Astronomers Association, *The Drama of the Skies* was launched in November 1935. Hans Christian Adamson, head of press relations for the American Museum of Natural History, edited the sixteen-page journal. Its publication presented an outlet for popular articles authored by Fisher's curators and legitimized the authority of those who lacked advanced degrees. A year later, Adamson shortened its title to *The Sky.* Charles A. Federer Jr. was hired as editorial assistant and his wife Helen Spence Federer managed its "Planetarium Page." On being relieved of the planetarium's administrative duties in 1937, Fisher assumed editorial responsibility for *The Sky.* But Fisher again demonstrated his lack of circumspection and sound managerial skills. His actions displayed an unbridled enthusiasm toward the communication of astronomical knowledge but little restraint for regulating the journal's growth within strict fiscal limits.

Fisher's inaugural issue of *The Sky* (March 1938) drew a strongly critical response from Yerkes Observatory director Otto Struve, whose research on the star Epsilon Aurigae was prominently featured. Struve objected to the sensationalized title that Fisher applied without his knowledge.[40] Apart from this corrective volley directed toward Fisher's editorial hand, Struve's comments perhaps reflected the opinions of professional astronomers toward Fisher and Stokley's adoption of high-entertainment planetarium programs. Struve characterized the majority of astronomers as being "essentially conservative in their attitudes toward the public," and "likely to be somewhat shocked by sensational headlines and exaggerated captions." He admitted that one of Fisher's primary tasks as an editor was "in making your publication attract public attention." But Struve advised Fisher that it was "not compatible with the modesty imposed upon a professional scientist by the character of his work to use sensational headlines." Fisher's reply seemed to express genuine remorse over the "liberties that we took with your excellent article" and promised to be more careful in the future. Fisher's apologetic tone evidently appeased Struve, who offered to submit additional copy. Struve remarked that he considered *The Sky* to be the "most progressive and interesting popular journal" then published on astronomy.[41]

Under Fisher's editorial control, *The Sky* doubled in size and the number of articles and use of illustrations expanded steadily, while a more elegant layout was devised. But these changes displayed little regard for cost-effectiveness and were not offset by adjustments to subscription fees. After an internal audit disclosed the reasons for its escalating costs, American Museum of Natural History officials elected to discontinue *The Sky.* In June 1939 assistant curator Dorothy A. Bennett, who had helped with the journal's production, resigned after accepting a position with the University of Minnesota Press. But a private ownership structure, announced later that year, prevented the journal from folding. Thereafter, its production was shifted to the newly founded Sky Publishing Corporation, created by the Federers. For nearly two years, this husband-and-wife team continued to manage *The Sky* from offices in the Hayden Planetarium. Larger opportunities arose in November 1941, when *The Sky* was merged with *The Telescope,* a bimonthly publication of Harvard College Observatory's Bond Astronomical Club.[42] The Hayden Planetarium's journal was removed from its birthplace and transported to Cambridge, Massachusetts, from which *Sky and Telescope* has grown into one of the largest-circulation astronomical magazines in the world, superseding the venerable *Popular Astronomy* after 1951.

Historian Thomas R. Williams has argued that G. Clyde Fisher was more a visionary builder than a tight fiscal manager. Fisher's administration of the Hayden Planetarium and its popular journal *The Sky* both appear to have failed for the same reason, namely his inability to oversee large projects within the scope of existing financial limitations. Without a doubt, Fisher succeeded in generating strong enthusiasm toward amateur astronomy in New York City and in promoting a diversity of educational programs using the Zeiss projector. As a "creator, motivator, entrepreneur, [and] showman," Williams notes, Fisher had few equals in the planetarium community. At the same time, however, he could "not accept the collateral responsibility for managing those activities within the framework of necessary controls and cost effectiveness" that remained vital to the institution's survival amidst troubling economic times.[43]

Under Alter and later others, *The Griffith Observer* retained its smaller format, was produced exclusively in-house, and came to establish its own specialized "niche" in the field of astronomical publications. By contrast, Fisher's vision of *The Sky* encompassed something far grander in scale. He sought to transform the journal into a large-format astronomical periodical that would achieve success in the business of astronomical journalism. To do so, however, required far greater resources (both human and financial) than Fisher's institution could reasonably provide. Continued production of *Sky and Telescope* (and later competitors such as *Astronomy*) has demanded the full-time services of professional journalists, astronomers, and numerous support staff that likely exceed even Fisher's highest aspirations.

Planetaria and World's Fairs

Chicago's Century of Progress Exposition

Celebrations of civic anniversaries have long provided occasions for American corporations to showcase the material and ideological products of national pride, patriotism, and prosperity. Modeled after the famous "Crystal Palace" Exhibition held in London in 1851, expositions of similar magnitude were hosted by the United States in 1876, 1893, and 1904. As Chicago's own centennial drew near, a world's fair honoring that city's transformation into a metropolis was conceived. This was the first Century of Progress Exposition.

A contextual analysis of these celebrations, exposing their deeper meanings and contradictions, has been provided by historian Robert W. Rydell. He notes how all such exhibitions "reflected profound concerns about the future," while deflecting criticism of the "established political and social order." The Century of Progress Expositions were no exception, being carefully orchestrated "exercises in cultural and ideological repair." Rydell argues that "[i]n the midst of America's worst crisis since the Civil War," these fairs "were designed to restore popular faith in the vitality of the nation's economic and political system and . . . the ability of leaders to [guide] the country out of the depression to a new . . . promised land." Spectators were offered glimpses of "an imperial dream world of abundance, consumption, and social hierarchy," a vision predicated upon the maintenance of existing power, race, and gender relations.[44]

Before the Adler Planetarium had opened its doors, Philip Fox was engaged in plans to involve his institution in the gala celebration. Fox convinced the Fair's Committee on Astronomy that exhibits should be centered at the planetarium, while additional displays were to be located in the Hall of Science.[45] Fox played a more pivotal role as master of ceremonies at the Fair's opening night on 27 May 1933. To illustrate the progress achieved by modern science and technology, Yerkes Observatory director Edwin B. Frost conceived the idea of tripping a switch by extraordinary and symbolic means. Light from the star Arcturus was gathered onto photoelectric cells at four remote astronomical observatories. Electrical impulses from these cells were then transmitted over telegraph lines to the Exposition. Selection of Arcturus was made because of its distance of forty light years. Starlight reaching telescopes in 1933 had begun their journey at the time of the World's Columbian Exposition, likewise hosted at Chicago in 1893.[46]

Fox offered a special planetarium demonstration titled "Drama of the Heavens" throughout the Exposition's two seasons. All planetarium visitors were charged a twenty-five-cent admission fee. Fox noted that receipts from "the two Fair periods were approximately three-quarters of the total to date" (a figure near $433,000).[47] From this estimate, Exposition attendance at the Adler Planetarium reached almost 1.3 million visitors. As a payback for the planetarium's assistance, Fox negotiated an agreement that approximately 3 percent of the Exposition's surplus proceeds would be donated to the Yerkes Observatory, a sum that amounted to roughly $4,800.[48]

New York World's Fair

The 1939 New York World's Fair embraced "The World of Tomorrow."[49] Personnel from the American Museum of Natural History and Hayden Planetarium made significant contributions toward the construction of two major exhibits. Chemist and science popularizer Gerald Wendt, chairman of the Fair's advisory committee on science, sought a means of opening the fair with "equal or superior dramatic value" to the method of starlight employed at Chicago's 1933 Exposition. American Museum of Natural History vice-director Wayne M. Faunce recommended a scheme involving "cosmic rays and ultra short wave radio" transmissions that were employed at the planetarium's own dedication ceremony. On opening night, 30 April 1939, a switch was thrown by Albert Einstein, whereupon the first ten cosmic rays lit successive portions of the fairgrounds.[50]

More educational in scope was a fifteen-minute audiovisual program designed for the "Theater of Time and Space," located within the Fair's amusement area. Brainchild of Alan R. Ferguson, secretary of Amusalon, Inc., this spectacle drew inspiration from the recent space travel programs hosted by the Hayden Planetarium. As envisioned by Ferguson, audiences would be taken on an imaginary journey into space, wherein a close inspection of the Moon, planets, and stars awaited them. Their "spaceship" would transport them to distant galaxies beyond the Milky Way, after which passengers were returned to the fairgrounds. Plans called for installation of a Zeiss planetarium projector in the Time and Space pavilion, but this device could not be secured and four stereopticons provided the substitute star fields. Construction costs and production expenses were underwritten by the Longines-Wittnauer Watch Company. G. Clyde Fisher acted as scientific consultant to the project.[51]

The Theater of Time and Space paved the way for creation of permanent theme parks such as Disneyland, which opened in 1955.[52] In the World's Fair performance, simulated spaceflights were made "scientifically accurate in every way," while "keyed to the lay mind" so that phenomena were "readily understood by those having little or no familiarity with astronomy." The pavilion itself, through which large numbers of people were shuttled, was so designed that its "contours melt[ed] into floor and side walls" to form "a limitless screen on which to project the celestial drama."[53] All of these characteristics have since become stock-in-trade features in audiovisual arcades. Despite its proximity to the motion picture capital of the world, it was not the Griffith Observatory of Los Angeles, but instead two East Coast planetaria and their prewar directors who pushed this educational medium toward its maximal use of entertainment and indirectly presaged future theme parks that have become so widely attended.

Revival of Popular Interest in Astronomy

Apart from Heber D. Curtis's address at the Fels Planetarium dedication and Frank Schlesinger's private communication to Stokley, two other contemporary accounts describe a revitalization of popular interest in astronomy during the 1930s. Both statements assign prominent roles to the development of planetaria and their widespread impact on public knowledge of the heavens. The first originated from Berwyn, Pennsylvania, amateur astronomer H. B. Rumrill, who operated the private Tredyffrin Observatory and was a former president of the Rittenhouse Astronomical Society. Rumrill declared that one of the "leading factors" behind renewed interest in astronomy was the "wonderful imitation of the sky as seen in a planetarium." Rumrill also cited an abundance of books, "written in non-technical style," along with a "prominence given in the newspapers and magazines" to astronomical discoveries, as important reasons behind this trend. Announcements regarding construction of the two-hundred-inch Palomar reflector, he noted, "aroused the most unusual interest."[54]

A more professional assessment of astronomy's relationship to the public came from the 1937 directive of the International Council of Scientific Unions and its Committee on Science and its Social Relations (CSSR). The CSSR's avowed purposes were "to consider the progress, interconnections, and new directions of advance" taken by the natural sciences "in order to survey . . . and to promote thought upon the development of the scientific world picture, and upon the social significance of the applications of science." Harvard astronomer

Bart J. Bok became principal author of the CSSR's study on astronomy. One of its primary goals was to evaluate "attention given by the general public to serious expositions of modern astronomical work."[55]

Section 3 of Bok's report focused on "Planetaria and Museums." In Bok's judgment, the Zeiss planetarium represented "one of the most important stimulants for the popularization of astronomy throughout the world." He noted how planetaria often formed "the nuclei of institutions with excellent museum exhibits," around which "active clubs of amateur astronomers and societies of interested laymen" had arisen. Adopting a more optimistic viewpoint than Stokley's 1936 analysis, Bok found no reason for excluding planetaria from cities whose populations reached five hundred thousand people. He characterized the greater popularity of American over European planetaria, from which "reports of steadily decreasing attendance" were gathered. Domestic institutions, Bok recounted, had adopted "a scheme by which the topic under discussion is changed monthly, thereby encouraging listeners to become 'repeaters' and eventually regular visitors." Finally, Bok sounded a conciliatory note toward the more high-entertainment programs staged by Stokley and his colleagues. He urged professional astronomers not to "frown habitually on the planetarium director who attracts crowds through spectacular demonstrations." The public itself, he argued, would not remain loyal to such demonstrations, unless "the fundamentals of scientific honesty [were] strictly observed."[56]

Summary

American planetarium directors devised a regular schedule of monthly programs that, if viewed in their entirety, constituted an introductory course in descriptive astronomy. This approach, dubbed the "American practice," attracted more repeat visitors than European planetaria. More than a million visitors were introduced to the Zeiss planetarium's artificial heavens during Chicago's Century of Progress Exposition.

Audience impressions of planetarium lessons attested to the unfamiliar beauty and majesty of the display. The exhibition of celestial phenomena and their law-like regularities occasionally drew forth spiritual responses and provided a backdrop against which the routines of daily living could be reexamined. Firsthand accounts, visitor surveys, and samples of poetry provide collective evidence that the ideals that guided planetarium donors, including intellectual enrichment, moral instruction, and the affirmation of supernatural purpose and design, were reflected among popular audiences.

Los Angeles's Griffith Observatory and New York's Hayden Planetarium became the seats from which two monthly popular astronomical journals were launched. Under the guidance of Dinsmore Alter (and his successors), *The Griffith Observer* has remained a successful, small-format niche periodical whose production is retained by its namesake. G. Clyde Fisher, by contrast, sought to transform *The Sky* into a large-format compendium of astronomical journalism.

Demanding far more resources than his institution could sustain, Fisher's vision only triumphed after the journal's operation was privatized and its production was removed from its birthplace.

Independent sources testify that planetaria became "one of the most important stimulants" behind the revitalization of public interest in astronomy. Aided by the appearance of popular books, magazines, and newspapers, American planetaria became foci within their communities for the establishment of groups of amateur astronomers, telescope makers, and enthusiastic sky watchers. In countless ways, planetaria conveyed the latest astronomical discoveries to audiences of every age and background, forging important linkages between the era's professional scientists and laypeople.

An enduring program tradition, pioneered by Fels Planetarium director James Stokley, explored leading astronomical theories of the Star of Bethlehem. By advocating a naturalistic explanation for the Star's appearance, planetaria have attempted to harmonize scientific and scriptural viewpoints in the Protestant natural theology tradition. Stokley raised the entertainment qualities of planetarium demonstrations to new heights by extending their subject matters and methods well beyond the strict lessons of descriptive astronomy. Stokley's dramatic showmanship, which spread to the Hayden Planetarium staff, portrayed scenarios depicting the "End of the World" along with simulated rocket flights to the Moon. Professional astronomers, however, objected to these sensationalized programs on grounds that scientific authenticity had been sacrificed at the expense of staging mass-entertainment spectacles.

These highly creative audiovisual productions demonstrated that planetaria could significantly aid popular understanding of future space travel at a time when scientists and engineers were taking preliminary steps to turn that fictionalized adventure into a reality. Planetaria left their marks on Depression-era audiences, which increasingly sought an escape from the realities of economic hardships. Related programs at the 1939 New York World's Fair helped to inspire permanent theme parks such as Disneyland.

Part III

The Postwar Period, 1947–1957

Chapter 5

Armand N. Spitz and Pinhole-Style Planetaria

The small planetarium . . . was conceived as an inexpensive educational device so that schools throughout the country could teach astronomy more effectively.
—Armand N. Spitz, 1964

The possibility of creating a much less expensive planetarium instrument by simplifying star projection techniques to the use of proportionally sized pinholes was chiefly realized by Philadelphia entrepreneur Armand N. Spitz (1904–1971). Spitz envisioned the establishment of smaller planetaria throughout the nation's public schools, museums, and colleges. He advocated cross-disciplinary instruction and subject-matter integration for planetarium audiences of all ages. Through a combination of technical innovations, effective marketing strategies, and altered social circumstances, Spitz distributed his projectors on an unprecedented scale. By the close of this second period of American planetarium development, more than a hundred permanent installations were in operation. In addition, Spitz labored to unite a new generation of planetarium directors who fashioned postwar careers around the pinhole-style projector.

Spitz's accomplishment enacted the single greatest transformation of the American planetarium community. In contrast to the prewar notions of Fox and Stokley, who urged that the number and kind of planetaria be limited to a handful of Zeiss installations found only in metropolitan centers, Spitz pursued a sharply different vision, that of a much larger, pluralistic community of smaller institutions that could present the basics of astronomy to children and adults alike. Not surprisingly, Spitz encountered opposition to the legitimacy of his projector's simplified design from Zeiss planetarium directors, actions clearly reminiscent of the "boundary-work" that accompanied introduction of Frank D. Korkosz's homebuilt projector at Springfield, Massachusetts (see chapter 2).

Spitz, however, was not the first person to conceive or manufacture a pinhole-style projector, although his system was the only one that spawned a notable impact on the American educational system. Neither of the two preceding cases (described below) appears to have influenced Spitz's own unique design. Before reconstructing his achievement, it is necessary to review those alternative solutions to the task of projecting star and constellation imagery without the sophistication and expense of a Zeiss planetarium instrument.

Alternatives to Zeiss Planetaria

Speaking at the 1933 dedication ceremony of the Franklin Institute's Fels Planetarium, astronomer Heber D. Curtis voiced the desire for a "small-size planetarium next to my classroom," where he could "instantly, and better than through hours of verbal explanation," depict "the mechanism of our motion through space, the configuration of the stars, [and] the changing heavens through the centuries and millennia."[1] Curtis's statement is the earliest known expression of an untapped potential awaiting the development of smaller planetaria. But whether Curtis's suggestion provided inspiration for Spitz to fashion his pinhole-style projector cannot be answered.

Unbeknownst to American astronomers or Zeiss planetarium directors, a portable projector that fully realized the aspirations set forth by Curtis was soon manufactured by the European engineering firm of E. Unglaube at Glogau (Glogow) in present-day Poland. The only surviving example of this device was installed at the senior high school at Falun, Sweden, in 1935 but not described until 1986.[2] Surprisingly sophisticated, this tabletop instrument featured a single starball, projected equatorial and ecliptic coordinates, and included motorized Sun, Moon, and planet projectors, plus interchangeable diurnal and annual motions. The number of Unglaube planetaria that were produced remains unknown. Had word of the projector's existence reached America at the time, the device might have become more widely used.

A second commercially marketed planetarium projector was originally designed as an astronomical teaching aid by W. Park, an instructor at Brantford, Ontario, during the 1943–44 school year. Later this device and auxiliary apparatus, including an orrery designed by Park, were moved to the Forest Hill Village School in Toronto.[3] In 1946 Park and several associates founded the Peerless Planetarium Company Ltd. in response to requests for duplicates of the projector installed at Montreal's Protestant High School. Several Peerless planetaria were employed at parochial and public schools, including Philadelphia's Central High School, whose Edmonds Planetarium was dedicated 11 December 1946.[4]

In January 1947, some nine months before Spitz's Model A projector was released, Science Associates, a Philadelphia supplier of "visual education aids," announced itself as "American representative" of the Peerless Planetarium Company. Science Associates was cofounded by Spitz and historian-turned-meteorologist David M. Ludlum (b. 1910). The company offered educational equipment for the study of astronomy, navigation, geography, meteorology, and climatology. Sales were pitched to instructors of the "Youth of our Air Age," and the claim was printed that now "Every School and College can afford a Planetarium." Historian Henry C. King has noted, however, that the $4,000 price of the Peerless projector "was not calculated to attract customers," nor did it offer a particularly realistic starfield.[5] Spitz was instrumental in securing a Peerless planetarium for Puerto Rico's Polytechnic Institute following his visit to the island as an educational consultant in 1945.[6]

Spitz and the Franklin Institute

Historian Brent P. Abbatantuono has furnished a detailed study of Spitz's life that answers many questions concerning his interest in astronomy and his development of the Model A projector.[7] Graduating from West Philadelphia High School in 1922, Spitz first matriculated at the University of Pennsylvania. Two years later he transferred to the University of Cincinnati, but left to pursue a career in journalism without receiving his degree. In 1928 Spitz fulfilled a personal ambition by becoming editor and later publisher of the suburban Philadelphia *Haverford Township News.* Economic depression forced the venture into bankruptcy in 1932 and changed the course of Spitz's life.

Attempting to find work as a foreign correspondent, Spitz traveled to France and was befriended by an officer on the ship who taught him the rudiments of celestial navigation and the construction of simplified astronomical instruments. His aspirations as a writer went unfulfilled and he returned to the United States, where he accepted a diminished role at the revitalized *Township News.* Spitz's former colleague, Vera Golden, was editor.[8] Their professional interests merged with personal ones and the couple was married.

Spitz was raised in the Quaker faith, a factor that might have influenced his attitude toward astronomy education.[9] Historian Frederick B. Tolles has studied the relationships between Quakers, capitalism, and early modern science that supports this claim. Tolles examined how this variant of Puritanism "looked upon the material world . . . as God's world in which men were called to do His will," and whose adherents' "hearts should not be set upon its evanescent goods but upon eternal treasures." Virtues of "industry and frugality," together with strict adherence to "[p]rudence, honesty, and a strong sense of order," were key ideological positions. In addition, historian Brooke Hindle notes that Quakers found a "positive reason for encouraging science in the prevailing expectation that it would ultimately improve the physical condition of man's life."[10] While generalizations of this kind cannot be applied uncritically to Spitz some three centuries later, these assertions nonetheless accord well with Spitz's enthusiasm toward astronomy and science education.

As the birth of his first child drew near, Spitz sought additional means of employment. He approached Richard W. Lloyd, executive vice president of the Franklin Institute, offering to prepare newspaper publicity for its events. Spitz's persuasiveness convinced Lloyd, who hired him on a part-time basis. This opportunity was expanded when Spitz took over editorial responsibilities for *The Institute News,* whose monthly publication began in May 1936. He inaugurated its biographical column "Who's Who in the Franklin Institute" and remained editor through December 1943. Spitz continued to parlay his resources and acquired a host of related duties, eventually becoming head of museum education. He organized the Institute's department of meteorology and taught courses in that subject, which proved significant for the development of his prototype planetarium.[11]

Spitz's principal acquaintance with astronomy began through lectures he attended at the Fels Planetarium as part of his "indoctrination" to the Institute's activities. Spitz recounted that because of an "unfortunate situation," he had been allowed to graduate from "one of the newest and most progressive high schools" in Philadelphia, "without having been introduced to the stars." This statement offers further evidence of the low status accorded to astronomy in so-called "progressive" education curricula. Spitz was impressed by the planetarium demonstrations and their impact on audiences of all ages. These performances fueled Spitz's desire to become a lecturer himself, but under James Stokley he was denied this opportunity. As Spitz expressed it, "I had asked for the privilege of lecturing in the planetarium . . . [but] was told that I did not know enough astronomy. I could not deny this because I had never studied any."[12] Spitz was perhaps excluded as much for his lack of a college degree as for the paucity of his astronomical training. He was not permitted to lecture until 1942, after Stokley had left the Fels Planetarium and Lloyd became the Institute's secretary-director.

Undaunted, Spitz set about to correct this shortcoming by researching and writing a popular book on astronomy! Reflecting his journalistic background, he argued that "the way to learn something for yourself is to write a book about it because you have to study [a subject] thoroughly enough to be able to interpret and explain [it]."[13] Spitz's literary effort enabled him to gain credibility as an author, much as G. Clyde Fisher's assistant curators had done at New York's Hayden Planetarium. He might have received added inspiration from Stokley's own labors in producing a popular-level textbook, *Stars and Telescopes* (1936).

Instead of reporting on current astronomical research in his book, titled *The Pinpoint Planetarium* (1940), Spitz chose classical descriptive astronomy as his subject. Several of Spitz's ideas might have arisen from an article, "Make Your Own Planetarium," that taught "junior astronomers" how to fashion a "shoebox" constellation viewer.[14] In this device, proportionally sized holes could be punched through flat sheets of paper to simulate the relative brightnesses of stars. Spitz's contribution was to provide appropriate instructions and star charts that, after being cut out and assembled, resembled the domed appearance of the night sky. Spitz's thoughts did not yet extend toward projection of these stellar images; his charts were to be illuminated from behind and viewed directly.

The rationale behind Spitz's book was an increased "popular interest in astronomy," although no explanation for its cause was given. Spitz praised the educational value of planetaria in offering the most "realistic" and "versatile" depiction of the heavens, but lamented that such experiences could not be made portable. Spitz intentionally reduced the number of stars on his charts so as to attract "anyone of any age" to the subject. He also limited their number to better match viewing conditions imposed by moonlit skies or the glare of city lights. Spitz's charts were styled an "appetizer" that could lead toward the "full feast" awaiting more advanced study of astronomy.[15]

Spitz's text was reviewed by Marian Lockwood, who praised his intention of helping beginners to overcome the "incomprehensibility" of ordinary star charts. While Lockwood regarded its descriptive materials as "well planned," she found the charts themselves "inadequate" for serious study of the heavens because so few stars had been plotted. Lockwood judged Spitz's effort as "original and novel," despite an execution which fell "somewhat short of its potential high destiny."[16] *The Pinpoint Planetarium* nonetheless served as an important conceptual link to Spitz's later pinhole-style projector.

Spitz and the Model A Projector

Following the birth of his daughter Verne, Spitz wished to project the stars upon her bedroom ceiling as a means of providing entertainment and instruction. At this stage, Spitz had no intention of marketing such a device.[17] His growing passion for astronomy instead was channeled into the construction of a Springfield-mounted reflecting telescope and three-foot-diameter replica of the Moon.[18] Spitz also began an eight-year association with the astronomy department of Haverford College and secured permission to use its telescope to observe double stars. More importantly, Stokley's negative attitude toward any non-Zeiss projection apparatus likely forestalled his plans. When Spitz first "announced his intention of trying to build [a projector] which could be sold for a modest amount," he was reportedly told that "there was no sense in trying to imitate a Zeiss [planetarium]. It just couldn't be done." Spitz responded that he only wished "to try and create a teaching aid that would help people to know the stars."[19] Not until Spitz was permitted to lecture at the Fels Planetarium did he again consider the design of an inexpensive projector.

Spitz's desire to construct such a projector might have been rekindled by a twenty-five-inch spherical device that projected 145 navigational stars onto a domed screen. It was manufactured by the Bausch and Lomb Optical Company for the U.S. Naval Air Station at Pensacola, Florida. A photograph showing servicemen training with the device graced the cover of the November 1942 *Sky and Telescope*.[20] This projector might also have inspired construction of the Peerless planetarium, whose twin hemispheres more nearly approximated the Bausch and Lomb instrument.

Spitz was reluctant to divulge information about his scheme to anyone at the Franklin Institute out of fear of ridicule for what he was trying to accomplish. As a result, he was forced to rely on the support of a wealthy Philadelphia family for the earliest phases of the Model A's development. Several members of the Wolff family had attended Spitz's wartime courses on meteorology. At dinner one evening the matriarch, identified only as Billie, mentioned that someone ought to make the subjects of astronomy and navigation more widely available to the public. Spitz expressed his desire to assemble an inexpensive planetarium, but a lack of resources had prevented its realization. Wolff then handed Spitz $200 out of her pocket and assured him that this gesture need not be repaid as a loan. Spitz promised her that he would make a start at bringing the device to reality.[21]

After rejecting both spherical and cylindrical surfaces as too difficult to shape with hand tools, Spitz began to explore the regular polyhedra, or Platonic solids, on which to plot his star field. He first surmised that a polyhedron bearing the largest number of identical sides (a regular icosahedron, consisting of twenty equilateral triangles) would most nearly approximate a sphere. During a year's spare time, Spitz labored through calculations needed to plot roughly one thousand stars onto its planar surfaces. Only after this task was accomplished did he conclude, albeit prematurely, that the figure's sharp, projecting corners would prove detrimental to his objective.[22] Consultation with various mathematical authorities, including Albert Einstein, then in Princeton, New Jersey, convinced Spitz that a twelve-sided dodecahedron would better serve his purposes.[23] Spitz unveiled the Model A projector at his Lansdowne, Pennsylvania, home in 1945.

Spitz's planetarium projected roughly one thousand stars down to the fourth magnitude. The unit stood three feet high and weighed twenty-five pounds. It was equipped with an electric motor that provided diurnal rotation in four minutes. The device could also be turned by hand. Its metal base was adjustable for a range of northern latitudes. The dodecahedron's twelve plates were fashioned from Vinylite plastic sheets. Below the instrument's polar axis, a series of stacked rings served as rudimentary Sun and planet projectors. Holes in the rims of individual rings projected solar and planetary images. By means of tabs and graduated scales, an operator could set the positions of these objects before each demonstration. A small arrow pointer was provided to aid the lecturer's presentation. Initial cost of the Model A projector was only $500. While unable to match the sophistication of the Unglaube projector, let alone the Zeiss Model II instrument, Spitz's device offered a realistic display of the stars and planets visible to the unaided eye. His long-standing goal of fashioning an inexpensive projector had at last been fulfilled.

To assemble his prototype, Spitz received added financial support from the Wolffs and tapped the mechanical expertise of two of his meteorology students. A loan of $600 enabled Spitz to purchase a lathe and drill press that were set up in the basement of his home. These were operated by the father-and-son machinist team of William Flood. In true Rube Goldberg style, Spitz's Model A projector emerged from this private workshop. Equipment was then moved to the cellar and garage attic of the Floods, where until 1948 the first production models of Spitz's device were fabricated.

Along with David M. Ludlum, who specialized in the sale of meteorological equipment, Spitz first marketed his planetaria through Science Associates. Ludlum was also cofounder with Spitz of the Amateur Weathermen of America (1946). Its journal, *Weatherwise,* launched in 1948, was edited by Charles A. Federer Jr. Spitz's Model A projector was not announced until October 1947, coinciding with its debut at Harvard College Observatory. Later that month Spitz hand-delivered the first production model to Eastern Mennonite College at Harrisonburg, Virginia. Planetarium director Maurice T. Brackbill reported to Federer, also editor of *Sky and Telescope,* that "the illusion of the night sky was striking, and the 'ah's' from the audience of faculty members and students attested . . . close approach to the excellence of the large planetariums."[24]

Available evidence suggests that Spitz entered the planetarium market only after he examined the competition posed by the Peerless projector, whose price he significantly undercut. When demands for Spitz's product rose, larger quarters were sought, especially when he was asked to supply twenty-foot-diameter domes to accompany projectors sold to the U.S. Naval Academy. Spitz Laboratories was incorporated in 1949, although its products were marketed by Science Associates through 1951. In that year Spitz resigned his duties at the Franklin Institute to devote full attention to the sale and utilization of planetaria. By the end of 1953 Spitz had sold his one-hundredth projector worldwide. His company experienced several relocations within the Philadelphia area before establishing its latest facility at Chadds Ford, Pennsylvania, in 1969.[25]

A final step that brought planetaria to the average consumer's reach was the advent of the Spitz Jr. projector. Manufactured by Harmonic Reed Corporation of Rosemont, Pennsylvania, its concept originated from a meeting between Spitz and Harmonic Reed's president, Thomas Liversidge. This toy planetarium, which retailed for $13.95, projected nearly 400 stars from its spherical design and was unveiled to prospective buyers on 8 March 1954.[26] It also brought closure to the effort begun almost twenty years before when Spitz first desired to project the stars upon his daughter's bedroom ceiling.

Brent P. Abbatantuono has reiterated "two key opinions" that guided Spitz's success in developing and marketing the pinhole-style projector. Quoting Spitz's associate Nigel O'Connor Wolff (b. 1916), he argued that Spitz "believed the planetarium to be 'the greatest teaching instrument ever invented,' and he felt it a shame a planetarium could be enjoyed only where some philanthropist donated a huge sum to purchase and house a Zeiss instrument."[27] What began as an almost whimsical desire on Spitz's part had now been transformed, through his dedicated efforts, into a valuable teaching aid that offered a low cost, realistic, and versatile depiction of the heavens.[28] His Model A projector was poised to redefine the American planetarium community by its provision of unprecedented career opportunities in astronomy education to instructors at museums, colleges, and public schools throughout the nation.

Acceptance of the Model A Projector

As word of Spitz's invention spread through the Franklin Institute, Stokley's protege Roy K. Marshall, who succeeded F. Wagner Schlesinger as director of the Fels Planetarium in 1945, reportedly issued a public statement that Spitz was planning to hoodwink the public and that his device was going to render a disservice to astronomy education.[29] In his 1943 sketch of Marshall's personality, Spitz asserted that "[t]hose who have been exposed to Dr. Marshall's vitriolic criticisms, or who have squirmed under his wrathful explosions at a job inadequately done, call him a stormy petrel. They may be right!"[30] Spitz thus expected to face Marshall's discomfiture over his invention of the pinhole-style projector. Shortly after Marshall's announcement, Spitz arranged an impromptu demonstration of the Model A in a darkened closet at the Institute. Marshall's attitude was significantly changed by the projector's performance, and he then published a supportive article suggesting that Spitz's equipment "may boom astronomy study."

Marshall described Spitz's projector as the least expensive and "most satisfactory substitute" available for Zeiss instruments. He characterized Spitz's product as being suitable for "smaller institutions," provided that it was not used in domes larger than thirty feet in diameter.[31] Also notable were Marshall's statements concerning the Zeiss firm's postwar "disintegration" and expiration of its original patents on the projection planetarium. This announcement effectively nullified Stokley's influential precaution regarding potential patent infringements attending the Springfield, Massachusetts, projector (see chapter 2). Such an opportunity was seized by Science Associates, which immediately advertised reproductions of the Korkosz instrument in December 1947.[32] Priced at $50,000, none of these instruments was ever remanufactured, however.

Marshall's initial attitude was symptomatic of the lingering prejudice among American planetarium directors against non-Zeiss projection instruments. But these opinions reflected only a portion of the negative sentiments that Spitz encountered as he campaigned for acceptance of the Model A projector. The remainder had been precipitated by a reversal of opinion among astronomers in response to the prewar entertainment spectacles staged by Stokley, Fisher, and Barton (see chapter 4). Spitz did not seem to be aware that a majority of astronomers had supported the Zeiss planetarium's introduction in the years surrounding its American debut.

Spitz claimed that astronomers' disapproval of planetaria stemmed from "the rather scornful accusation that . . . [these] were just oh and ah instruments."[33] Elsewhere, Spitz clarified that this expression signified "demonstrations . . . [which] became shows . . . theater . . . spectacle."[34] To try to counteract this criticism,

Spitz recalled the involuntary reaction of audiences when first seeing the planetarium's stars projected onto the domed ceiling. He wrote, "The infallible aspirated 'ah-h-h-h' which comes with the switching on of the stars . . . bears testimony to the astonishment felt by most people who are seeing the show for the first time."[35] Under those circumstances, he implied, its occurrence was a valuable and expected reward, having nothing to do with the staging of imaginary trips to the Moon or the end of the world. Spitz was gratified when "no less earnest" a response came from local school children who witnessed demonstrations at his company's factory. It was from the imbalance struck between education and entertainment, Spitz argued, that many professional astronomers remained "anti-planetarium" for years. Only gradually did they come to recognize, in his judgment, "what we were trying to do educationally."[36] Spitz's call for the renewed teaching of astronomy was exactly the prescription needed for planetaria to regain credibility in the eyes of most astronomers.

A significant example of the latter took place on 10–11 October 1947, as the first production units of Spitz's projector were readied for sale. Through the influence of Charles A. Federer Jr., Harvard College Observatory director Harlow Shapley invited Spitz to exhibit the Model A before a joint meeting of the Bond Astronomical Club and the American Association of Variable Star Observers (AAVSO). This proved an ideal opportunity for Spitz's device to be critiqued by a host of skillful observers. On the night before the demonstration (which occurred inside the observatory dome of Harvard's fifteen-inch refractor), Spitz and Federer made final adjustments to the positions and brightnesses of stars, using a needle file and black wax pencil. When groups assembled for the demonstration, Federer gave the lecture with the instrument mounted atop the observing ladder. A second demonstration was repeated the following night.[37] As retold later by Grace S. Spitz, audience reactions were "quite warm indeed," even among those who "had notably been lacking in enthusiasm."[38] This unofficial endorsement of the Model A served to reinforce Marshall's revised opinion that Spitz's projector could succeed at its appointed task and bring the artificial heavens to newfound audiences around the globe.

Spitz and the Postwar Planetarium Market

One of the most important factors that allowed Spitz to gain a secure foothold in the planetarium market was the collapse and restructuring of the Zeiss firm following the war. During the spring of 1945 General Patton's Third Army occupied much of central Germany while reconstruction work began on the Jena factory. After the signing of the Yalta Agreement, Jena was slated to join the Soviet-dominated East German state. In June 1945, only days before the Red Army moved in, Allied forces ordered an evacuation of Zeiss's leading scientists, engineers, and instrument makers, along with important company files. More than a hundred executives and their families were conveyed westward to

Heidenheim, where an abandoned factory building served as temporary headquarters. Operations were later transferred to Oberkochen, where the West German division of Zeiss was slowly reborn.[39]

Soviet troops, meanwhile, oversaw a resumption of operations at the depopulated Jena plant through October 1946, when a majority of personnel and equipment were removed to Moscow. Dismantling of the Jena factory was ordered but its employees clung to a minimal system of production. In June 1948 the company's assets were confiscated by Soviet officials and the Jena firm was reorganized as a Volkseigener Betreib (VEB) or "People's Property Concern."

The effect of these actions was to remove the sale of all Zeiss planetaria from the market for a number of years. None had been manufactured since the last Model II instrument was delivered to Pittsburgh's Buhl Planetarium in 1939. Apart from difficulties experienced in restarting their operations, Zeiss personnel anticipated that the demand for projectors would remain so low that the line was temporarily discontinued.[40] The VEB-Jena factory succeeded in producing its next planetarium instrument for Stalingrad in 1954, while the Oberkochen firm supplied its first new model to São Paulo, Brazil, in 1957.[41]

Roughly a decade elapsed during which Spitz faced very little, if any, free-market competition in the fabrication of planetaria. Production of the Unglaube projector was never resumed after the war and, as noted, Spitz sharply undercut the market price of his nearest competitor, Peerless Planetarium Ltd. of Canada. These factors enabled Spitz to become the world's largest manufacturer of planetaria. As Brent P. Abbatantuono has argued, Spitz's Model A projector "revolutionized the availability of artificial skies just as Henry Ford's Model T had done for the automobile."[42] A more detailed analysis of the institutional contexts in which Spitz planetaria were installed and those personnel that made up the postwar American planetarium community is presented later in this chapter.

Postwar Planetarium Installations

Morehead Planetarium

Despite the impossibility of securing any new projectors from either Zeiss division, independent efforts yielded two more major planetarium installations during the postwar period. The first involved purchase of a used Zeiss Model II instrument by Union Carbide Company president (and North Carolina native) John Motley Morehead. Morehead had earned a degree in chemical engineering from the University of North Carolina at Chapel Hill in 1891. He and his father invented the electrical arc furnace, which the former used to perfect the manufacture of calcium carbide. Morehead joined the Union Carbide Company in 1905 and eventually became its president. From 1930–33, he was appointed special diplomatic envoy to Sweden by President Herbert Hoover. It was there that he gained an appreciation for the Zeiss star projector. Following his wife's death in 1945, Morehead established a foundation that supplied funds to erect a planetarium on his alma mater's campus.[43]

Morehead's projector was originally installed at Stockholm, Sweden, in 1930. Twice relocated, the instrument was dismantled and stored at Gothenberg during the Second World War. In 1947 it was purchased by Morehead, crated for shipping, and reassembled on the Chapel Hill campus of the University of North Carolina. The Morehead Planetarium sported the first such instrument to be erected in a rural American city and operated by an institution of higher education. Morehead had also served on the advisory committee to New York's Hayden Planetarium during the 1930s. This role proved influential in his subsequent gift of a large Copernican planetarium, manufactured by Fecker of Pittsburgh, to complement the Zeiss projector. Morehead was the only other major American planetarium to install a large ceiling orrery. In addition, the planetarium houses a small collection of astronomical instruments purchased by university president Joseph Caldwell in 1824. On a much-reduced scale, the Morehead collection resembles the Adler Planetarium's Mensing Collection.

Roy K. Marshall became the Morehead Planetarium's director when that facility opened its doors on 10 May 1949. Marshall departed after two years to pursue a more lucrative career in the production of television commercials.[44] He was succeeded in 1951 by Morehead technician Anthony F. Jenzano (b. 1919), who had reassembled the Swedish projector for Marshall. His institution was to host a gathering of "planetarium executives" during this second developmental period. Jenzano also played a crucial role in Morehead's selection by the National Aeronautics and Space Administration (NASA) to conduct its astronaut training program after 1958 (see chapter 7).

Morrison Planetarium

A more difficult pathway toward acquisition of a major planetarium was followed by the California Academy of Sciences in San Francisco. Through bequests from the May T. Morrison estate, a sum of $350,000 was reserved for a planetarium to honor her late husband, attorney Alexander F. Morrison, who had provided legal counsel for the Academy. Had it not been for the wartime optical and mechanical experiences of two Academy employees, G. Dallas Hanna and Albert S. Getten, however, construction of a large planetarium instrument might not have been attempted. A preliminary sketch of the Morrison projector was prepared by Russell W. Porter, hired as a consultant only a year before his death.[45]

Work on the project commenced in 1948 and required four years to complete. The Morrison Planetarium, whose custom instrument featured an inversion of the star hemispheres and planetary "cages," was opened 8 November 1952 as part of the Academy's Hall of Science erected in Golden Gate Park. George W. Bunton (1910–1995), chief technician at Los Angeles's Griffith Observatory, became Morrison's first director.[46] Notable planetarium "firsts" achieved under Bunton's tenure were the use of automated programs (controlled by magnetic tape recordings) and the adoption of experimental programs (1957–59) that featured abstract lighting effects and electronic music produced by the local artistic duo Vortex (see chapter 7). Bunton assumed a leadership role among planetarium executives and was appointed chairman of the 1955 American Association of Museums planetariums section.

Richard H. Emmons and Planetarium Education

Introduction of Spitz's pinhole-style planetaria inspired other attempts to project constellation imagery using homebuilt devices. One such projector was constructed by students enrolled in an introductory physical science course taught at the Canton, Ohio, campus of Kent State University in 1949. Instructor Richard H. Emmons used this experience to earn a master's degree in education and prepare the first graduate-level thesis examining the roles of smaller planetaria for instructional and community relations purposes.[47]

Emmons's project was undertaken following a visit to Pittsburgh's Buhl Planetarium. While Emmons regarded the Buhl's presentation as "excellent," he noted the lack of opportunity for audience members to ask questions during the lecture. In Emmons's opinion, "high maintenance expenses [only] partly defrayed by paid admissions" dictated the necessity of programs being changed frequently at all major planetaria. Cost effectiveness required that "[p]atrons must be induced to return," with the result that these facilities "had not been used at the maximum educational efficiency." Emmons suggested that "the ideal demonstration for the adult novice" should remain unchanged.[48] In Emmons's view, smaller planetaria held a strong potential for education, despite their lack of technical sophistication.

Demonstrations of the Canton planetarium began in December 1949. Over the next six months, some 1,500 people viewed its program. Emmons argued that planetarium instruction could be readily incorporated with the teaching of "physics, mathematics, geography, navigation, engineering, mythology, history, and philosophy." He added that children often fared better than adults whenever new concepts were introduced because of the latter's "erroneous attitudes [which] have to be unlearned."[49] Emmons's study did not measure student learning in relation to the planetarium, although it pointed out the significance of smaller facilities for visitors of all ages.

The planetarium's popularity was used in an election asking voters to approve municipal bonds to secure continued funding for the Canton campus of Kent State University. The issue was defeated and the planetarium became Emmons's personal property. It was later reerected behind his home and used for school and public demonstrations during the 1950s.[50] Perhaps its most significant influence was on Emmons's daughter, Jeanne, who literally grew up in association with it. After earning her Ph.D. in astronomy education, Jeanne (Emmons) Bishop went on to become the first woman president of the International Planetarium Society (1983–84), the largest association of planetarium professionals, founded in 1970 (see chapter 8).

Prelude to the Space Age

An important factor that enabled planetaria to gain significant popularity in the postwar period was the growing anticipation that spaceflight would soon become a reality. Rapid progress toward that objective was being made. The promise of spaceflight gave a tremendous impetus to the acquisition of planetaria, especially by smaller public museums and science centers, which kept their visitors informed about the latest technical developments. The engineering steps toward spaceflight reached their culmination in well-publicized efforts to launch the first artificial earth satellites in conjunction with the International Geophysical Year (1957–58).

V–2 Rocket Program

The postwar capture and subsequent launch of German V–2 missiles at White Sands, New Mexico, after being refitted for high-altitude research constituted an important step towards opening the space frontier.[51] Coordination of competing civilian science and defense research needs by the unofficial V–2 Panel yielded notable advances in atmospheric, solar, and cosmic ray physics. As historian David H. DeVorkin has demonstrated, the V–2 Panel's momentum was then "redirected to[ward] planning committees for the International Geophysical Year" and "deciding which experiments should be supported during Project Vanguard," the U.S. artificial satellite program.[52]

Apart from gathering abundant physical science data, the V–2 missions provided an unexpected bonus whose meaning could be readily grasped by the average citizen. A montage of photographs, taken from an altitude of 100 miles by a rocket launched 7 March 1947, was reproduced on the June 1948 cover of *Sky and Telescope.* Captioned "the curved earth," this image was pondered by editor Federer, who asked whether it might not represent a "picture of the future?"

Hayden Planetarium Symposia on Spaceflight

In October 1951 Hayden Planetarium chairman Robert R. Coles convened the first of three public symposia devoted to the possibility of spaceflight. German Rocket Society émigré Willy Ley was appointed to coordinate its panel of speakers. Afterward, editors at *Collier's* magazine commissioned an eight-part series devoted to the topic, featuring illustrations by the nation's foremost astronomical artist, Chesley Bonestell. Four symposium speakers, including Harvard University astronomer Fred L. Whipple, contributed articles to the series's inaugural issue (22 March 1952), whose cover proclaimed "Man Will Conquer Space Soon."[53] The *Collier's* series, and remaining symposia convened in 1952 and 1954, provided a forum in which rocket engineer Wernher von Braun became a leading spokesperson for the advocacy of spaceflight.

When the permanent theme park Disneyland opened in Anaheim, California (1955), its attraction "Tomorrowland" depicted simulated views from an earth-orbiting space station and a rocket to the Moon ride. To develop and promote these concepts, Walt Disney produced a one-hour television program, "Man in Space," which was broadcast on 9 March 1955. Ley, von Braun, and others assisted in the technical aspects of production. As historian Howard E. McCurdy notes, "Disney and *Collier's* exposed millions of Americans to the possibility of spaceflight . . . in a manner that allowed people to visualize how spaceflight would actually appear."[54] McCurdy's study of such popular attitudes confirms and explains historian Walter A. McDougall's assertion that the "space age" was long preceded by a period of "cultural anticipation," stretching back to the writings of science fiction novelists.[55]

McCurdy also demonstrates that historical analogies have long played "an important role in the promotion of space exploration." Of these, the most widely adopted is a view of space as the ultimate frontier, a variation of the "frontier thesis" first annunciated by historian Frederick Jackson Turner in his classic paper "The Significance of the Frontier in American History" (1893). Despite historians' recognition that the "popular image of the American frontier contains more myth than substance," this deeply ingrained stereotype has been little altered among the public. McCurdy argues that "to advocates of space exploration, the Turner doctrine, however dimly understood, became the basis for a new adventure."[56] Opening of the space frontier became a prominent rationale for the construction of American planetaria in the postwar period.

The power of visual imagery to shape opinions, ideas, and attitudes is a well-known social phenomenon. Hayden Planetarium director Neil de Grasse Tyson has reaffirmed the symbolic meanings of photographs taken from space during the 1960s. In Tyson's judgment, those publicized views of Earth reminded us that "landmasses did not have political boundaries drawn upon them." Apollo missions to the Moon depicted Earth as "small, fragile, and distant," the only habitable planet within our solar system.[57] A new metaphor, "spaceship Earth," symbolized the unity and interdependence of humankind and reshaped public consciousness concerning our planet's finite natural resources. From compelling images such as these, the U.S. space program unwittingly boosted the fledgling environmental movement of the 1970s.[58]

Whether stimulated through verbal, textual, or electronic media, public perceptions concerning the approach of a new age, the space age, blossomed during the late 1940s and early 1950s. What had seemed little more than escapist fantasies of science fiction writers (and planetarium demonstrations) in the late 1930s appeared with dramatic confirmation to be lurking around the corner. As the Cold War intensified, public awareness of rockets, missiles, and the prospects of spaceflight grew commonplace.

International Geophysical Year

By far the largest and most carefully planned scientific investigation of the 1950s was the International Geophysical Year (IGY), whose eighteen-month campaign officially lasted from 1 July 1957 to 31 December 1958. This endeavor accomplished what *New York Times* science journalist Walter Sullivan termed an "assault on the unknown." The IGY incorporated scientists from more than sixty nations and combined the labors of some 60,000 people, whose attention was directed toward problems stretching "literally from pole to pole."[59] Never before had such a degree of international cooperation been undertaken for scientific purposes, and the spirit of collaboration remained largely undeterred by Cold War tensions. The scope of activities pursued on behalf of the IGY poses challenging questions about the growth of large-scale research practices in twentieth century science and technology. Nonetheless, the IGY's diffusion of efforts across a host of geophysical problems exhibited a horizontal, rather than vertical, integration of practitioners' skills and resources.[60]

Coordination of this enterprise began at a dinner party hosted 5 April 1950 at the home of James A. Van Allen, head of high-altitude atmospheric research at the Applied Physics Laboratory of Johns Hopkins University. One of Van Allen's guests, geophysicist Lloyd V. Berkner of the Carnegie Institution of Washington, D.C., suggested that the scientific community organize a third International Polar Year on the twenty-fifth anniversary of the second one (1932–33). This timeframe was expected to coincide with the next sunspot maximum in 1957–58. Berkner's suggestion found ready support among those present and from a variety of colleagues elsewhere.

Discussions in support of this venture were held the following year by the Comité Spécial de l'Annee Géophysique Internationale (CSAGI), one of many committees appointed under the auspices of the International Council of Scientific Unions (ICSU), which ratified the IGY proposal in 1952. Two years later, CSAGI members broadened the scope of its plans to include exploration of the Antarctic continent and the environment beyond Earth's atmosphere. Geophysicist S. Fred Singer recommended that an artificial earth satellite be launched in culmination of these activities.[61] Singer's proposal unwittingly dealt a significant impact to U.S. astronomy education and fostered the development of American planetaria in unprecedented fashion.

Plans to orbit an artificial satellite captured public imagination as no other IGY experiment. The bulk of geophysical investigations were carried out in remote corners of the world and generated little public awareness. But because the satellite launches were to be conducted publicly, a much higher media profile ensued. While containing only small instrumented probes, the IGY satellites nonetheless promised to fulfill the long-awaited dream of achieving spaceflight. In turn, each held the prospect of a tangible end product—a man-made point of light rapidly crossing the sky—that under the right conditions could be glimpsed by anyone. Behind these expectations, however, lay the critical assumption that the first satellites launched would be of American origin.

Soviet participation in the IGY was not announced until April 1954. The following year, a Soviet IGY committee formally registered with CSAGI. Historian Rip Bulkeley has argued that because of their late entry, Soviet scientists and administrators were "never in a position to influence the basic programme" of the IGY, which was dominated by "American scientists and their political allies."[62] U.S. officials ignored an April 1955 Moscow newspaper article describing Soviet plans to launch an artificial satellite. Western experts regarded the communist regime as incapable of achieving such an ambitious goal within the allotted timeframe. On 2 August 1955, however, academician Leonid I. Sedov, chairman of the newly created Interdepartmental Commission on Interplanetary Communications (ICIC), formally disclosed the possibility of a Soviet earth satellite. Another year elapsed before it was clear that Soviet efforts were to be conducted under the IGY banner.

American involvement with the artificial satellite program gained public endorsement from the White House on 29 July 1955. Funding allocations were provided by the National Academy of Sciences, and responsibility for the Vanguard satellite was awarded to the Naval Research Laboratory, under the direction of John P. Hagen. Despite American reliance on its armed forces to deliver the satellite into orbit, the Vanguard mission was hailed as a peacetime activity executed under civilian control, whose objective was scientific rather than political or military.[63] But recent studies have shown that this was a cover story constructed by national security officials for purposes of surmounting the perceived obstacle of satellite overflight of other nations and establishing the principle of "freedom of space." Overflight, however, was never an issue among members of the Jet Propulsion Laboratory's Homer J. Stewart Committee, whose 3 August 1955 selection of Vanguard over the competing Army Ballistic Missile Agency's Orbiter proposal attempted to insure the best scientific return from an IGY satellite. The psychological importance of being first into space was not lost upon security analysts, but was dismissed by the Eisenhower administration, which displayed no propensity to engage in a nationalistic race with the Soviet Union.[64]

Spitz and Project Moonwatch

One of the vitally important aspects of the Vanguard program was to determine the orbital parameters of satellites.[65] While the satellite's radio transmissions were to be used in deriving a preliminary orbit, a completely redundant optical tracking program was envisioned in the event of transmitter failure. Under the direction of Fred L. Whipple, the Smithsonian Astrophysical Observatory (SAO), which had relocated in 1955–56 to Cambridge, Massachusetts, was assigned the task of establishing a network of twelve optical tracking stations placed strategically around the globe.[66] Each station was to operate a wide-field photographic instrument known as a Baker-Nunn camera for the capture of satellite images. Measured positions were to be relayed to the SAO's computation facility for derivation of the satellite's orbit and calculation of future ephemerides.[67]

One important problem, however, remained: the Baker-Nunn cameras could not obtain satellite images without the establishment of preliminary orbits. Thus, it became necessary to recruit experienced amateur astronomers who could record satellite observations during evening or morning twilight. Project Moonwatch, as this effort was dubbed, provided amateur scientists with a unique opportunity to participate in the data collection program of the IGY. To make these satellite observations, widespread observing teams or "stations" were equipped with low-power monoscopes, adjusted to view selected altitudes along the local meridian. Each monoscope was mounted on a tabletop and incorporated a small flat mirror that reflected a narrow strip of sky. Shortwave radios provided accurate time signals that were recorded on audio tape. An observer called out "Time" when a satellite crossed the center of his or her monoscope's field of view. Moonwatch teams supplied SAO scientists with positional satellite data long after the IGY officially ended.[68]

In February 1956 Armand N. Spitz was selected as national coordinator of visual satellite observations for the U.S. Moonwatch effort. He was chosen on the strength of his reputation as a popularizer of astronomy and the influence that his Model A planetarium was exerting throughout the nation. Spitz was ideally suited for the initial phases of this task, which involved national lecture tours to recruit potential volunteers. Observing teams were drawn from a host of national associations, including the Astronomical League, American Association of Variable Star Observers, Association of Lunar and Planetary Observers, Western Amateur Astronomers, and members of the Civil Defense Ground Observer Corps.[69] But Spitz's lack of formal astronomical training prevented him from being able to guide the reductions of Moonwatch data. Spitz remarked, "I am not a mathematical astronomer. I don't get along with mathematical equations. I am not very much of a scientist. You can call me an interpreter of science if you want to."[70] Consequently, veteran observer Leon Campbell Jr. was chosen as supervisor of station operations.

As a measure of Spitz's success in promoting this opportunity, at least seventy-nine U.S. stations, comprising over 1,200 individuals, had officially registered by the start of the IGY. Dozens of other Moonwatch groups were established around the world, especially in Japan. Two nationwide alerts, held on 17 May and 19 July 1957, tested the readiness of teams to observe mock satellites towed by Civil Air Patrol planes and to communicate their findings to Cambridge.[71] SAO director Whipple dismissed initial skepticism expressed toward the use of observations made by amateur astronomers: "Time, I'm happy to say, has proved these fears to be groundless. The amateurs joined up in droves, all around the globe. They did a splendid job for Project Vanguard, and they have been doing an increasingly more effective one for the American space effort ever since."[72]

Spitz's labors on behalf of the IGY satellite program brought him an unexpected measure of success. On 3 June 1956 he was awarded an honorary Doctor of Science degree by Otterbein College at Westerville, Ohio, in recognition of his achievements in the field of astronomy education. This honor was bestowed following the installation of a planetarium and observatory donated by alumnus Alfred H. Weitkamp.[73] For Spitz, who had left the University of Cincinnati without receiving a baccalaureate degree, this symbolic recognition stood as the culmination of his twenty-year devotion to the study and teaching of astronomy. That same year, he was named a special consultant to the National Science Foundation (NSF). Following the launch of *Sputnik,* Spitz used NSF resources to help organize and support the first nationwide symposia on planetarium education in 1958 and 1960 (see chapter 8).

Coordination of the satellite program brought noted personal as well as professional changes to Spitz's life. In recruiting teams for Project Moonwatch, Spitz addressed groups of amateur astronomers and participated in extensive committee work. Among the leading organizations that supplied members for this effort was the Astronomical League (founded in 1947), the nation's largest federation of amateur astronomers.[74] The League's executive secretary (1949–54) and later president (1955–57) was Grace C. Scholz (b. 1912), a medical statistician by training.[75] As League president, Scholz was appointed to the artificial satellite program's National Advisory Committee, which met at SAO headquarters on 1–2 February 1957.[76] Through these and other contacts, a personal relationship developed between Scholz and Spitz that led to Spitz's divorce and marriage to Scholz in 1958.

That Spitz and other IGY officials had been informed about the possibility of a Soviet-launched satellite is a matter of record. Spitz reported to Moonwatch participants that during its 1956 Barcelona conference, CSAGI president Sydney Chapman described cooperative plans between the United States and the USSR for tracking each other's satellites.[77] Soviet IGY committee members described, in consecutive issues of the journal *Radio* between June and September 1957, the frequencies on which the Soviet satellite was to communicate by telemetry. Written announcement of the same was sent to CSAGI vice president Berkner, but the letter appears to have miscarried.[78] E. Nelson Hayes's account of the events leading up to and immediately following the launch of *Sputnik I* on 4 October 1957 discloses that "considerable thought" was given to this issue by SAO director Whipple in memoranda prepared earlier that summer. Whipple recommended that all satellite-tracking personnel be put on general-alert status starting 1 July.[79]

One of the ironic statements that emerged from this period was uttered by Spitz only a short time before *Sputnik's* launch. On account of their lasting impression, Spitz's words were recalled years later by John C. Rosemergy, then science department head at Ann Arbor (Michigan) High School. Spitz had traveled there under the dual purposes of inspecting the school's newly installed Argus Planetarium and establishing a Moonwatch team among its students. At the close of a conversation that had drifted toward the artificial satellite program, Spitz remarked to Rosemergy, "I just hope that the first one we see is ours."[80] That Spitz's wish was not fulfilled became the factor by which the third and largest period of growth (1958–70) in the American planetarium community, chiefly affecting the nation's public schools, was ushered in.

Postwar Professionalization

AAM Planetariums Section

The nation's smaller museums and science centers comprised the second-largest category of institutions (after universities and colleges) that acquired planetaria before 1958. It is not surprising, therefore, that the earliest means of collaboration among planetarium directors emerged and flourished within the museum discipline. Since hosting its first annual meeting in 1906, the American Association of Museums (AAM) had grown into the largest and most prestigious organization of its kind. Its leaders recognized the need to support a variety of programs in research, education, and exhibition among public institutions of all sizes. Within the AAM, various sections were created to address a diversity of specialized interests among the arts, sciences, and humanities. From this body of museum educators, and not the ranks of academic astronomers, came the first viable association of planetarium directors.

As the head of museum education at the Franklin Institute, Armand N. Spitz was familiar with the AAM. In turn, its officers and members were to become acquainted with his Model A planetarium. Spitz demonstrated the pinhole-style projector before the children's museums section at the 1949 AAM conference in Chicago.[81] The following year, he was appointed to a six-member panel that discussed contemporary issues in museum education.[82] In 1951 Spitz explored whether an organization of planetarium directors might be established within the AAM. On the day preceding the association's annual conference in Philadelphia, Spitz hosted an informal session on the topic of "planetarium problems," which was followed by a demonstration at the Fels Planetarium. Delegates were invited to inspect Spitz Laboratories (then located in west Philadelphia) and raise questions about current issues in planetarium programming. Spitz later recalled that "only ten or twelve" delegates had attended the organizational meeting, but "found that we could help each other, and even enjoy doing so."[83] The decision to organize a "planetariums section" gained support within the AAM.

Maxine L. Begin (1922–1985), curator of education at the Minneapolis Public Library's Science Museum, hosted the section's first official session at the 1952 AAM conference. Ernest T. Luhde, director of the Stamford (Connecticut) Museum, presided. A five-member panel that included hostess Begin offered discussion on the topic "Planetariums, Their Use as a Community Service."[84] The adopted format allowed institutional hosts and moderators to be changed yearly and spelled long-term success for the inauspicious group, whose meetings lasted through the late 1960s (or perhaps beyond).

While self serving toward his own business interests, Spitz's aspirations in fostering the planetariums section proved insightful. From a variety of perspectives, his proposal must be regarded as a success. Designed to attract the nation's smaller planetaria, the AAM section soon welcomed larger institutions and their directors into its proceedings. Spitz correctly surmised that personnel from smaller facilities would be inadequately prepared or supported to create and sustain an independent organization. Because of geographic isolation, there was little chance for planetarium directors to attend professional meetings, apart from those in which other museum employees could participate. Thus, it made sense to dovetail the planetariums section into an established association. Choice of the AAM as an umbrella organization in which to nurture a contingent of planetarium directors was perhaps the wisest choice that Spitz could have made.

As an alternative, Spitz might have attempted to organize the nation's university and college planetarium directors within the auspices of the American Astronomical Society (AAS). But that strategy almost certainly would have presented formidable obstacles to Spitz's objectives. Lacking both formalized training in astronomy and a baccalaureate degree, Spitz would have found it difficult to establish his credentials before this body of professional astronomers. Furthermore, the AAS had long exhibited an "ambivalent relationship with the world of astronomy education."[85] Attention directed toward education was chiefly driven by the job market's cyclical demands for researchers. Seldom were concerns for undergraduate, let alone pre-collegiate education, taken seriously. Nonetheless, Yerkes Observatory astronomer Thornton Page advocated use of a simplified "Tin Can Planetarium" for elementary classroom instruction in 1950.[86] But any attempt by Spitz to unite the discipline's academic planetarium directors likely would have met with failure.

By contrast, AAM officials who approved the planetariums section realized that their organization stood to acquire new members as the number of American planetaria increased. Spitz was not blind to the sales potential that might accrue from genesis of the planetariums section, although his primary goal remained the diffusion of astronomical knowledge. Spitz was once described by Joseph M. Chamberlain (b. 1923), then chairman of the (renamed) American Museum-Hayden Planetarium, New York, as a "man with an endless stream of ideas."[87] Spitz's business interests formed an integral part of his approach to living, but they never appear to have dominated what he valued most—communicating the joy of learning about the heavens to listeners of all ages.

At the 1953 AAM meeting in Buffalo, New York, delegates learned of San Francisco's Morrison Planetarium from lecturer Leon E. Salanave and examined programming needs for specialized audiences. The 1954 Santa Barbara, California, conference attracted representatives from the Griffith Observatory and Morrison Planetarium. Dinsmore Alter and George W. Bunton considered the planetarium's role in terms of "Education, Entertainment, [and] Culture." The meeting was chaired by John Patterson, newly appointed director of Boston's future Hayden Planetarium, then under construction. For two years (1951–53), Patterson oversaw the traveling planetarium of Boston's Museum of Science. During the five years it was operated, Boston's portable planetarium gave some 3,000 demonstrations to over 125,000 visitors.[88]

Collectively, these actions demonstrated a newfound collaboration between the host of smaller institutions and the nation's major planetaria. They revealed an awakened interest within the latter's personnel to learn from and about their more numerous colleagues, whose actions they were soon to emulate. Those directors whom Spitz had united recognized the importance of his vision and embraced the challenge of fashioning a viable planetarium association within the AAM.

Besides founding the AAM planetariums section, Spitz attempted to improve communications among its members by creating a newsletter, which he named *The Pointer,* after the lecturer's projection arrow used to identify stars and constellations. This venture, however, was shorter lived and less successful than the association it was designed to serve. Spitz convinced Robert R. Coles (1908–1985), who succeeded Gordon A. Atwater as chairman of New York's Hayden Planetarium in 1951, to assume its editorship. Previously, Coles had supported Spitz's efforts to construct his first Model B projector.[89]

The Pointer's significance lay in its being the first regular publication to originate within the American planetarium community. Its purposes were to inform and advise all members of the discipline, though it was aimed primarily at newcomers who were most in need of suggestions from experienced colleagues. Because the AAM did not publish its own proceedings, Spitz might have wished *The Pointer* to serve as a journal of record for the planetariums section. Such hopes faded, however, when Coles was replaced at the Hayden Planetarium after two years by Joseph M. Chamberlain, who declined to assume the editorial task.[90] This turn of events served to mute the AAM's valuable communications tool. It is unclear when and by whom *The Pointer* was revived, although its continued existence was referred to as late as 1960.[91]

Planetarium Executives

A second, informal association of directors came into being in 1952, possibly sparked by the growth of smaller planetaria and creation of the AAM planetariums section. Morrison Planetarium director George W. Bunton recalled, "It may have been the introduction of the Spitz instrument that prompted the men of the major institutions to establish an annual executives' conference to compare notes and exchange ideas among themselves."[92] These planetarium executives purposefully drew members only from major institutions and focused their attention on problems associated with metropolitan locations. This was hardly surprising, as executives faced administrative tasks that were materially distinct and occurred on significantly different scales from those encountered by directors of smaller planetaria. In contrast to the AAM's mixed-gender association, these executives were composed almost entirely of Caucasian men. Their small numbers and privileged status lent a somewhat distinctive atmosphere to meetings and activities, which evoked descriptions of fine dining and elegant accommodations.[93]

Discussions at meetings encompassed administrative concerns over promotion and publicity of programs, attendance figures, and the nature, purpose, and impact of exhibits, along with technical and production-related matters. Among the issues raised at the 1954 Pittsburgh meeting was the possibility of creating an "[e]xchange of sky shows on a 'package' basis," including script, music, and audio tape. The use of projected skylines or panoramas, replacing traditional cove-mounted silhouettes, encouraged the sharing of original artwork for customized photography.[94] The design and use of special effects projectors, along with motion pictures, sound effects, and music, all signified that techniques of showmanship pioneered by James Stokley during the 1930s had since become routine, sophisticated, and expected audience fare among major planetaria of the 1950s.

Reminiscent of visits exchanged during the 1930s by Stokley and Fisher, a congenial atmosphere of collaboration was apparent among postwar planetarium executives. Previous attempts to fashion an informal association of personnel after that of the Astronomical Neighbors had failed some fifteen years earlier from a lack of institutional diversity. Conditions had changed in several important respects in the postwar period, however. Rapid means of transportation, especially on airline flights, more readily linked institutions formerly separated by continental distances. Of greatest importance, however, was the larger number of major planetaria (including those under construction), plus their cadre of newer directors. Gone from the scene were Fox, Stokley, Fisher, Barton, and Marshall, along with Bennot and Lockwood. Directors remaining from prewar service were F. Wagner Schlesinger (Chicago), Dinsmore Alter (Los Angeles), and Arthur L. Draper along with C. V. Starrett (Pittsburgh). In addition, social and economic changes stemming from the Second World War had turned the United States into a much stronger and more prosperous nation. Particularly evident was the significant role that science and technology would play in shaping the future of society. Few executives, however, took seriously the suggestion of Dinsmore Alter that college freshmen be required to take a "science orientation course," two-thirds of which might be taught by means of the planetarium.[95]

At least some planetarium executives came to view their institutions (and the strengthening of their discipline) in a rather different light compared to their prewar colleagues. In particular, John Patterson, third chairman of the AAM planetariums section, urged executives to work toward common goals alongside directors from smaller planetaria. While acknowledging that significant differences marked the large and small facilities, Patterson ventured that "[a]ll planetariums can and should contribute to the refining of the [public's] moral and intellectual faculties," and prove themselves to be "true cultural aids" to their communities. Though regarding the educational value of smaller planetaria as "completely satisfactory," Patterson nonetheless clung to conventional wisdom by citing their inabilities to match the "dramatic entertainment features" staged by major facilities.[96] When Charles A. Federer Jr. chose not to leave Sky Publishing Corporation to direct the Boston Museum's new Hayden Planetarium, Patterson became one of the first directors from a small planetarium to be admitted to the executives' ranks. He strove to convince skeptical colleagues to join forces with, rather than remain united against, their smaller allies. Joseph M. Chamberlain, another executive who shared Patterson's experience with smaller planetaria, had lectured using the Merchant Marine Academy's Spitz planetarium at King's Point, Long Island. Chamberlain succeeded Coles as general manager of New York's Hayden Planetarium in 1953 and became its chairman three years later.

One of the most revealing discussions occurred at the May 1957 meeting hosted by Israel M. Levitt of Philadelphia's Fels Planetarium. Debates concerned the education vs. entertainment issue and degree of a star instrument's use during public programs. Reflecting Chicago's conservative programming style, Albert Schatzel indicated that their projector was featured during 45 minutes of every hour-long presentation and served to put the "accent on astronomy." Starrett concurred with this percentage but suggested that trends had moved "away from straight astronomy to more dramatic shows." Levitt interjected that a planetarium's mission was no longer to teach astronomy, because it had become necessary "for [public] shows to compete with the theatre." Anthony Jenzano of Chapel Hill seconded Levitt's remark that planetaria "do not teach—just spark interest." Alter dissented from previous speakers and argued that "astronomy should be taught accurately but simply" with the planetarium. He noted that while staging "spectacular shows" (i.e., "A Trip to the Moon") at his own institution, all scripts were subject to prior "approval of the observatories," a move seemingly intended to reestablish the planetarium's credibility in the eyes of professional astronomers.[97] Jenzano planned to offer an "End of the World" show as Morehead's "1958 spectacle," though it was to be offset by a more educational program titled "Land, Sea, and Sky," billed as a "tribute to the International Geophysical Year."[98]

These opinions reveal an almost complete departure from the "American practice" of monthly program rotations pioneered by Fox and Bennot (see chapter 4). Few if any institutions still regarded the elements of descriptive astronomy as the foundation of their programming styles. Remarkably little consensus remained about the planetarium's role as a pedagogical device for teaching astronomical concepts. Executives were becoming convinced of their audience's greater visual sophistication, resulting from media-driven forms of mass entertainment. Visual sophistication implied that audience members would no longer be content with prior levels of performance and would find over-reliance upon the star instrument to be inadequate. To avoid declining attendance, program content at major planetaria required adaptation to evolving audience needs (a fact recognized by Richard H. Emmons). What had been viewed as an "audacious" programming style during the 1930s had since become the normative practice for a major planetarium's survival in the late 1950s. This dilemma, in turn, forced the drafting of a defensible programming philosophy by one of the executives' leading spokesmen (see chapter 8).

On more than one occasion, planetarium executives debated whether or not to establish a formal organization. That issue surfaced during the 1955 Chapel Hill meeting, at which an "Association of Planetarium Officials" was proposed. No precise credit can be assigned to anyone for advocating this scheme, although circumstantial evidence suggests that Jenzano (as host) brought this matter before the group.[99] No conclusion was reached after preliminary discussion; the proposal was tabled until the following year. When the identical question again was raised, executives voted against its adoption. Executives' meetings had proven successful without the adoption of a formalized structure. Few needs were required beyond those of a gracious hosting institution that changed hands annually. Election of officers and drafting of a constitution were likely viewed as unnecessary and constraining acts for so small and specialized a group. Rather than formalize domestic arrangements, executives looked with anticipation toward a proposal from Joseph M. Chamberlain to open the 1959 executives' conference to an international contingent of planetarium directors (see chapter 8).

Gender and Planetaria

While excluded as directors of major planetaria for years to come, women gained postwar readmission to the American planetarium community through advent of Spitz's Model A projector. The educational institutions that most readily accommodated women as planetarium directors were small, regional museums that interpreted their area's local history or natural resources. Unlike school districts, museums were able to offset the purchase price of a planetarium, dome, and auxiliary equipment through modest admission fees. Significant fund-raising efforts and charitable donations came from community service groups such as the Junior League. Such lower-prestige positions, offering a modest salary, were often filled by women educators. Within this setting, however, their careers became self empowering. Several women museum directors procured and managed Spitz planetaria—a dual responsibility not shared at major Zeiss installations. These opportunities seemingly offered women all of the professional responsibilities and rewards that they sought.

At least seven women became planetarium directors before the space age had begun; some remained active in the following decade, when the number of women astronomers increased sharply. Maribelle B. Cormack (1902–1984), curator of the Roger Williams Park Museum in Providence, Rhode Island, directed its planetarium after 1953. Cormack earned a bachelor's degree from Cornell University and a master's degree from Brown University. She participated in the two national symposia on planetarium education held in 1958 and 1960. Cormack wrote extensively in the fields of natural science and children's literature. She was awarded an honorary doctorate by Rhode Island College in 1966.

Genevieve B. R. Woodbridge, who earned a bachelor's degree from the University of Minnesota, directed the Grout Museum of History and Science in Waterloo, Iowa, starting in 1955. She opened its planetarium the following year and retained both directorships past 1970. Louise L. Morlang and Mrs. R. L. Sullivan[100] successively directed the Townsend Planetarium at Charleston, West Virginia's public library and children's museum, while Claudia Robinson supervised the Dallas, Texas, Health and Science Museum's planetarium.

Miss Charlie M. Noble, professor of astronomy at Texas Christian University, acquired a Spitz planetarium in 1949 and gave impromptu demonstrations until permanent quarters could be erected at the Fort Worth Children's Museum in 1955. Noble was perhaps the first woman college instructor to employ the pinhole-style projector, although her teaching extended to "junior astronomers." In 1956 Noble received the Astronomical League's national award in recognition for her extensive youth mentoring program. Noble's name was subsequently affixed to the Fort Worth planetarium, making her possibly the first woman to be so honored.[101]

The woman who achieved the highest recognition from planetarium colleagues was Maxine (Begin) Haarstick, curator of education at the Minneapolis Public Library's Science Museum. Haarstick's institution hosted the first official meeting of the AAM planetariums section in 1952. She had earned both bachelor's and master's degrees from the University of Minnesota. Haarstick's active role in the planetariums section, gauged by almost yearly presentations, led to her selection as chairperson of its 1957 meeting. She became the first woman to attain such recognition from male-dominated colleagues and was re-elected chair of the section in 1964. Her most notable paper, "How to Succeed in the Planetarium," appeared in *Museum News.*[102] Haarstick's career was capped by assuming both planetarium and museum director's posts, which she held concurrently after 1959.

During prime years of the "feminine mystique," women planetarium directors were few in number, discouraged from pursuing scientific or technical careers believed to be reserved for men. The typical career pathway open to men, which led to the directorship of a major planetarium, remained virtually closed to women. Only within the nation's smaller museums and exceptional universities were the barriers to women's participation as planetarium directors gradually removed. While their number and status remained far short of the mark attained by male colleagues, a few stereotype-breaking women attained both museum and planetarium directorships. While not all inclusive, this pattern reveals the largest gender-specific differences in postwar planetarium management. No men seemed to have served in this dual capacity. Whether by constraint or desire, these women evidently looked no further toward careers in the nation's major planetaria.

In steadily growing fashion, women directors began to act as a positive force for change within the American planetarium community. Their presence under the dome served as an important role model and demonstrated that girls as well as boys could learn about the heavens. For those few women who embraced this challenge, neither size of dome nor type of planetarium instrument mattered. Regarding the supposed distinction between large and small planetaria, Haarstick replied that the dome's diameter "tells us nothing except how to figure its circumference."[103] What counted most was being able to present astronomy lessons to children and adults in an enthusiastic and entertaining fashion. It was this objective that male colleagues came to realize they shared with women directors: a desire to promote the study and teaching of astronomy at all age levels. But the fewer and lesser rewards these women received, along with the motivations that guided them, remained primarily intrinsic.

Planetarium Growth and Diversity

To more fully understand the changes that swept the American planetarium community during the postwar period, consideration must be given to the discipline's rapidly evolving size, composition, and diversity. By October 1957, exactly one decade after Spitz first marketed the pinhole-style projector, at least 116 planetaria of a permanent nature had been established in the United States, along with 6 located elsewhere in North America (4 in Canada and 2 in Puerto Rico).[104] This quantity represents more than a fifteen-fold increase over the prewar census. As noted above, all but two of these institutions (Chapel Hill's Morehead Planetarium and San Francisco's Morrison Planetarium) were smaller facilities that featured non-Zeiss projection apparatus. Of greater interest, however, are the nature of these institutions and their principal clienteles.

Among the 116 pre-*Sputnik* U.S. planetaria, the largest institutional category (52, or 45%) was that operated by universities and colleges. Many of these facilities were established by physics or astronomy departments to present instruction in descriptive astronomy. Only limited public outreach programs seem to have been offered on a regular basis. A majority of their directors were likely faculty members who possessed the Ph.D. degree. Responsibility for the planetarium's operation was sometimes attached to the department chair's duties or occasionally shared among colleagues. Despite the availability of support for postwar scientific research and teaching, none of these institutions reported receiving federal dollars for their planetarium purchase or construction.[105]

The next largest category of planetaria was that acquired by public museums and science centers. These 41 institutions, composing 35% of the total, obtained planetaria as additions to existing facilities or else incorporated them into the design and construction of new facilities. In contrast to prewar financial hardships, postwar economic prosperity and the baby boom ushered in a new era of museum building. Children and their families became the leading audiences that attended museums as leisure time grew commonplace.[106] Awareness of the approaching space age, coupled with sales of inexpensive planetaria, fueled the desire to incorporate astronomy education into the repertoire of many smaller museums.

Unlike its academic counterpart, the museum profession had long recognized the importance of a titled "director" as one who was chiefly responsible for overseeing an institution's mission. Few directors of smaller museum planetaria, however, possessed advanced degrees in astronomy. Such qualifications were only expected among larger facilities in metropolitan centers. Some directors were strictly volunteers who received no financial remuneration for their services. Nor were they expected to conduct or publish any form of research.

Public education was the overriding purpose of a smaller museum. Many directors acquired their astronomical skills through informal means, often in conjunction with previous wartime services.

A third category of planetaria, which comprised but 15 institutions (13%), was that operated by the armed forces. Celestial navigation was the principal form of instruction taught in these facilities, for which Spitz designed an astronomical triangle, geocentric Earth, and other specialized projectors that later became standard planetarium accessories.[107] Several training institutions, like the U.S. Naval Academy, were among Spitz's earliest customers. Naval Air stations and Air Force bases, as well as Merchant Marine academies, added to the list. Instructors were chiefly noncommissioned officers who had gained extensive experience during the recent war. Because of frequent turnover, few of these directors retained their posts for very long.

Somewhat ironically, the smallest category of American planetaria was that of private or public schools and districts. Only 8 institutions (7%) procured these devices for student instruction, an indication of the straitened circumstances under which postwar schools were operated. Why was this the case? Historian Arthur Zilversmit has noted that "[a]fter ten years of depression, followed by four and one-half years of total war," American schools were completely unprepared to meet the demographic challenges posed by the coming decade.[108] At the same time, Butler University professor Harry E. Crull lamented that "we have acquired a frame of mind which encourages the belief that astronomy has become an area of esoteric knowledge for the elect."[109] Regardless of whether the Model A's price was too high, that sufficient room to erect the instrument and dome was lacking in overcrowded buildings, or that astronomy was perceived as an overly technical, impractical science within a dominant life-adjustment curriculum (see chapter 6), Spitz's desire to see school rooms transformed into astronomy classrooms went largely unfulfilled. Not until after the launch of *Sputnik* would American education come to embrace the ideals of an earth- and space-science curriculum. Only then would sufficient funds from a new sector, the federal government, become available to turn Spitz's dream into a reality and planetarium instruction into a professionally organized discipline chiefly centered within U.S. public schools.

Summary

Armand N. Spitz transformed the American planetarium community through his introduction of the dodecahedron-shaped, pinhole-style projector that bore his name. By adopting a favorable pricing structure that undercut his nearest competitor, Spitz succeeded in having his product installed at more than a hundred educational institutions across the United States and abroad. He considerably widened the planetarium's sphere of influence by enlarging its accessibility, reaching audiences that might never have attended the demonstrations in large urban centers. Spitz marketed his device at the time when a new era of national prosperity, social stability, and child-oriented leisure time was enveloped by the dawning space age. Lack of free-market competition enabled Spitz to fill a nearly vacant manufacturing niche and become the world's largest producer of planetaria.

Nonetheless, Spitz fought against lingering prejudice held toward his planetaria, from both professional astronomers and Zeiss planetarium directors. Credibility of the Model A projector was boosted following a notable demonstration given before a host of experienced observers. By arguing for a return to the teaching of basic astronomical concepts, Spitz gradually won acceptance for his device and its realistic depiction of the stars and their apparent motions.

Spitz expected that his projector would be installed in public schools throughout the nation. Ironically, this intended market generated the least number of subscribers. Planetarium sales came chiefly from collegiate physics or astronomy departments, regional museums and science centers, and armed forces schools. The sale of Spitz projectors was significantly aided by *Sky and Telescope* editor Federer, whose feature articles constituted word-of-mouth advertising of the most valuable kind. Spitz had accurately foreseen a potential opportunity, exhibited a willingness to take risks, and ignored remarks from those who were regarded as his intellectual superiors. His determination in bringing large projects to completion were added factors in explaining Spitz's success.

By launching his simplified projection device, Spitz strongly reshaped the American planetarium community's postwar growth and diversity. In contrast to the prewar domination of Zeiss installations, there developed a much larger and more pluralistic community of smaller institutions and their directors. Dozens of individuals, including a handful of women educators, fashioned successful careers around Spitz's planetaria without having earned advanced degrees in astronomy. The efficacy of the smaller planetarium as a cross-disciplinary instructional tool was independently demonstrated by Richard H. Emmons.

Two groups of American planetarium directors met regularly to exchange interests, experiences, and concerns. Both groups undertook steps toward professionalization by placing disciplinary goals above those of personal or institutional gain. Despite marked differences in their respective facilities, each recognized that it held a number of objectives in common with the other, the foremost being the delivery of programs and exhibits to planetarium visitors of all ages.

The AAM planetariums section matured into the largest and most diverse body of American planetarium directors yet assembled. Its personnel came chiefly from smaller, regional museums and science centers, although representatives from major planetaria were welcomed into the section. Without Spitz's advocacy of the AAM section as an umbrella organization, it is unlikely that the directors would have achieved the same degree of success in so rapid a fashion. Efforts to sustain a newsletter within the discipline encountered more limited success, however.

A group of "planetarium executives," whose prewar ranks had been thinned by resignation, retirement, or death, drew members chiefly from large metropolitan institutions. A few executives attained positions through prior experience with smaller, and even portable, planetaria. Boston's John Patterson urged cooperation with colleagues from smaller-domed facilities. Executives confronted the growing visual sophistication of audiences and embraced the entertainment spectacles once pioneered by James Stokley. Significant levels of institutional cooperation were achieved in spite of widely divergent programming philosophies. Executives resisted the adoption of any formal organization because it might impede the versatility of their gatherings.

The rise in popularity of Spitz planetaria coincided with civilian research in exploring the upper reaches of Earth's atmosphere. Captured V–2 missiles were converted into scientific instruments that collected geophysical and astronomical data. Photographs taken from aloft first revealed the Earth's curvature and strongly hinted that the space age would soon become a reality. Dreams of the exploration and conquest of space were no longer confined to the literature of science fiction or the fleeting exhibits of a World's Fair. New York's Hayden Planetarium hosted symposia on the prospects of spaceflight, while subsequent distribution of these ideas by magazine and television coverage reached millions of citizens. A revision of the "frontier thesis" replaced earlier emphases on natural theology and moral instruction that characterized planetarium programs during the formative period.

Success of the V–2 program generated proposals to launch the first artificial satellites in culmination of the International Geophysical Year (1957–58). Those plans created unprecedented opportunities for amateur participation via Project Moonwatch. It is impossible, however, to separate the influences of Spitz's Model A projector, the satellite program, and mass media representations of space exploration on renewed popular interest in astronomy. A third distinct period of planetarium growth was to emerge in response to the Soviet launch of *Sputnik I* on 4 October 1957.

Part IV

Planetaria in the Space Age, 1958–1970

Chapter 6

Sputnik *and Federal Aid to Education*

Except for the Japanese attack on Pearl Harbor, probably no event [took] the American people quite so much by surprise as the launching of the first Soviet earth satellite.
—Walter Sullivan, 1961

Launch of the first *Sputnik* satellite on 4 October 1957 caught Americans off guard as dramatically as news of Pearl Harbor.[1] Soviet technological prowess was confirmed when a much heavier satellite and its canine passenger were boosted into orbit the following month. Project Vanguard, by contrast, suffered embarrassing failures; the U.S. Army was directed to loft *Explorer I* into space on 31 January 1958. Political historian Walter A. McDougall has argued that by not anticipating *Sputnik*'s extraordinary propaganda value, the Eisenhower administration was dealt its "greatest defeat."[2] Understanding how and why Americans reacted with such alarm to *Sputnik* is essential for explaining the support provided for the planetarium community's third period of growth.[3]

Sputnik, the Media, and the Public

Historians agree that the launch of *Sputnik* delivered a sudden jolt that redefined national goals and objectives concerning the exploration of space. James R. Killian Jr., then MIT president and newly appointed chairman of Eisenhower's Presidential Science Advisory Committee, wrote, "As it beeped in the sky, *Sputnik I* created a crisis of confidence that swept the country like a windblown forest fire. Overnight there developed a widespread fear that the country lay at the mercy of the Russian military machine and that our own government and its military arm had abruptly lost power to defend the homeland itself."[4] *New York Times* science journalist Walter Sullivan regarded *Sputnik's* impact as "profound," having triggered a "shudder [felt] through large parts of the world." In his judgment, much of the populace was thrown into "dismay and confusion." Sentiments of wonder and excitement, which might otherwise have accompanied humanity's first venture into space, became swallowed in fear.

Accusations were spread that "the Russians had not played the IGY game according to the rules."[5] But a longer view has demonstrated that it was "not Soviet foul play so much as an administrative blunder" that garnered the United States its second place finish.[6] Sullivan's remarks were echoed by J. Allen Hynek, associate director of the Smithsonian Astrophysical Observatory's (SAO) satellite tracking program, who reported "a strange mixture of awe, admiration, and fear," the last being amplified because "there had been no warning."[7] Almost unanimously, scholars have asserted that Americans reacted toward *Sputnik* with "shocked disbelief." As details of the Soviet feat were disclosed, public outcry grew into a "torrent of anxious commentary" that "came close to hysteria."[8]

To what extent were public attitudes shaped by media coverage of the event? That question was posed by Pulitzer Prize-winning author Walter A. McDougall, whose account of the "space technological revolution" offered this explicit claim: "The *press* assumed Sputnik meant Soviet superiority, and the *press* pushed the panic button."[9] McDougall's contention has been critically examined by Lehigh University professor of journalism Jack Lule. His analysis of the language employed in October 1957 newspaper coverage points toward three primary dramas that defined *Sputnik's* immediate impact: "a drama of defeat; a drama of mortification and national self-reflection; and a drama invoking the dream and dread of space." In weighing McDougall's assertion that media influence constructed the public's response, Lule's assessment supports such a view. "In the final analysis," he writes, "the press *selected* its primary themes, its main actors, its own dramas to enact."[10]

Informed contemporaries, however, saw beyond the superficial rhetoric being circulated in media reports. One was Canadian geophysicist J. Tuzo Wilson, who played a significant role in the IGY's scientific programs. Wilson refused to accept the notion of Americans being dealt a "technical defeat" by *Sputnik.* If there was a defeat, he argued, "it was a political defeat, arising from failure to appreciate the psychological impact of the first satellite and failure to support unconventional technical ideas soon enough."[11]

In McDougall's judgment, the devastating blows that *Sputnik* delivered shattered two major premises: (1) "the evident superiority of American liberal institutions," not only in the "spiritual realm of freedom" but in the "material realm of prosperity," and (2) "the overwhelming American superiority in the technology of mass destruction."[12] *Sputnik* stood as an affront to the nation's most cherished values and beliefs, namely that Western democracy and capitalism sustained the world's premiere socioeconomic system. The satellite offered dramatic proof of the existence of Soviet hardware capable of launching an intercontinental ballistic missile (ICBM), which fueled the myth of an American "missile gap." The shock with which the space age began was amplified by strong discomfort that the United States had unthinkably lost its lead in this important enterprise.

Further insights on public reactions toward *Sputnik* were presented by SAO's editor-in-chief E. Nelson Hayes. By laying preparations for tracking all IGY satellites, the SAO unwittingly became a national clearinghouse (and sounding board) on the Soviet satellites. The SAO reportedly received "hundreds of telephone calls" and "thousands of letters" in response to the national "emergency." From a superficial analysis of this unsolicited testimony, Hayes concluded that "[n]ot a few" people expressed belief that "the scientist was once again meddling in cosmic affairs that were not his business."[13] From this viewpoint, *Sputnik* symbolized the potentially evil creation of a none-too-fictional Dr. Frankenstein. Its sudden appearance threw an ominous shadow as threatening as the mythical figure crafted in Mary Shelley's famous novel. By opening the space age, humanity had accepted social and moral responsibilities for all risks encountered beyond the Earth's atmosphere. These sentiments likely underscored a significant fraction of the cultural ambivalence identified by Lule as the "dream and dread" of outer space.

Despite repeated assurances from President Eisenhower that the American satellite program had "never been conducted as a race with other nations," the public mood remained unaltered. Eisenhower's apparent lack of concern drew sharp criticism and charges of complacency from political adversaries. His response toward *Sputnik* was forged during the first two months of the crisis; afterward, no major deviation from this approach was taken. Framed within the context of an enduring Cold War struggle and pressures to restore a balanced budget, Eisenhower declared, "We face, not a temporary emergency, such as war, but a long-term responsibility." For the rest of his term, he remained genuinely puzzled over the degree of anxiety triggered by *Sputnik.*[14] As a consequence, the administration's failure to restore public confidence on issues of national security, pride, American power, and world leadership served to define the platform of the Democratic National Party and helped to insure its success in the 1960 campaign for the White House.[15]

In the aftermath of *Sputnik,* critics blamed the nation's public schools and identified perceived failures of progressive education as a principal reason that Americans had fallen behind the Soviets in space technology.[16] Whether these charges were accurate or not, such claims touched off a "bitter orgy of pedagogical soul-searching."[17] Progressive education's postwar incarnation was known as life-adjustment education, whose objectives had received support from the U.S. Office of Education.[18] Among contemporary critics of life-adjustment education was historian Arthur Bestor, who decried the subversion of education's main purpose, "the deliberate cultivation of the ability to think."[19] Likewise, historian Scott L. Montgomery has argued that this "well-meaning, literal-minded, anti-intellectualism" hoped to instill social and cultural values through an emphasis on "health, safety consciousness, orientation to work, the promotion of intelligent consumership, maturation of interpersonal relationships," and so forth. Life-adjustment education was directed toward "what a student might possibly do, or how he or she might labor and behave," rather than "who[m] they might be or might become."[20]

Sputnik brought this succession of debates to a sudden climax and produced changes of lasting significance. Studies reported that the average Soviet high school graduate had been exposed to a much heavier training in mathematics, science, and foreign language instruction than his or her American counterpart.[21] Such findings were seen as "evidence that we had fallen far behind not only in teaching basic skills but in science and technology," lending apparent confirmation to previous attacks on life-adjustment education.[22] Unless American schools were to reverse their emphases on social skills development, critics maintained, the United States would lose an escalating military-technological race.[23] One of the first organizations to counteract such prevailing thinking was the National Science Foundation (NSF). Through its support of major new science curricula, initiated by the Physical Science Study Committee (PSSC) in 1956, a more rigorous educational program took shape that aimed to make "more understandable to all students the world in which we live and to prepare better those who will do advanced work."[24]

Satellites, Congressmen, and Presidents

National Defense Education Act of 1958

After the launch of *Sputnik I,* prominent scientists and engineers urged the Eisenhower administration to undertake a concerted effort to strengthen American science education. The president, however, was convinced that the Soviet satellite had not upset the military balance of power and turned a deaf ear toward such proposals. Stung again by *Sputnik II* on 3 November, Eisenhower appointed MIT president James R. Killian Jr. as his science advisor. Killian shared Eisenhower's fears of "heat[ing] up the cold war, accelerat[ing] the arms race, and encourag[ing] technological excess." But through Killian's influence, Eisenhower became convinced that training future scientists and engineers had become the nation's "most critical problem."[25] Along with Marion B. Folsom, secretary of the Department of Health, Education, and Welfare (HEW), and Elliot L. Richardson, assistant secretary of HEW's Task Force on Higher Education, the administration drafted recommendations for expanded educational support. Eisenhower strongly "disliked excessive government spending and the possibility of an unbalanced budget," but these measures were undertaken to restore public confidence and counter Democratic challenges to his leadership.[26]

Concurrently, a pair of bills were introduced into Congress by Representative Carl Elliott (Dem., Ala.) and Senator Lister Hill (Dem., Ala.), following congressional hearings conducted by Senator Lyndon B. Johnson. Debates surrounding this legislation, its passage by the Eighty-fifth Congress, and its signing into law on 2 September 1958 are recounted by historian Barbara Barksdale Clowse. She contends that creation of the National Defense Education Act (NDEA) of 1958 "probably owed as much to . . . strenuous bipartisan effort as it did to the national-security crisis."[27] Eisenhower gave his signature only because of the NDEA's avowed intention of serving as "an emergency undertaking to be terminated after four years."[28]

NDEA legislation broke long-standing resistance to federal support of education and brought sweeping changes to U.S. science, mathematics, and foreign language instruction. Because of strict-constructionist interpretations of the Constitution, federal lawmakers had required public school districts to seek their own means of support (usually derived from local property taxes). While the issue of federal aid "had been raised periodically in the Congress since the 1870s," it had "consistently failed to pass, no matter how compelling the demonstration of educational calamity." During the twentieth century, this deadlock had remained polarized around three sociopolitical concerns: race, religion, and the fear of federal control. Racial segregation only began to weaken following the Supreme Court's ruling on Brown vs. The Board of Education in 1954, while aid for parochial schools was denied on the grounds of separation of church and state. Even the "awesome new fact of the atomic bomb" had done little to shake prevailing attitudes regarding federal involvement with education.[29] Passage of the NDEA symbolized the "triumph of the arguments of the scientific establishment over those of professional educational groups" and was viewed as a "defeat for general federal aid and the formula for local control proposed by the National Education Association."[30]

The NDEA's significance lay in establishment of an important legislative precedent. By repeated extensions applied during the 1960s, NDEA funding paved the way toward passage of the Elementary and Secondary Education Act (ESEA) of 1965, whereupon general federal support at last became a reality.[31] Further, NDEA funding brought to fruition arguments expressed by Harvard University president James B. Conant for the consolidation of resources at district-wide levels and the establishment of comprehensive high schools. Blueprints for these institutions were sketched in Conant's influential report *The American High School Today* (1959). Scott L. Montgomery indicates that high school consolidation had become a "widespread reality" by 1965.[32] Historian John A. Douglass contends that in the wake of *Sputnik,* the Cold War American research university was redirected as a tool for national defense. Federal control of higher education was avoided through decentralization and adoption of a multi-agency approach in the disbursement of funds.[33]

NDEA legislation triggered the third and largest phase of U.S. planetarium development, whereby the nation's public schools became the chief recipients of planetaria during the 1960s. Federal support not only fulfilled Armand N. Spitz's intention of placing pinhole-style projectors in public school classrooms, it resulted in the creation of new professional associations whose members formed allegiances along regional, and finally international, parameters.

Title III Funds: Support for School Planetaria

The National Defense Education Act of 1958 was composed of ten subchapters or "titles," which specified how "existing imbalances in our educational programs" were to be remedied "as rapidly as possible."[34] According to its Title III provisions, appropriations of $70 million per year were authorized over the next four years. Section 441 stated that these amounts were to be distributed "to State educational agencies . . . for the acquisition of equipment . . . and for minor remodeling." Nowhere in this title, however, were exact specifications given about what types of equipment or remodeling were considered eligible.[35] But it was under the heading of new equipment or the retrofitting of an educational space that planetarium projectors and domes were suited for NDEA assistance.

Under "Allotments to States," section 442, (a), (2), it was noted that funds were to be assigned on a matching basis, although the "allotment ratio" itself could range between one-third and two-thirds of a project's total cost. The principal stipulation, section 443, (a), regarding disbursement of NDEA funds was the submission, through a state's educational agency, of a "State plan" satisfying requirements listed elsewhere in the Act. Responsibilities fell on state departments of public instruction to insure that locally initiated projects submitted for NDEA approval were compatible with existing goals and objectives. In one particular state (Pennsylvania), enthusiasm for NDEA assistance ran so high among officials that its newly implemented earth- and space-science curricula became models for education reforms initiated elsewhere. Not coincidentally, that attitude enabled Pennsylvania to acquire more school planetaria than any other state, outstripping New York and California combined.

Designed to serve as an "emergency" measure, the NDEA was renewed and amended by later Congresses, which enrolled other academic subjects under its auspices. Annual appropriations for Title III provisions grew to as much as $130 million during fiscal year 1971. NDEA expenditures were later merged with titles created under the Elementary and Secondary Education Act (ESEA) of 1965, whose Subchapter II provisions specified annual appropriations of $100 million for the creation of "Supplementary Educational Centers and Services." Under ESEA directives, "planetariums" became one of a number of "[c]ultural resources" bolstered by this Act.[36]

When school districts submitted proposals for NDEA assistance, it became necessary for administrators to adopt guidelines for determining the eligibility of science equipment purchases and space remodeling. Formal specifications therein allowed "Planetariums" to be regarded as "Eligible Science Equipment for Teacher Demonstration." At the same time, conversion of "existing space in a completed building into a planetarium"[37] legitimized the procurement and incorporation of these facilities in the nation's public schools.[38]

Few public or private schools were fortunate enough to install planetaria before the space age had begun. Yet, schools and districts were to become the largest institutional category of the American planetarium community in the post-*Sputnik* era. Federal assistance at last topped the support of Zeiss planetaria furnished by the Weimar Republic during the interwar period (see chapter 1). Of greater importance to U.S. astronomy education was reintroduction of that subject into national curricula after an absence of nearly sixty years. Significant differences were readily apparent, however, because astronomy's role no longer reflected the satisfaction of college-entrance requirements nor its value as a "mental discipline" subject, two prominent rationales exercised in pre-1900 curricula. After 1957 astronomy and space science were looked upon as academic disciplines to be mastered in the nationwide rush to catch up with the Soviets. Public schools were not the only beneficiaries to reap the rewards of a national defense ideology. Colleges and universities were being readied for an influx of the first baby boomers.

Higher Education Facilities Act of 1963

Five years after NDEA legislation gained approval, a similar measure was passed by the Eighty-eighth Congress, which extended assistance toward "instruction or research in the natural or physical sciences, mathematics, modern foreign languages, or engineering," to the nation's colleges and universities.[39] The Higher Education Facilities Act (HEFA) of 1963 addressed problems facing the "security and welfare of the United States." Both of these conditions, it was argued, might become "jeopardized" unless "this and future generations of American youth" were assured of opportunities to attain the "fullest development of their intellectual capacities." Such actions were deemed necessary to accommodate "rapidly growing numbers" of young adults who "aspire[d] to a higher education."[40]

Composed of four subchapters or titles, HEFA legislation established criteria by which federal grants or loans were to be used for the construction of undergraduate and graduate academic facilities. Subchapter I appropriations allocated $230 million annually for fiscal years 1964–1966. Under Subchapter IV, the term "academic facilities" was inferred to mean "structures suitable for use as classrooms, laboratories, libraries, and related facilities necessary or appropriate for instruction of students."[41] Two years later, generalized support, including payments and loans for student tuition, became available upon passage of the comprehensive Higher Education Act (HEA) of 1965. The HEA subsumed earlier HEFA legislation in much the same way that ESEA assistance encompassed the more narrowly conceived NDEA measure. The ESEA and HEA of 1965 have been called "truly pivotal" legislative actions by Barbara Barksdale Clowse, who argues that they established federal support "for all manner of educational undertakings."[42]

That planetaria were eligible for federal assistance under HEFA provisions is shown by the case of Clarke College in Dubuque, Iowa, which received a loan of $1.6 million for construction of a new "[s]cience building and planetarium."[43] Unlike school planetaria, however, which served a district's student population, many collegiate-owned planetaria were operated for the benefit of the public. These institutions reached broader clienteles than the undergraduate students for whom they were designed. Monthly programs were delivered as a form of payback for local taxpayer support. An unforeseen consequence of HEFA legislation was that many university and college planetaria offered astronomy instruction in the form of adult and continuing education. This informal education held the potential to influence public attitudes regarding federal support for continued space exploration (see chapter 7).

The Spitz A3P Projector and the Space Science Classroom

The advent of federal assistance to U.S. science education exerted profound changes within Spitz Laboratories, Inc. According to Herbert N. Williams, who joined the company as sales representative in 1952,[44] reconceptualizations of its product line and marketing strategies began soon after the 1958 symposium on planetarium education, hosted at the Cranbrook Institute of Science in Bloomfield Hills, Michigan (see chapter 8).[45] That meeting occurred less than one week after NDEA legislation was signed into law by President Eisenhower. Delegates needed no encouragement in teaching astronomical concepts pertinent to the space age. Quoting delegate Charles G. Wilder, editor Miriam Jagger expressed how attendees "could scarcely fail to be aware that the 'cold scientific war of the Sputnik era' has had far-reaching consequences for planetaria."[46] In Williams's estimation, "two things . . . had to happen: [i]nstrumentation . . . had to change, and our approach to education . . . had to take on an entirely different complexion."[47]

The first of these initiatives resulted in the design of a more sophisticated, versatile projector that met Williams's expectations for the "planetarium of the future." Designated the A3P, Spitz Laboratories' new device was to become the most widely manufactured planetarium instrument in the world. More importantly, a new marketing strategy was conceived that satisfied demands for space science education heard at state and national levels. Before the impact of federal assistance on the American planetarium community can be documented, it is necessary to describe this post-*Sputnik* entrepreneurial thrust.

The A3P embodied the largest innovation in projector design since the Model A itself was introduced (see chapter 5).[48] In place of its awkward-looking dodecahedron, a spherical starball was substituted, giving the instrument a sleek, professional appearance. Techniques of miniaturization yielded a fully integrated system of planet projectors. This allowed synchronous movements of the Sun, Moon, and planets to be controlled by a single switch, and individual planet projectors could easily be reset. The A3P's circular planetary "analogs" were derived from Tychonic representations of the solar system.[49] While not as accurate as the linkages employed in Zeiss projectors, the A3P's capability of showing a full range of planetary phenomena represented a milestone in the technical development of smaller planetarium instruments. This achievement put the A3P almost on a par with larger, more expensive foreign-made instruments.[50]

Wallace E. Frank (b. 1921), an MIT graduate and former employee of the Franklin Institute, served as principal designer of the A3P. An early shareholder in Spitz's venture, Frank had worked in the bioengineering field before joining the company as executive vice president. In 1959 Frank was elected president instead of Armand N. Spitz. This transition removed the elder patriarch and

FIGURE 1. Artist's conception (circa 1917) of a Ptolemaic (i.e., projection) planetarium, designed for the Deutsches Museum, Munich. *From "Der Aufbau der Astronomie im Deutschen Museum (1905–1925)," by Franz Fuchs.* Deutsches Museum Abhandlungen und Berichte *23, no. 1 (1955), 62. Reproduced by permission from Deutsches Museum, Munich.*

FIGURE 2. Zeiss Model II planetarium projector and lecturer with arrow pointer, circa 1930. *From "Canned Astronomy: What the New Planetariums for Chicago and Philadelphia Will be Like," by Albert G. Ingalls.* Scientific American, *September 1929, p. 200.*

FIGURE 3. Rosicrucian planetarium projector, circa 1936: the first American-built projection device. *Courtesy Rosicrucian Egyptian Museum, San Jose, California.*

FIGURE 4. Advertisement for Spitz pinhole-style planetarium projector, *Sky and Telescope,* October 1947. *From* Sky and Telescope, *October 1947, p. 27. Reproduced by permission from Sky Publishing Corporation and Spitz, Inc.*

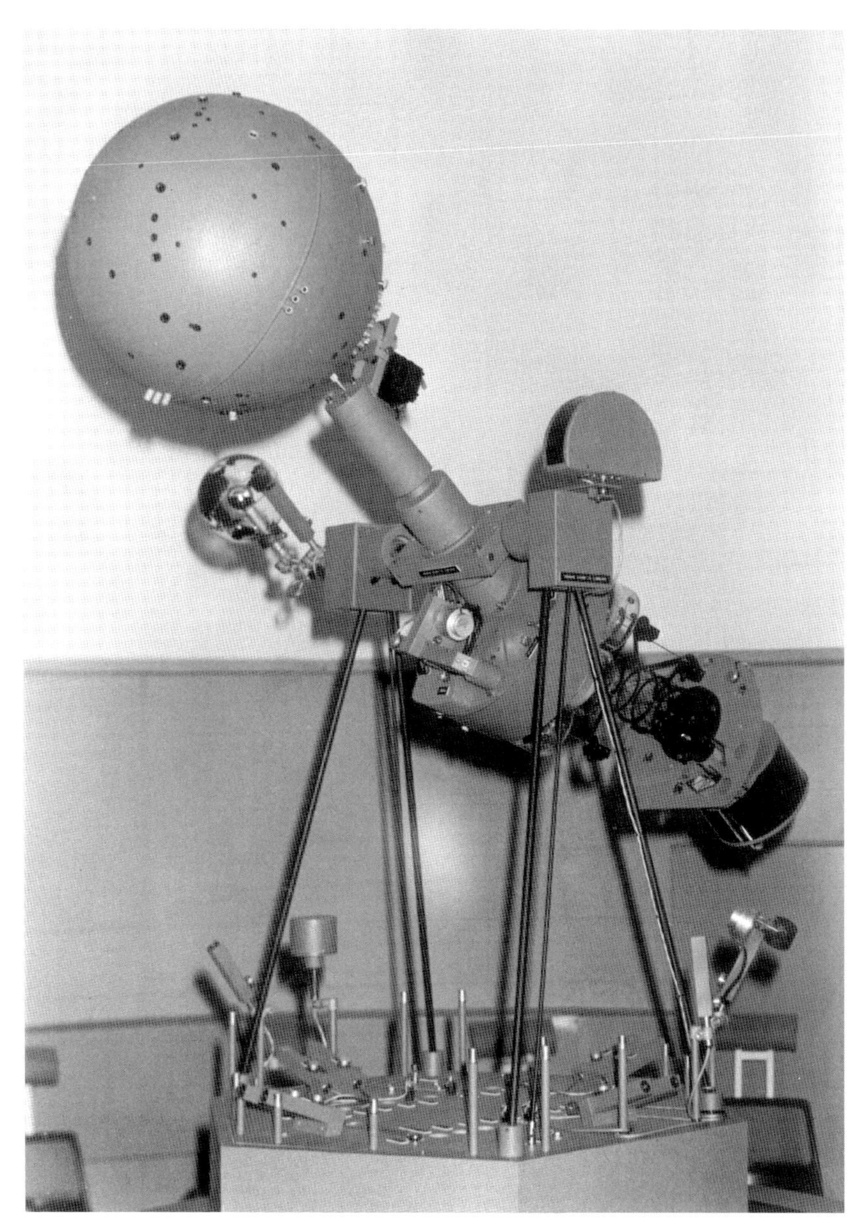

FIGURE 5. Spitz A3P planetarium projector, designed by Wallace E. Frank. *Photo by Jordan D. Marché II.*

FIGURE 6. Gemini 4 astronauts Edward H. White (l.) and James A. McDivitt practice celestial navigation skills inside the Morehead Planetarium, Chapel Hill, North Carolina (1965). *NASA photograph. Courtesy Morehead Planetarium, Chapel Hill, North Carolina.*

FIGURE 7. A single frame from a time-lapse motion picture of cumulus cloud development, taken with a 180-degree fisheye camera lens at the University of Nevada-Reno's Fleischmann Atmospherium-Planetarium (1960s). *From "Techniques of Extreme Wide-Angle Motion-Picture Photography and Projection," by O. Richard Norton.* Journal of the Society of Motion Picture and Television Engineers, *February 1969, 83. Reproduced by permission from SMPTE.*

FIGURE 8. Experimental setup for studying the behaviors of indigo buntings inside the Robert T. Longway Planetarium, Flint, Michigan (1960s). *From "Migratory Orientation in the Indigo Bunting,* Passerina cyanea,*" part 1, by Stephen T. Emlen.* The Auk, *July 1967, 316. Reproduced by permission from the American Ornithologists' Union.*

FIGURE 9. Conference of American Planetarium Educators (CAPE), Michigan State University, 22 October 1970. Host Von Del Chamberlain and keynote speaker George O. Abell are visible in lower right corner. *Courtesy Michigan State University Archives and Historical Collections. Negative no. 703198–1.*

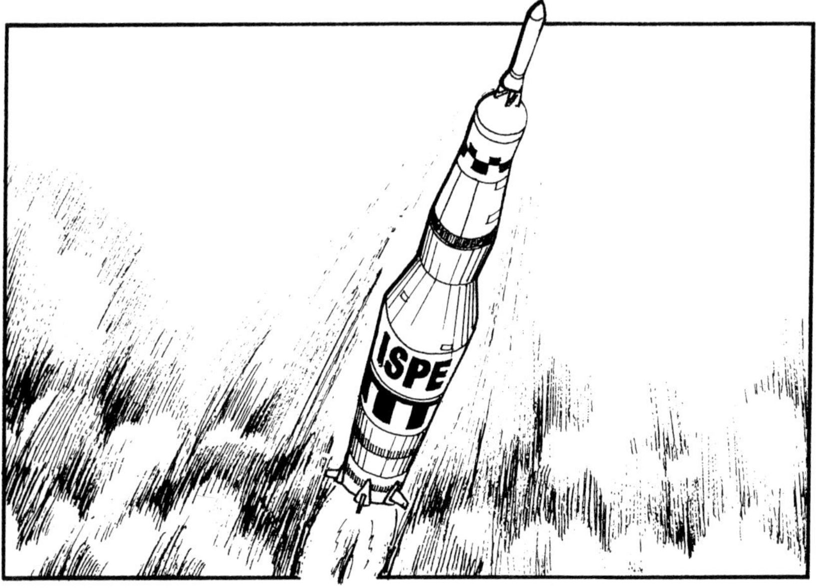

Vol. 1, No. 1
June 21, 1972

FIGURE 10. Cover of the inaugural issue of the ISPE journal *The Planetarian*, 21 June 1972. *Courtesy International Planetarium Society.*

closed the era presided over by the planetarium community's leading prophet. While establishing the feasibility and acceptance of small, pinhole-style projectors, Spitz's ideas were likely regarded as antiquated for the coming era defined by federal assistance.[51] It was under Frank, and not Spitz, that the American planetarium community came of age during the 1960s.

An important marketing strategy framed the A3P within a larger "package" designated a "space science classroom." When Armand N. Spitz had sold his first Model A projectors, provisions for makeshift domes and installations were shouldered by customers. Nothing of this sort, however, could be left to chance in the space age. Complete architectural services and professional installations became the norm. With the advent of NDEA legislation, planetaria moved to the forefront of educational theory and practice. Development of a fully integrated learning system, encompassing both projector and domed classroom facility, enabled these devices to find their way into schools and colleges in record numbers. By means of this approach, the A3P's potential as an instructional tool capable of delivering a space science curriculum was fully realized.

A Space Science Classroom is the title of a sixteen-page booklet produced by Spitz Laboratories in about 1961, with the twin purposes of promoting the A3P and redefining the planetarium's educational mission.[52] Of central importance was the adoption of a holistic instructional approach, derived from the latest pedagogical theories. *A Space Science Classroom* posited the birth of a "space science revolution," which was compared to the "industrial revolution" of an earlier era. "[T]he last five years," the report argued, had produced a type of mental stress that was later diagnosed in Alvin Toffler's *Future Shock* (1972): "We have been projected into the future with a suddenness that has completely outmoded our past ways of thought and action." The dawning space age, it was claimed, necessitated a "basic understanding of space" among "all citizens." To meet this daunting challenge, a profound "curriculum revision" was called for at "all levels."[53]

Full attention was paid to school-age children, for whom it was argued that, "[b]y starting astronomy early in the child's education," a "better comprehension and retention" of "concepts and relationships," along with a "more rapid assimilation of knowledge," became possible.[54] Presentation of scientific ideas to children of almost any age was derived from the educational theories advanced by Harvard University psychologist Jerome S. Bruner in his widely circulated report *The Process of Education* (1960). Bruner's disciplinary approach to curriculum design stood as the antithesis of all "progressive" and "life-adjustment" models.[55] Further evidence of Bruner's influence can be found in statements that planetaria were "ideal place[s] to establish the principles of the scientific method." Through repeated exposures to this "laboratory," planetarium-trained students, it was suggested, would emerge better prepared to begin their "scientific career[s]."[56] These assertions affirmed the inquiry-based approach that underpinned the latest mathematics and science curricula produced through NSF sponsorship.[57]

The "space science classroom" approach likewise targeted colleges and universities. Along with physics and astronomy departments, institutions that offered teacher training were expected to become major users of planetaria. "As more public schools initiate courses in the earth-space sciences," it was argued, correspondingly "more teachers will have to be qualified in the new courses" through planetarium instruction.[58] Collegiate planetaria were needed to train the next generation of astronomy instructors, who were to take their places in the space science classrooms being erected throughout the country. This initiative served to reestablish the cycle of astronomy teaching and learning that had been largely absent from the nation's schools and colleges since the turn of the century. Thus, at no time since the formative period was the educational role of planetaria as thoroughly envisioned, or closely integrated with leading curriculum theories, than in the pages of *A Space Science Classroom*.

A second design innovation, which bore immense practical significance to the space science classroom, prompted traditional concentric seating patterns in planetaria to be abandoned. This development arose from a compromise struck between the planetarium's circular architecture and the pedagogical features of a standard classroom that emphasized a teacher-directed focus. As former Spitz employee E. Jack Spoehr has explained, under the former regime, "[t]here was no practical way to introduce a science demonstration table, chalk board," or other projection screen into the planetarium when half of the audience's back was turned towards it. The circular seating pattern, "so firmly entrenched in classic planetarium design[,] was increasingly recognized as a handicap to good instructional geometry and was replaced with a succession of 'unidirectional' seating patterns," which allowed all students to face in roughly the same direction.[59]

Another Spitz Laboratories' promotional effort, a sixty-six-page manual *The Use of the Planetarium in the Teaching of Earth and Space Sciences*, was published in 1960.[60] Rather than offering a sequence of programs geared toward particular grade levels, the manual presented a collection of "lecture suggestions" for imparting contents to pupils of all ages. By stressing "observational study" over the memorization of facts, students might acquire a firmer grasp of "<u>why</u> and <u>how</u>" our knowledge of the heavens had been obtained. Use of the planetarium to teach confirmation of the heliocentric theory (Oskar von Miller's original intention for the projection planetarium) was explicitly encouraged: "the exquisite deductive reasoning process which led to the heliocentric system can be really appreciated." Integration of astronomy with other subjects (a practice long advocated by Armand N. Spitz) was retained. Planetarium instruction could enjoin "all the basic sciences" and "demonstrate the interrelationships between them."[61] *Use of the Planetarium* served as a template for the assembly and delivery of complete planetarium lessons, including pre- and post-visit activities. This latter objective was not realized, however, until chapters of Spitz's *Planetarium Director's Handbook* were issued sequentially by education director Michael A. Bennett, starting in 1971.[62]

Availability of federal funds spurred the first significant competition within the small to mid-sized planetarium market. During the 1960s both foreign and domestic manufacturers began producing comparably priced planetaria. An M–1 projector made by the Goto Optical Manufacturing Company of Japan was installed at Bridgeport, Connecticut's Museum of Art, Science, and Industry in 1962.[63] Bridgeport's director Phillip D. Stern went on to establish his own corporation, Planetariums Unlimited, Inc., of Holbrook, New York, which marketed its Viewlex series of projectors. Harmonic Reed Corporation of Rosemont, Pennsylvania, makers of Spitz Jr. projectors, upgraded their product line with a series of NOVA planetariums. By undercutting the A3P's cost, Harmonic Reed captured a portion of the lower-end school planetarium market, installing projectors in classrooms suitable for single-class demonstrations. All of these manufacturers benefited from acceptance of the space science classroom approach pioneered by Spitz Laboratories.

NASA's 1966 Planetarium Survey

From the humiliation caused by *Sputnik I* and *II,* American political leaders became painfully aware that formulation of a national space policy transcending IGY scientific experiments was an imperative responsibility. Legislation creating the National Aeronautics and Space Administration (NASA) as a civilian agency was signed into law by President Eisenhower on 29 July 1958. The agency's existence became official on 1 October, when it subsumed the older National Advisory Committee on Aeronautics.[64]

By the mid–1960s,[65] officials in NASA's Educational Programs Division recognized that planetaria had acquired a "new significance" because of substantial contributions made through "educational and informational programs." They noted a "rapid increase" in the number of existing or planned installations, which had begun to foster new regional associations of planetarium directors. To collect information about this growing body of facilities and instructors, a comprehensive survey of all "permanent-type planetarium installations" in the United States was implemented. With the collaboration of directors, manufacturers, and state education officials, a database of over four hundred institutions was compiled. NASA's survey questionnaire was mailed in November and December 1966, from which a response of "97 or 98 percent" was reported.[66]

In a table titled "Distribution by States of American Planetariums," the regional distribution of planetaria was already apparent. NASA's list, however, was far from accurate in defining state memberships among future regional associations. "New England States," for example, never composed an independent association, but joined that created by other "Mideast States." Among these "incipient" regions, the latter (including Pennsylvania's 69 installations) claimed the largest number of American planetaria (127). "Rocky Mountain States," by contrast, reported the least number (11). This table exhibited the earliest tri-fold classification of planetaria on the basis of institutional categories. Among them,

171 were operated by elementary and secondary schools, 162 by colleges and universities, and 88 by museums and observatories. Museum planetaria were outnumbered almost two to one by both public school and collegiate installations. From this data alone, it is apparent why a new type of professional association, whose principal clientele was drawn from the rapidly expanding school and collegiate facilities, came to dominate the community's social structure and subsumed the "planetariums section" of the AAM (see chapter 8).

The most significant aspects of the 1966 NASA planetarium survey are its tabulations of instrument types and funding sources. Respondents were asked in "Question 2a. Equipment" to "Describe basic planetarium projector," including "Name of manufacturer" and "Model." In "Question 4a. Financing," respondents were asked to "List major sources of funds for purchase and installation of equipment and physical facilities." One of five responses that could be checked was "U.S. government."[67] No other query of planetarium installations preceding or following NASA's survey addressed both fundamental questions.[68]

Among 421 installations surveyed, 152 (36%) indicated receipt (or expected allocation) of federal assistance. Of that number, 6 had been operated by armed forces divisions before 1958. Thus, 146 new planetaria (35%) received federal support in the post-*Sputnik* era.[69] While that last percentage is not particularly high, it must be recognized that those installations represented a gain of 126% over the total number of permanent planetaria operated before 1958 (116 institutions; see chapter 5). Among the 152 planetaria receiving federal assistance, nearly two-thirds (99) were operated by elementary and secondary schools. More than one-quarter (42) belonged to colleges and universities, while the remainder (11) belonged to museums and observatories.

Out of 152 planetaria receiving federal aid, 107 installations (70%) either obtained or intended to purchase Spitz A3P instruments. Those installations reflected a gain of 92% over the total number of planetaria operated before 1958. Of the 99 elementary and secondary schools receiving federal support, 67 selected A3P instruments, while 34 out of 42 colleges and universities and 6 out of 11 museums and observatories reached the same decision. Among states, Pennsylvania led the elementary and secondary schools with 19 A3Ps, while California topped the colleges and universities list with 16.

Information from NASA's survey regarding the nature of federal funding is not available. Categories of school and collegiate institutions presumably reflected the support obtained from NDEA and HEFA legislation (and their ESEA and HEA successors). These data are indicative of the funding sources operating near their peak allocations, and hint of continued planetarium growth into the early 1970s. They offer the most compelling evidence in support of the arguments expressed above, namely that in response to federal assistance, introduction of Spitz A3P projectors and "space science classrooms" established the third major phase of American planetarium development. As a result, the planetarium community adopted new measures of collaboration that were to (a) subsume its earlier locus of activity within the museum discipline, and (b) unite its members into an international professional association by decade's end.

Pennsylvania's Earth- and Space-Science Curricula

Still to be explained among NASA's survey results was the extraordinary number of planetaria (69, or roughly 16% of 421 institutions) then established in Pennsylvania. Before 1958 only 9 installations, including Zeiss-equipped facilities at Philadelphia and Pittsburgh, were found in that state. Other pre-*Sputnik* planetaria were divided among 3 colleges and universities, 2 secondary schools, 1 public museum, and 1 armed forces installation. But by the end of 1966, 51 elementary and secondary schools, 12 colleges and universities, and 6 museums and observatories had acquired or were expecting to install planetaria. Of these, 31 out of 51 schools, 2 out of 12 colleges and universities, and 1 museum (operated by its local school district) received or anticipated federal assistance. At least 19 planetaria would house Spitz A3P instruments. How did Pennsylvania come to lead the nation in establishing school planetaria?

Pennsylvania's accomplishment arose in large measure from the dynamic leadership of officials at the Department of Public Instruction (DPI) in Harrisburg, who designed and implemented the nation's first comprehensive earth- and space-science curriculum for secondary students. Two years later, a similar course of study was launched for elementary-level students. Upon the advent of federal assistance, districts throughout the state were urged to install planetaria in new secondary school buildings. Pennsylvania's school planetaria represent the culmination of the DPI's instructional strategies. Still another factor responsible for the state's record planetarium growth was its large number of independent city and borough school districts. During the 1963–64 school year, a total of 657 districts operated secondary schools.[70] Surrounding states such as Maryland, by contrast, operated consolidated school districts that pooled educational resources on a county-wide scale. Such states contained perhaps an order of magnitude-fewer school districts than Pennsylvania. Even if every one of a state's consolidated districts had acquired planetaria, their numbers could not have matched those obtained within the Commonwealth of Pennsylvania.

In 1958 Charles H. Boehm, Pennsylvania's superintendent of public instruction, attended an Air Force Association Convention and Aerospace Panorama in Dallas, Texas. Boehm later described the event as "a preview of the kind of world in which today's youth would live the greatest period of their lives."[71] He returned to organize a committee of fourteen educators drawn from the fields of astronomy, geology, meteorology, and oceanography. Members of this advisory committee included Arthur L. Draper and Israel M. Levitt, directors of the Buhl and Fels planetaria in Pittsburgh and Philadelphia. An earth- and space-science curriculum was drafted and introduced throughout the state during the 1959–60 school year. This curriculum received a citation of merit from the Air

Force Association for "an outstanding contribution [made] to space age education."[72] To promote its curriculum and recruit potential teachers, some 20,000 copies of *A Suggested Teaching Guide for the Earth and Space Science Course* were distributed. Three years after its introduction, the number of Pennsylvania schools teaching earth and space science annually had climbed from 240 to more than 400, while the estimated number of students enrolled had soared from 25,000 to 38,000, far exceeding the expectations of its creators.[73]

According to John E. Kosoloski, DPI's earth- and space-science coordinator at the time, the curriculum's success largely stemmed from its origination at "the state level, from the capitol, from the area where leadership is expected." Yet Kosoloski claimed that the DPI never pushed the curriculum in a top-down, "autocratic" fashion. Regardless of how it was perceived, the course design and materials were offered as a state-level recommendation, but never as a legal mandate. "We recommended and suggested it," he writes, "but . . . we did not order it."[74] Another DPI initiative, appearing in a manual on school construction standards, recommended that new secondary schools be equipped with planetaria, or observatories, or earth- and space-science laboratories.[75] To the lasting benefit of its pupils, ambiguities surrounding the legal import of these statements strongly aided new planetarium construction. By the end of 1969 the DPI had witnessed the establishment of planetaria in 85 school districts, with 25 others in the planning or completion stages.[76]

DPI influence likewise extended to the state's colleges and universities. Boehm recognized that for an earth- and space-science curriculum to be adopted among Commonwealth schools, a corresponding teacher certification program must be implemented. Among the skills expected of prospective teachers were "a working knowledge of the use of the planetarium for laboratory work." As a result, planetaria were to be installed at each of the fourteen state college campuses. A minimum of twenty-four semester hours of coursework in geology, meteorology, and astronomy was required for teacher certification. At the elementary level, NDEA funds were used for the purchase of instructional equipment and training of more than 6,000 teachers in general science education, a "gigantic undertaking . . . done at no cost to local authorities."[77]

In utilizing NDEA funds, Kosoloski and other officials considered it their "responsibility as citizens" to maximize the potential afforded by this unprecedented funding opportunity. Pennsylvania's earth- and space-science curricula satisfied the condition of a "State plan" mandated by NDEA lawmakers. Kosoloski described his enthusiasm for this task as being "not [simply] a job; it is a calling."[78]

The 1959 *Teaching Guide* argued that "[e]vents of the past several years demand that every citizen of the Commonwealth be well acquainted with the Earth and the Space that surrounds it." The course's explicit purpose was "to broaden a student's understanding of his physical environment both on Earth and in Space." A sequence of "Topics" was prepared for each unit of study. Specific educational objectives, or "Understandings," provided the relevant terminology and concepts to be mastered by students. "Suggestions to Teachers" offered potential lesson strategies for achieving the objectives. Four curricular units and their allotted lengths of study were: "The Changing Earth" (11 weeks), "The Earth in Space" (9 weeks), "Weather and Climate" (9 weeks), and "The Oceans" (3 weeks). When an expanded version of the *Teaching Guide* was issued in 1963, a closer integration of its contents with traditional instruction in biology, chemistry, and physics was displayed.[79]

DPI authors revealed an adherence to the curricular theories of Jerome Bruner in their claim that, as a "discipline," earth- and space-science constituted a "distinct meaningful pursuit of learning." Mindful of the recent IGY success, committee members argued that the Earth, its atmosphere, oceans, and surrounding space were not "separate unrelated subjects," but instead "closely related parts of Man's total physical environment." Astronomical charts reflected the *Guide's* emphasis on an understanding of maps used to record and interpret earth- and space-science data. A portable telescope was among the types of equipment "strongly advised" for purchase. A class "trip to an observatory or planetarium" was thought to "unquestionably stimulate interest" among students.[80]

How did an earth- and space-science curriculum differ from that of a traditional astronomy course? Eight of nine subunits within "The Earth in Space" contained objectives and lessons routinely employed by astronomy instructors.[81] Lesson outlines bore no sign of adaptation to school planetaria. Only within the final subunit, titled "Space Travel," were nontraditional subjects encountered. Concepts of orbital and escape velocities, atmospheric drag, and rocket propulsion were introduced. The launch of artificial satellites and the hazards of manned spaceflight were discussed. Aspects of long-duration space travel, closed ecological systems, and alternative propulsion systems rounded out this subunit.[82] In a 1963 revision of the *Guide,* "Space Travel" was redefined as "Astronautics," while greater attention was paid to the "Advantages of Man in Space Vehicles." An expanded account of satellites launched for communication, navigation, and meteorological purposes, along with Pioneer V's flight to Venus, received mention. A new subunit, titled "Man in Space," featured graphics of a Mercury space capsule, descriptions of weightlessness, and the G-forces experienced during launch and reentry. A full-page illustration depicted the problem of "Re-entry from the Moon." Study of closed ecological systems was elevated to a single-page subunit entitled, "Space Biology."[83]

Creation of the 1961 *Earth and Space Guide for Elementary Teachers* was sparked when instructors from the Deep Run Valley Joint Elementary Schools at Blooming Glen, Pennsylvania, asked students what questions they might have about Earth and space in preparation for teaching those subjects in the following year. Such questions, it was argued, enabled "the natural curiosity of children" to be directed "toward a desire to find their own answers." This problem-solving approach was dubbed the method of "scientific inquiry—making an effort to find out." Occasional failures to reach a "final, satisfying conclusion" were not to be disparaged, but rather were regarded as a "highly desirable situation." "As the child begins to adopt this attitude," it was argued, "he himself is becoming 'a scientist.'"[84]

The *Guide for Elementary Teachers* consisted of four sections: "The Earth, Primary and Intermediate," and "The Universe, Primary and Intermediate." Its format began with "Children's Questions," which it was argued "should be given and discussed first," without reference to the *Guide.* Simplified answers were furnished under "Related Concepts and Information," with an explicit caveat that "[c]oncepts cannot be taught, as such." Instead, the child "must gradually accumulate the knowledge which becomes a concept," as illustrated by an example, "The earth is very large." "Experiences and Activities" encouraged students to investigate or model natural phenomena (such as phases of the Moon) and to grasp principles revealed by observations. Neither a one-time visitation, nor repeated access, to a projection planetarium was assumed. An "umbrella planetarium," decorated with the Big and Little Dippers, was used to show why circumpolar star patterns remained visible all year long. Integration with mathematics and other disciplines was advocated: "[a]lmost all aspects of social studies are related to man's way of adapting himself to, and making use of, his earth and space environment."[85]

Again, only subsections devoted to "Space and Space Travel," at both primary and intermediate levels, revealed significant departures from traditional instruction in elementary astronomy. Constellation study was not emphasized but activities demonstrating how a three-dimensional arrangement of stars could yield the apparent two-dimensional shape of the Big Dipper were provided. The nature of rocket transportation and travel to the Moon posed the most pressing educational issues. One of the most important questions addressed at the intermediate level was "Why doesn't the moon fall from the sky?" Two pages of text and diagrams explained the situation as resulting from "two forces in opposition." Such an explanation did not invoke the misleading concept of centrifugal force. Notions of rocket staging, guidance, weightlessness, and reentry, along with space stations and the possibilities of life beyond Earth, rounded out this approach. Despite the obstacles that were to be overcome, it was confidently asserted that "Man will explore space."[86]

Pennsylvania's earth- and space-science curricula were predicated on a combination of factors put into motion by the IGY and *Sputnik.* State education officials understood the significance of these events and crafted new teaching guides as broadly integrative and timely as the range of IGY experiments themselves. Renewed allocation of NDEA funds enabled Boehm's curricula to acquire an unexpected significance.[87] In Kosoloski's hands, they became important tools for leveraging the greatest amount of federal assistance awarded to any state for the construction of school planetaria. After ten years of effort, Kosoloski's vision of delivering an earth- and space-science education to Pennsylvania students had situated *one-fifth* of the nation's planetaria (and employment opportunities) within his own state.

Summary

The American planetarium community's third and largest period of growth arose from the "crisis of confidence" triggered by *Sputnik I* and *II*. Long-standing resistance to federal support of education was overturned by passage of the National Defense Education Act (NDEA) of 1958. Availability of matching Title III funds triggered the first large-scale construction of planetaria in schools and districts, thereby fulfilling Armand N. Spitz's original intentions. Federal assistance at last exceeded the stimulus to European planetarium growth furnished in the interwar period by the Weimar Republic.

Legislation further allocated federal resources to the nation's colleges and universities under the premises of improving national security and preparing for an influx of baby boomers. Teacher certification programs reestablished the cycle of astronomy teaching and learning that had been absent from the nation's schools and colleges for roughly sixty years. As a result, U.S. astronomy education witnessed an unprecedented renaissance, strongly aided by the superpowers' race to the Moon.

Launch of the *Sputnik* satellites was taken as evidence of the apparent superiority of the Soviet educational system. American schools, which had been dominated by "progressive" and "life-adjustment" curricula, finally abandoned these approaches and sought to instill new forms of instruction that would enable citizens to meet the challenges of an escalating military-technological race. Planetaria thus acquired a new and far-reaching significance, further aided by the approach to curriculum theory advocated by Harvard University psychologist Jerome S. Bruner.

Suddenly, the earth and space sciences were viewed as academic disciplines to be mastered in the nationwide rush to catch up with the Soviets. The necessity of offering a space-science education to rapidly growing numbers of young people justified the construction of hundreds of school and collegiate planetaria during the 1960s. A lesser but significant impact occurred in the context of informal instruction, whereby these same planetaria furnished their communities with adult and continuing education, often free of charge.

Advent of federal assistance triggered competition among planetarium manufacturers, both domestic and foreign. Spitz Laboratories responded by designing its highly successful A3P projector and marketing the device under its "space science classroom" approach. Elevation of the financial stakes stimulated management initiatives within the company that displaced the elder Spitz into retirement.

By the mid–1960s, a comprehensive (NASA-sponsored) survey affirmed the impact of federal assistance on the acquisition of U.S. planetaria and disclosed their geographic distribution. Within a few more years, planetarium directors would pursue the formation of regional associations as a further step towards disciplinary professionalization (see chapter 8).

Dynamic leadership within Pennsylvania's Department of Public Instruction envisioned planetaria (and astronomical observatories) as vehicles for the delivery of its earth- and space-science curricula. Officials leveraged the greatest amount of federal assistance awarded to any state, which enabled the Commonwealth to acquire one-fifth of the nation's planetaria by decade's end.

CHAPTER 7

New Horizons in Planetarium Utilization

A number of new initiatives in American planetaria arose during the third developmental period. These ranged from almost purely instructional functions (involving the training of NASA's Mercury, Gemini, and Apollo astronauts) to the strictly theatrical (experimental stagings of the first planetarium light shows). "Fisheye" cinematography, a new "immersive" projection technology, made its debut in the planetarium community. While the first two initiatives were confined to major facilities, the third was pioneered in a smaller collegiate installation.

In addition, planetaria were transformed into research laboratories whose artificial skies were employed to study the star-recognition abilities of nocturnally migrating birds. More importantly, educational research was conducted on audiences of children and adults, with hopes of determining the most successful instructional strategies for presenting astronomical concepts. Many of these initiatives enhanced the career trajectories of planetarium apprentices and directors alike, thereby assisting the community's process of professionalization (see chapter 8).

Morehead Planetarium's Astronaut Training Program

Planetaria had been used for teaching principles of star recognition and celestial navigation to U.S. servicemen, but they acquired a new significance with the coming of manned spaceflight. As part of their training for Earth orbital and lunar landing missions, all NASA astronauts who flew in space between 1961 and 1975 received specialized instruction at Morehead Planetarium at the University of North Carolina in Chapel Hill. Anthony F. Jenzano, Morehead's director, submitted the winning proposal for hosting NASA's original Project Mercury astronauts. The duration and sophistication of training exercises exceeded Jenzano's expectations, and numerous personal and professional rewards accrued to him, his staff, and the institution.

Reasons behind Morehead's selection had much to do with Jenzano's technical and conceptual prowess, along with the planetarium's proximity to Cape Canaveral in Florida. Astronauts first began training at NASA's Langley Research Center at Langley Field in Virginia. Jenzano addressed his proposal to this institution in 1959.[1] One of Jenzano's leading suggestions for enhancing the "effect of reality" was his conception of "one (or more) lightweight, inexpensive dummy capsule(s)" in which astronauts could be positioned. As the planetarium sky was moved "past the window(s) along a simulated 'flight path,'" astronauts' views corresponded precisely to those objects seen in accordance with the capsule's orientation.[2] Jenzano thus described the first use of a planetarium as an analog spaceflight simulator. His proposed training exercises were judged superior to ordinary studies of the heavens performed under a planetarium's domed ceiling.

Apart from Jenzano's suggestion of the capsule mockup, other factors were responsible for NASA's selection of the Morehead facility. Earlier that year Jenzano performed an extensive overhaul of the Zeiss Model II instrument. Its star field was significantly enhanced by the installation of two collars, manufactured by Zeiss-Oberkochen, for depicting the forty-two brightest stars. This technical upgrade left the Morehead instrument unsurpassed among U.S. planetaria and bestowed praise on the facility for possessing "the highest qualifications of any available."[3] Further, Morehead Planetarium was situated in a rural community whose 25,000 inhabitants placed much smaller demands on its services than any metropolitan installation. Morehead's flexibility of accommodation was repeatedly called upon when astronauts needed to make last-minute reviews of the sky.

John Motley Morehead's donation of the planetarium had initially drawn criticism from local residents who regarded the $3 million facility as "rather impractical." They asked, "Why didn't he give us something useful?"[4] But with the advent of manned spaceflight, Morehead's gift became very practical and such criticism gave way to newfound pride.

Training sessions for Mercury astronauts began at Morehead in February 1960. Each astronaut and his backup spent two days practicing star recognition and introductory celestial mechanics. One of the principal instructors for this task was James W. Batten, a doctoral candidate in earth- and space-science education at the University of North Carolina.[5] On the first day of training, zodiacal stars and constellations were introduced under the full planetarium sky; patterns of association and sequential recognition were emphasized. Activities on the second day centered on star identification along anticipated orbital paths.

To demonstrate this effect, technicians installed two "orbital line projectors" whose inclinations were adjusted to match preselected flight patterns. Astronauts took turns sitting inside a modified Link trainer, whose canopy contained an opening that simulated the view seen through a Mercury capsule's window.[6] The trainer was provided with electric yaw controls to simulate the actions of rocket thrusters. Initially, the line projectors were illuminated and star fields along their paths identified from inside the trainer. Astronauts practiced recognition of those fields, and then performed simulated course corrections without the projector's aid.

What had led NASA to initiate astronaut training exercises at Morehead? Manned spaceflight posed many new risks and uncertainties; political pressures mounted as the Soviet Union strove to achieve that same objective. To forestall a possible disaster, NASA took every precaution that might insure the safety of its crews and the success of its missions. As James W. Batten remarked, "It is most important that [the astronaut] knows where he is, because if for some reason the remote control reentry mechanism fails, he must manually reenter the atmosphere at the correct place."[7] Raymond Zedekar, a training officer in NASA's Flight Crew Operations Division, expressed the matter rhetorically: "We're going into the unknown. . . . What are the requirements, how much training will be needed?"[8] Project Mercury astronauts spent an average of sixty hours in planetarium training sessions.

Star recognition offered an alternative method of determining the capsule's orientation. As part of NASA's effort to build redundancy into spacecraft component systems, celestial navigation was a skill that could be effectively taught. As explained by Project Gemini astronaut Walter Cunningham, "Our intention in studying here at the planetarium is to be able to locate ourselves at all times in space. If all else fails, we will use the stars as our only reference."[9] Stellar positions also provided the foundation on which other, more sophisticated, guidance systems depended.

During Project Mercury, the occasion arose whereby an astronaut *did* make use of Morehead's training to successfully reorient his spacecraft before reentry. On L. Gordon Cooper's flight aboard *Faith 7,* an equipment malfunction forced him to determine and correct the capsule's orientation with a minimum expenditure of fuel. Cooper's mission might have ended in disaster had this maneuver not been satisfactorily accomplished.[10] Three years earlier, Cooper attested to the change in attitude which Morehead's training provided: "I thought I knew a lot about the stars, but I found there was more to know than I had anticipated."[11] Cooper's actions vindicated the training program's effectiveness and solidified NASA's commitment to the program through its Gemini and Apollo missions. Even before Cooper's flight, astronaut Frank Borman voiced gratitude for the training he received and found it "hard to express how much we learned from the planetarium."[12] Borman's words were echoed by Neil Armstrong, who opined, "We didn't realize how little we knew about the beyond until we came here."[13]

Following Project Mercury, a second round of nine astronauts began training at Morehead in January 1963, while a third round of fourteen was initiated in March 1964. Spacecraft hardware and software had become more reliable; thus, Morehead's training came less and less to resemble the emergency function it once assumed. NASA's Gemini missions were designed to perform critical rendezvous and docking procedures necessary for executing a future lunar landing. These objectives demanded more sophisticated spacecraft maneuvers by each two-man crew, and correspondingly greater familiarization with the stars. Accordingly, Jenzano and his staff constructed a two-seat Gemini capsule simulator with electric yaw controls, while a "distance and ranging simulator" depicted the Agena target vehicle. Additional personnel were hired to instruct the new astronauts. Donald S. Hall, an educator under Jenzano, became Morehead's assistant director in 1963. Richard S. Knapp then joined the staff as an instructor and succeeded Hall in 1968. Both men worked closely with each astronaut team, including the *Apollo 11* trio commanded by Neil Armstrong. As Knapp later recalled, "We knew their mission was the one slated to land on the moon, so it was particularly exciting to work with them."[14] Hall and Knapp's experiences were instrumental in their becoming directors of major American planetaria after 1969.

Training for Apollo lunar landing missions placed new demands on the Morehead team because astronauts were to depart from and return to Earth along predetermined flight corridors. Complex spacecraft maneuvers, requiring precise stellar orientations, were necessary to accomplish these objectives. An Apollo simulator, sporting wide- and narrow-field sighting devices, was fashioned by the Kollsman Instrument Corporation. Seating only one astronaut, the simulator's design provided its occupant with remote control over the Zeiss instrument. These actions allowed trainees to refine vehicle orientation changes by economically shifting from one planetarium star field to another.[15] Before setting the lunar module down onto the Moon's surface, Neil Armstrong spent some 130 hours of training at the Morehead facility. Emergency procedures again became essential to insure the safe return of the disabled *Apollo 13*. Its command module pilot, John L. Swigert, performed visual star field alignments only hours before their critical reentry.[16]

The last groups of spacefarers to be trained at Morehead were twenty-one astronauts chosen in 1966 and another eleven recruited in November 1967 who participated in later Apollo, Skylab, and Apollo-Soyuz missions. While the latter two programs were confined to earth-orbital missions, stellar recognition and navigational skills acquired during Apollo flights were relevant to the performance of astronomical observations conducted with Skylab instruments.[17] One of the original Mercury astronauts, Donald K. "Deke" Slayton, was grounded in 1962 after being diagnosed with a minor heart condition. But in July 1975 Slayton became part of the three-man American crew aboard the Apollo-Soyuz mission. He was the last astronaut to have been trained at Morehead.

Fruitful and rewarding relationships were maintained between NASA and the Morehead Planetarium under Anthony F. Jenzano's leadership. Between 1960 and 1975, sixty-two astronauts undertook training exercises, spending thousands of man-hours beneath its starfield.[18] Morehead's astronaut training program doubled as a superb public relations vehicle for the planetarium, university, and surrounding community, even though most astronaut visits were announced after the fact in order to minimize security risks and the appearance of autograph seekers. Astronaut training demonstrated the usefulness of Morehead's donation by helping the nation to achieve a politically sanctioned endeavor, thus erasing former criticism of the institution's purported impracticality.

Why, then, did astronaut training at Morehead come to an end, along with the Apollo era? First, manned space exploration came to a temporary halt as NASA concentrated on sending unmanned spacecraft to the planets. Second, orbital missions conducted by NASA's Space Shuttle program established a division of labor among astronauts as either pilots or mission specialists. The latter group required little training in stellar recognition that was mandatory of earlier pilots. Finally, the sophistication of digital computer simulations, along with advanced electronic guidance systems, rendered the analog planetarium sky and its accessory hardware obsolete. These developments were not foreseen by Jenzano, who once predicted that planetarium training would become "even more important during the era of the space shuttle."[19]

Morrison Planetarium Light Shows

Before America's Mercury astronauts had been selected, San Francisco's Morrison Planetarium hosted a series of experimental programs (1957–59) in which compositions of electronic music were accompanied by projections of abstract visual imagery. These performances represented a strictly theatrical use of the planetarium's domed environment and shifted programming entirely towards the entertainment end of its spectrum. Ticket revenues from these performances far exceeded initial expectations. But core administrative and philosophical issues emerged that brought the short-lived programs to a halt. Staged some fifteen years before the introduction of laser light shows in planetaria, these concerts foreshadowed the ambivalence that surrounded the light show phenomenon of the 1970s and beyond.

In 1957 Morrison Planetarium director George W. Bunton was approached by two Bay-area artists, musician Henry Jacobs and filmmaker Jordan Belson, who proposed staging a performance of "sound experiments" within the planetarium's acoustical and visual spaces. The duo chose Morrison in part because of its high fidelity sound system, already used to create the first automated public programs. Bunton was initially receptive and regarded the effort as a "legitimate form of art," featuring sounds "created and composed by technological means." This aspect of the performance strengthened its legitimacy before the planetarium's overseer, the California Academy of Sciences. Bunton characterized the electronic music as having an "eerie, 'out of this world' texture." By means of a rotary switch (which would later be termed a joystick), the audio could be made to swirl rapidly around the room in either direction, an effect that gave rise to the performance's name, Vortex.[20]

Vortex performances are the earliest known examples of abstract visual imagery being projected inside a planetarium dome. A generation earlier, however, American composer, conductor, and inventor Thomas Wilfred conceived a new lighting concept, termed Lumia, that expressed the "relationship of light not only to color but to form and texture as well."[21] In 1928, Wilfred designed a silent visual carillon, a tall domed building on whose exterior surface Lumia effects would be projected (by means of Wilfred's earlier invention, the clavilux). Wilfred's carillon was never realized, although he accepted an artist's residency at New York's Grand Central Palace and directed an experimental theater devoted to performances of music and light.[22]

Vortex's first performance was given 28 May 1957 and offered free of charge. In response, a large contingent of listeners to a local FM radio station, KPFA, appeared. What most disturbed Bunton was the composition of this audience, roughly one-half of whom he described as "Bohemian," or belonging to the beat generation and widely regarded as "denying themselves nothing in the way of worldly pleasures and sensations." On this occasion, it appeared to Bunton as if the planetarium had been "invaded."

Admission fees were charged at all subsequent performances. Still, gross receipts doubled for each concert series; Morrison received a fee plus half the net proceeds.[23] According to Bunton, the "Bohemian" element gradually faded and was "almost totally absent" by the fourth program series. Preparations for Vortex performances nonetheless conflicted with ordinary planetarium functions and generated tension among his staff. While Bunton described Jacobs and Belson as "very competent artists," he complained of their indiscriminate use of the star instrument so as to deliver its "maximum performance." Bunton voiced objections to the ways in which the pair had "take[n] possession, move[d] in, [and] intrude[d] upon the domain that is all but sacred to our technical staff (including myself)."

Vortex drew praise among media critics as an important cultural asset of the Bay area. But after the third concert series, Bunton reported that certain Academy members had grown "extremely antagonistic" toward the art form. To try and dispel the "general nuisance" raised by Vortex, Bunton appealed to Academy director Robert C. Miller, who submitted the matter to the council. Bunton "strongly recommend[ed]" that the programs be discontinued on account of internal "friction and discord" spawned. Administrative approval (which evidently favored ticket revenues over harmonious staff relationships) granted two additional performances. Vortex 5 ended its run in January 1959. While never stating his personal opinion of the concerts, Bunton noted that he stood "accused of approving of 'that junk'" as long as he permitted it to continue. Subsequent appeals from Jacobs and Belson to perform again were rejected.[24]

Vortex programs posed the most serious challenges to matters of identity, purpose, and audience satisfaction that planetaria had faced since their inception, raising the education-versus-entertainment debate to a new level. Vortex revealed conflicting attitudes and opinions inside major planetaria concerning support for technical and artistic innovation versus preservation of the status quo. The economic incentives behind such program decisions could not be ignored. A generation before, James Stokley's dramatic programming style had transformed planetarium demonstrations away from an emphasis on pedagogy and toward the realm of high-entertainment spectacles. With Vortex (and future) planetarium light shows, however, that movement was accelerated towards a purely theatrical, entertainment function.

In the year that Vortex was terminated, Bunton published a cogent essay that addressed some, but not all, of these issues. Vortex performances were never mentioned by name, but their influence was clearly discernable. Science museums and planetaria, Bunton argued, were increasingly hard pressed to fulfill their obligations "to interpret science for the layman." Familiar problems of seeking an appropriate "balance" between education and entertainment grew more acute when "success has come to be measured in terms of attendance." Planetarium (and museum) executives were subjected to pressures compelling them "to cater to the interests of the public." Executives were torn, Bunton argued, between "extremes of policy" that reflected desires to stage profitable, highly popular programs and to maintain a "sense of obligation" to restrict their contents to a "high scholarly level." It was pointless to debate matters of attendance and revenue, Bunton urged, unless an institution's aims and objectives were clarified—a conclusion identical to that reached by Joseph M. Chamberlain (see chapter 8).[25]

Lessons gained at Morrison Planetarium were never circulated among the broader planetarium community (beyond its small coterie of executives) and were subsequently forgotten. Bunton departed Morrison in 1962 to direct the smaller Bishop Museum Planetarium in Honolulu, Hawaii, perhaps in response to administrative tensions generated by Vortex performances. Planetarium light shows did not reappear until the advent of Laserium at Los Angeles's Griffith Observatory in 1973.[26] Both there and elsewhere, related issues resurfaced as planetaria again weighed the box office success of laser light shows against philosophical and administrative problems. Bunton's experiences with Vortex were to be relearned time and again by every institution that repeated his little-known experiment.

UNR's Fleischmann Atmospherium-Planetarium

For almost forty years, planetaria accurately reproduced the night sky, illustrating a full range of astronomical phenomena. Little attention was paid, however, to the realistic simulation of meteorological phenomena occurring in the "other half of nature's sky," its daytime complement.[27] Through a combination of technical innovations and imaginative insight, that barrier was overcome when the world's first "weather theater" or "atmospherium" was opened on the University of Nevada–Reno (UNR) campus.

In 1960 the Max C. Fleischmann Foundation slated $190,000 for construction of a planetarium. It was to be managed by the Desert Research Institute (DRI), the largest research arm of the university, which specialized in Great Basin studies. Concurrently, atmospheric physicist Wendall A. Mordy was appointed the DRI's director. Mordy had conducted research by taking time-lapse motion pictures of cloud development, especially wave-cloud phenomena. When he learned of the Fleischmann Foundation's gift of a planetarium, Mordy recognized the potential for expanding his studies to include coverage of the entire daytime sky.

Mordy's opportunity arose from the advent of "fisheye" camera lenses and motion picture projection systems. The first exhibition of fisheye cinematography occurred at the 1960 Seattle World's Fair in a production titled "Journey to the Stars."[28] Independent development of the concept was pursued by the Jam Handy Corporation of Detroit, Michigan, whose "wrap-around motion picture projection system" had been designed to train U.S. Navy jet pilots.[29]

Mordy recommended an "atmospherium-planetarium" to university president Charles Armstrong, who in turn convinced the Fleischmann Foundation of its value for teaching and research. The foundation's gift thereby was raised to $480,000 to complete Mordy's enlarged conception. Spitz Laboratories installed its A3P planetarium instrument, featuring the first "prime-sky" xenon arc lamp, and the atmospherium (a standard 35 mm film projector equipped with a customized fisheye lens manufactured by the Jam Handy Corporation). O. Richard Norton, formerly assistant director of San Francisco's Morrison Planetarium, was chosen curator. UNR's Fleischmann Atmospherium-Planetarium was dedicated 15 November 1963.[30]

Despite the atmospherium's avowed purpose of replicating "time-lapse motion pictures of the *daytime* sky," the new medium's full potential was only starting to be realized.[31] One of Norton's "high priority" tasks was to integrate the atmospherium's capabilities with the planetarium's astronomical programs. Realistic simulations of spacecraft encounters with planets were filmed by driving the fisheye camera along a motorized track toward model globes. Results were judged superior to those of ordinary zoom lenses because of the extraordinary range of magnifications possible. Stunning visual effects were also achieved by projecting standard-format film through the fisheye lens. Eruptive solar prominences leapt from one end of the domed screen to the other above the Sun's immense limb, lending an element of realism that was impossible to achieve by conventional techniques.

On-location filming soon demonstrated the atmospherium's "almost unlimited" theatrical capabilities. Norton installed the fisheye camera on the floor of Arizona's Meteor Crater and recorded the panoramic vista under changing illumination. Resulting scenes attested to the atmospherium's versatility in generating realistic, dynamic "skylines." Launch of an unmanned Saturn I rocket was filmed by the remotely operated camera, which yielded unmatched footage for the topic of space exploration.

Turning his camera downward, Norton began experiments in underwater fisheye cinematography. A watertight Plexiglas box, containing a hemispherical window, was constructed to house the camera. By suspending the device from a stationary platform, activities occurring within a shallow tide pool were filmed.

During playback on the domed screen, "sea anemones 6 ft in diameter and brilliantly colored echinoderms of equal size" could be seen performing their life processes.[32] Later a flotation device was fabricated, enabling the partially submerged camera to be towed over a Hawaiian coral reef. Possibilities opened to research and education in marine biology seemed almost limitless.

The atmospherium's capabilities were realized at a pace comparable to those of program developments within America's first planetaria. Pathways leading from research toward education and entertainment were bridged in less than ten years. Such a transition reiterated the shift from pedagogical lessons to spectacular audiovisual programs staged by prewar Zeiss planetarium directors (see chapter 4). An important difference, however, was that 35 mm fisheye cinematography remained confined to a single institution. More than a decade elapsed before the University of Arizona committed unrestricted funds for the construction of a planetarium and fisheye projection system to honor deceased benefactor Grace H. Flandrau. Norton was subsequently appointed director of the Arizona facility.[33] Until then, Reno's atmospherium-planetarium remained unique. How might this delay in the acceptance of fisheye cinematography be accounted for?

A partial explanation may be reached through comparison of Nevada's atmospherium projector with the Zeiss planetarium. While the Model II star instrument embodied a fully developed pedagogical tool, the atmospherium itself remained conceptually and technically experimental. No market for the duplication of its products or services was immediately envisioned. More importantly, the atmospherium did not represent a stand-alone teaching unit; without development of appropriate "software" (i.e., films), its utility remained inert. In turn, radically different skills and equipment were needed to produce original films—techniques that were only laboriously and expensively acquired, as Norton's experience demonstrated. Few planetarium directors possessed such skills, or the time and budgets necessary to obtain them. As a result, fisheye cinematography in planetaria did not break away from this isolated context until the production and distribution of feature-length films displaying high entertainment values was achieved more than a decade later.

The significance of Reno's atmospherium-planetarium was not to be demonstrated for years to come. Two important outcomes, however, are attributed to realization of a "frameless picture which places the viewer in the center of the activity."[34] The first sprang from a consortium of similarly equipped planetaria that became organized in the late 1970s. Seeking a better label than "atmospherium," producers rechristened it "Cinema–360."[35] A second, more notable influence was its extension to a higher-resolution 70 mm film and a "rolling-loop" projection system, fabricated by Canada's IMAX Systems Corporation. Concurrent development of the tilted-dome or "space theater" concept gave rise to San Diego's Reuben Fleet Space Theater, which premiered the first so-called OMNIMAX film in 1973.[36] Commercial success of both IMAX (flat-screen) and OMNIMAX (hemispherical) films carried enormous implications for museum and planetarium administrators in the final decades of the twentieth century.[37]

Planetaria as Research Laboratories

Celestial Orientation of Migratory Birds

One of the most intriguing applications of planetaria has been in studying the abilities of migratory birds to use celestial cues for navigational purposes. These investigations were undertaken in West Germany on old-world warblers of the family *Sylviidae,*[38] but were subsequently applied by American ornithologists, especially toward an understanding of the orientation system of the indigo bunting (*Passerina cyanea*).

In his doctoral research, E. G. Franz Sauer of the University of Freiburg undertook nocturnal orientation experiments with several species of old-world warblers, including blackcaps (*Sylvia atricapilla*), garden warblers (*S. borin*), and lesser whitethroats (*S. curruca*). Sauer's populations of birds had been raised in captivity and possessed no direct experience of the natural sky. But when exposed to the stars after exhibiting seasonal migration readiness, the warblers oriented themselves along their species-specific migratory directions. Cloudy skies left the birds disoriented.[39]

To refine his understanding of these birds' behaviors, Sauer performed experiments on the warblers at the Olbers Planetarium in Bremen, West Germany. When its artificial stars matched the time and date of the natural sky, the warblers oriented themselves as before, supporting hypotheses that they possessed internal clocks and innate stellar recognition abilities. By adjusting the planetarium to depict skies viewed from lower latitudes, Sauer recorded deviations in their orientation that he claimed corresponded to birds' migratory flight paths. Finally, Sauer explored the birds' behaviors under conditions simulating diurnal and seasonal rotation of the sky through 360 degrees.

Sauer concluded that old-world warblers relied on a two-coordinate navigation system, aided by an internal clock that sensed both diurnal and annual passages of time.[40] Additional experiments on populations of captured warblers lent support to Sauer's findings, although certain behaviors were only weakly confirmed.[41] But Sauer was unable to determine which stellar patterns, if any, were most significant to the warblers. Sauer's research was later criticized for his methods of data recording by direct observation. Nonetheless, his conclusion that birds relied "heavily on information [gathered] from the stars" appears justified.[42]

In the United States a series of experiments was undertaken by Stephen T. Emlen, a doctoral candidate at the University of Michigan and later member of Cornell University's Division of Biological Sciences. In 1964–65, Emlen studied a population of twenty-four captive indigo buntings, of which a majority exhibited seasonal migratory orientations when exposed to the natural sky. Emlen performed extensive tests on the buntings at the Robert T. Longway Planetarium in Flint, Michigan. Emlen discovered that buntings maintained their directional orientations after the planetarium sky had been offset through positive or negative rotations corresponding to three, six, and twelve hours of time.

Emlen's findings posed a number of challenges to Sauer's claims regarding old-world warblers by implying that buntings possessed no internal clocks and did not rely on a two-coordinate navigational system. Through selective obstruction of the planetarium's star field, Emlen discovered that stars in the northern circumpolar region were most important in allowing buntings to maintain their migratory orientations. Emlen hypothesized that buntings navigate according to a "star-compass," which involves gestalt recognition of stellar patterns in the vicinity of the celestial pole.[43]

By further experiments performed at Cornell University's planetarium, Emlen established the importance of celestial rotation as the means by which juvenile buntings acquire knowledge of circumpolar star patterns. To confirm the rotation hypothesis, Emlen modified the planetarium instrument by selecting an arbitrary bright star (Betelgeuse in the constellation Orion) as a fictitious celestial pole. Under these conditions, the stars of Orion formed a prominent "circumpolar" grouping that revolved around Betelgeuse every twenty-four hours. Juvenile buntings were reared under these artificial skies, and on reaching maturity, exhibited seasonal migratory orientations corresponding to the planetarium's modified axis of rotation. Emlen concluded that buntings learn to perceive the north-south axis of celestial rotation from observations of stars that are visible throughout the night and exhibit the least diurnal movements.

Emlen's results demonstrate the selective advantages of this type of star recognition ability over the course of evolutionary history. Because precession of the Earth's axis gradually shifts circumpolar star patterns through a cycle of 26,000 years, it is difficult to imagine how birds could retain a hereditary star map without compensation occurring over thousands of generations. Avian recognition of stellar configurations near the celestial pole eliminates that difficulty. Buntings appear to use this technique as one of their means of nocturnal migration from the east-central United States to regions of Central America or the Caribbean islands and back again.[44]

Sauer and Emlen's findings demonstrate the importance of planetaria as research tools in helping to answer intriguing questions found in the biological sciences. While designed as instructional aids for teaching the confirmation of scientific theories, planetaria were employed in the generation of new scientific knowledge. These studies signify a continuing realization of the untapped potential residing within this pedagogical device.

Planetaria and Educational Research

Educational research was not conducted in planetaria before the 1960s. Much of this delay is perhaps attributed to two unfavorable conditions:

(a) astronomy was not widely taught in elementary or secondary schools before *Sputnik's* launch and (b) the majority of planetaria serving children and young adults were located in public museums, where their educational potential was ignored by researchers. This myopia changed dramatically with the proliferation of school planetaria resulting from NDEA legislation and the introduction of "space science classrooms." Only after the space age had begun did educators consider it important to understand what types of learning occurred in planetaria, as compared with traditional classroom instruction.

Doctoral research in planetarium education was conducted by six men and two women at American universities through 1970. Apart from three institutions being located in one state (Michigan), no distribution pattern is apparent within this sample; none of the universities or faculty advisors produced more than a single planetarium-related dissertation. Several researchers had acquired extensive planetarium experience before enrolling in doctoral studies. Only half remained active in the planetarium community afterward.

These studies employed two distinct methods of inquiry: (a) demographic surveys administered to assess current usage of planetarium teaching strategies and (b) quantitative experimental analyses measuring the effectiveness of planetarium versus classroom instruction. Chronologically, the demographic surveys preceded the experimental comparisons, and revealed a progressive sophistication of researcher questions that addressed the planetarium's educational functions.

Three researchers assessed planetarium utilization within elementary, secondary, and adult and continuing educational contexts, while a fourth performed content analyses on selected elementary- and secondary-level programs. Ruth A. Korey examined the "scope and nature of programs in relation to elementary education" at nearly two hundred planetaria nationwide. Organized class visits constituted the services most frequently requested; programs were descriptive in nature and subject to constraints of dome and staff size. Little formal evaluation was conducted by instructors and limited opportunities existed for student participation.[45]

John T. Curtin's analyses of program contents revealed no significant differences among programs delivered to elementary and junior high school audiences. Constellation study occupied roughly half of instructors' allotted time. Questions posed by lecturers were classified according to Benjamin S. Bloom's *Taxonomy of Educational Objectives.*[46]

Demographic survey data permitted Dale E. McDonald to draw comparisons between the instructional use of planetaria and observatories in secondary schools. Concepts of the celestial sphere and geocentric motions were commonly associated with the former, while studies of solar system objects and astronomical instrumentation dominated the latter. Planetaria received higher usage for astronomical instruction than authorities recommended, while the opposite condition prevailed regarding observatory tutelage.[47] These findings, however, were chiefly applicable to district administrators or instructors attempting to maximize the planetarium's educational potential. Such data became obsolete as the number of facilities expanded rapidly through the 1960s and beyond.

Selected characteristics of adults enrolled in continuing education classes were examined by Maurice G. Moore, director of the Longway Planetarium in Flint, Michigan. Moore analyzed the word-recognition abilities, media habits, and attitudes toward expenditure of funds for space research among those who attended planetarium programs and those who did not. Attendance was correlated with higher word-recognition abilities and more positive attitudes regarding space exploration.[48] Moore's research provided evidence for a link between the popularization of science and increased levels of support for space science and engineering policy decisions. His findings expose a common weakness in simplified models of science communication, which posit a strict, one-way flow of information from expert knowledge producers to passive knowledge consumers.[49] That tools such as planetaria, designed for the popularization of science, have helped to shape public support of funds for continued space exploration highlights the necessity of employing more sophisticated communication models employing feedback loops.

An important advance accompanied the first experimental comparisons of planetarium and classroom lecture-demonstrations. It was expected that planetaria would provide the most effective means of instruction because they were capable of producing the most versatile and realistic simulations of the night skies. Educational theory predicted that "[t]he closer the vicarious experience is to the actual objective experience, the stronger and more accurate [will be] the perception of the learner."[50] By this argument, three-dimensional representations of the sky shown by planetaria were thought to offer superior instruction compared to the two-dimensional figures sketched on chalkboards. Yet four independent tests of this hypothesis reported disconfirming evidence. Instead, classroom lecture-demonstrations yielded results that differed little from, or were judged superior to, planetarium demonstrations.[51] These findings delivered the first major blow to instructors and manufacturers by revealing the apparent inadequacy of the planetarium as an instructional device.

Researchers suspected, however, that the problem lay not with the planetarium or its utilization, but rather with the means employed to assess student performance. Instructors unwittingly uncovered the difficulties experienced by students of all ages in attempting to transfer acquired knowledge from the context of instruction to the context of evaluation. Glenn E. Warneking, former director of the Earth and Space Laboratory of the Frederick County, Maryland, school district, called attention to a well-known educational phenomenon (dubbed the "cue" theory) that recognizes that "a class tested for recall in the same surroundings as that in which the instruction was initially received will do better than one tested in a different setting."[52]

George F. Reed examined several reasons why the lighted classroom environment offered a "decided advantage" over the planetarium chamber. In the former setting, he noted, the chalkboard diagrams were "unavoidably similar" to those portrayed on tests. Reed also observed that in a well-lit classroom, students engaged in note taking, asked spontaneous questions, and their facial expressions revealed whether knowledge was being assimilated. In the darkened planetarium, by contrast, note taking became impossible, student questions were held until the presentation's conclusion, and facial expressions could not be monitored. Thus, the planetarium's instructional effectiveness, Reed argued, was "dependent upon more than superiority in terms of an audio-visual device" and the interest and motivation of students.[53]

From these studies, the planetarium's measured lack of effectiveness was reinterpreted as a shortcoming of the research process itself. Today, student learning abilities are recognized as being influenced by a much greater number of factors than educators had once believed. In coming decades, research in planetarium education examined many of these other factors, especially the effects of student participation on the acquisition and retention of astronomical knowledge (see Epilogue).

Summary

New horizons in planetarium usage emerged with the appearance of analog flight simulators for training NASA astronauts, experimental stagings of electronic music concerts, and the development of 35 mm fisheye cinematography. More than sixty astronauts that flew on Mercury, Gemini, and Apollo space missions were trained at the University of North Carolina's Morehead Planetarium under the direction of Anthony F. Jenzano. Star recognition offered a backup to other, more sophisticated vehicle guidance systems. Apart from providing routine and life-saving navigational skills, the astronaut training program became an excellent public relations vehicle for the planetarium, university, and surrounding community.

San Francisco's Morrison Planetarium witnessed administrative flare-ups in response to hosting the first experimental light shows. George W. Bunton disclosed the conflicts encountered when financial incentives and audience satisfaction goals clashed with deeper institutional identities and purposes. Planetarium light shows represented one extreme along the continuum that lay between education and entertainment. Bunton's little-known experiment foreshadowed the problems that accompanied the laser light show phenomenon of the 1970s (and beyond). That experience may also have influenced his decision to undertake a reverse career move to a smaller planetarium facility.

At the University of Nevada-Reno's Fleischmann Atmospherium-Planetarium, a decade of research and innovation by director O. Richard Norton demonstrated the enormous potential of 35 mm fisheye cinematography. His development of this new "immersive" technology roughly paralleled the movement from pedagogy to entertainment that characterized James Stokley's programming style during the 1930s. The lack of available fisheye motion pictures, and the technical skills needed to produce them, long prevented this medium from being replicated at other institutions. The wrap-around motion picture concept was later adapted to higher-resolution OMNIMAX formats.

Planetaria were transformed into laboratories for conducting biological and educational research. Their artificial skies provided ideal environments for testing the abilities of migratory birds to recognize celestial cues. The importance of stellar rotation by circumpolar star patterns was experimentally confirmed as the indigo bunting's means of nocturnal migration. Such research signified that planetaria, as tools designed for pedagogical confirmation of scientific theories, could be applied to the production of new scientific knowledge—possibilities never imagined by Oskar von Miller.

Significant educational research in American planetaria only began in the 1960s. Not until "space science classrooms" became widespread fixtures among public schools did educators recognize the importance of determining their most effective teaching strategies. Demographic studies, assessing patterns of usage, gave way to experimental comparisons between planetarium and classroom teaching practices. Independent findings reported that planetarium methods were either indistinguishable from, or else judged inferior to, classroom treatments.

Researchers suspected, however, that this surprising conclusion arose from an incomplete transfer of student knowledge from the context of instruction to the context of evaluation. Among adult and continuing education students, evidence was gathered of positive reinforcements between the popularization of astronomy and increased levels of public support for science and engineering policy decisions. In turn, these findings challenge simplistic, top-down science communication models.

CHAPTER 8

New Routes to Professionalization

The time may not be far distant when planetariums will be as numerous as museums. In this age of emphasis on science, such a trend is not only welcome, but almost mandatory.
—Joseph M. Chamberlain, 1957

. . . I am quite confident that now Planetarium is a profession and that we're going at it in the proper way.
—Armand N. Spitz, 1970

Fundamental steps toward disciplinary professionalization were achieved among American planetaria by the close of the third developmental period. These stages reflected the tremendous institutional growth that resulted from federal legislation enacted in the wake of *Sputnik,* in which hundreds of new planetaria, chiefly in public schools and districts, were created (see chapter 6).

Recruitment and training of new planetarium instructors became one of the most important issues faced by planetarium leaders. A host of strategies, including NSF-sponsored summer institute programs, planetarium internships, and graduate degree programs, were created to alleviate this manpower shortage.

Starting in the mid–1960s, planetarium educators organized themselves into regional associations. By decade's end, a viable, continent-wide professional association, which adopted formal bylaws and launched its quarterly journal, had been established. Forty years after Chicago's Adler Planetarium had opened its doors, the North American planetarium community had come of age.

NSF-Sponsored Planetarium Symposia

On 7–10 September 1958, the first nationwide symposium on planetarium education was convened at the Robert R. McMath Planetarium of the Cranbrook Institute of Science in Bloomfield Hills, Michigan. The Cranbrook meeting was organized by curator of education James A. Fowler and director Robert T. Hatt. Support originated from the National Science Foundation (NSF), which presumably underwrote delegates' travel and accommodation expenses. Its sponsorship almost certainly resulted from Armand N. Spitz's role as a special consultant to the agency from 1956 to 1960.[1]

The conference attracted some 101 delegates from sixty-seven institutions. As Hatt remarked, the meeting concerned itself "particularly with the problems of the smaller planetaria," although representatives from five major facilities played an active part. Donald H. Menzel, director of the Harvard College Observatory, delivered the keynote address "Observatories in Space." Delegates explored topics ranging from "Astronomical Subjects and Applied Science" to "Presentation Methods and Special Effect Techniques." The Cranbrook symposium's proceedings were published under NSF sponsorship as *Planetaria and Their Use for Education,* constituting the American planetarium community's first professional monograph.[2]

Symposium speakers addressed issues concerning audiences of all ages. Integration of planetarium lessons with school curricula received close attention, although Raymond J. Howe, curator of education at the Kansas City Museum, noted an absence of scientific data on "what constitutes the most effective learning experience in a planetarium."[3] Howe's remark pinpointed the lack of formal research conducted in planetarium education, a shortcoming remedied in the coming decade (see chapter 7). John C. Rosemergy of Ann Arbor High School's Argus Planetarium encapsulated the nation's yearlong introspection on its educational weaknesses with the following remarks: "Sputnik made the American secondary school, and particularly its science programs, the subjects of a suddenly magnified and panicky concern." High school planetaria, Rosemergy argued, "are a resounding answer to some of the questions directed to our schools because of Sputnik."[4]

The diversity of opinions expressed was matched by that of delegates in attendance. Reflecting an increased role for women in American planetaria, four of the thirty-two presentations were delivered by women directors, including Maxine B. Haarstick (Minneapolis Public Library) and Margaret K. Noble (Washington, D.C., public schools). Disagreements arose over the use of "live" versus "recorded" programs and whether planetaria should embrace science fiction topics or restrict themselves to delivering fact-based lectures. Nonetheless, a consensus of opinion emerged that techniques of "good theater" were essential for success, no matter what subject or audience was addressed. Participants concurred on two broad purposes under which most planetaria were operated: "to stimulate interest in astronomy and related sciences" and "to increase the layman's understanding and knowledge of the universe." Delegates returned to their institutions with hopes of developing new programs capable of "teach[ing] and inspir[ing] every audience with maximum effect."[5]

The Cranbrook symposium exerted a strongly unifying effect on those in attendance, and its impact was carried to the broader astronomical community.[6] The nationwide meeting helped to solidify the recognition, integrity, and authority of smaller planetaria and their educational missions, which had first coalesced under the American Association of Museums planetariums section. A short-lived measure of cooperation was attained between large and small planetaria, whose personnel were to meet again in two years. In a culminating gesture, host James A. Fowler was elected chairman of a committee appointed to investigate the formation of a nationwide planetarium association.

American planetarium directors again convened under NSF sponsorship at the Cleveland Museum of Natural History's Ralph Mueller Planetarium on 28–31 August 1960.[7] Planetarium director Dan Snow, along with museum director William E. Scheele, hosted at least ninety-three delegates. The symposium's proceedings were published as volume 2 of *Planetariums and Their Use for Education.* However, notable differences stand out between this gathering and that hosted by the Cranbrook Institute. Fewer sessions allowed delegates to present more in-depth assessments on specific topics. Editor Richard H. Roche noted that the "pragmatic nature" of many talks reflected their context as "professionals addressing a group of professionals."[8] Representatives from major planetaria declined to participate, having recently attended an international conference of planetarium executives in New York City. No account of the Cleveland meeting was reported in *Sky and Telescope,* suggesting that the novelty had already worn off. Ironically, one of the symposium's most important decisions—to organize a national association of planetarium operators—received no mention from the proceeding's editor and was touched on by a single delegate's remarks.

Two of the five sessions, "The Planetarium and the Science Curriculum" and "Public Services and Public Programs," considered the social roles and audiences served by planetaria, whether large or small. As humankind stood poised on the brink of manned space flight, contributions of planetaria "to the culture of our time" were endorsed by editor Roche, who regarded the symposium's objectives as improving "school and planetarium cooperation." Delegates perceived that the "education of today's children" remained strongly dependent upon "better training of teachers."[9] Science journalist David Dietz explained how planetaria could secure improved publicity for their programs. Means of sharpening communication skills and developing standards of performance were topics that had few counterparts at the Cranbrook meeting. Like the nation's fledgling space program, the American planetarium community itself was evolving rapidly.

The symposium's most introspective session was titled "Staffing a Planetarium." There, the recruitment, training, and compensation of personnel were candidly discussed for perhaps the first time. Citing results derived from a previously administered questionnaire, James A. Fowler sketched a composite portrait of the planetarium director's most desirable traits. Delegates could only agree, however, that such qualifications depended strongly upon the institutional contexts and audiences served.[10] As planetaria faced critical manpower shortages, a supply of well-trained personnel remained one of the community's most pressing needs.

On 30 August the delegates accepted recommendations from Fowler's standing committee to establish a national professional association, electing to call itself the American Association of Planetarium Operators (AAPO).[11] From this title, it is apparent that delegates sided against adoption of the word "director," perhaps because of its exclusionary connotation. An organization comprised of personnel from smaller institutions, the AAPO sought to enroll all people associated with planetaria, regardless of formal title. Creation of a job information and referral service was suggested by Wofford College Planetarium director Howard Pegram.[12] Despite an auspicious beginning, however, the AAPO remained an organization that existed in name only. No further meetings were convened under its auspices; its effect upon American planetaria was negligible. How, then, can this dramatic turnaround be explained?

The AAPO appears to have suffered a leadership vacuum following the Cleveland symposium. No other institution or director emerged to organize and host a third biennial meeting. Especially puzzling is the fact that, after spearheading the AAPO's formation, James A. Fowler ceased to play an active role in its future. From the number of delegates who had gathered at Cranbrook and Cleveland, no apparent shortage of qualified personnel can account for this anomaly.

One factor that hindered sponsorship of future symposia was the withdrawal of NSF funding. Termination of this revenue source coincided with the expiration of Armand N. Spitz's consultancy role in 1960, which further coincided with his replacement as Spitz Laboratories president by Wallace E. Frank. The waning of NSF support might also have stemmed from newer perceptions that the national security crisis triggered by *Sputnik's* launch had abated. Another factor that might have kept the AAPO from becoming a viable organization was a sense of loyalty among members of the flourishing AAM planetariums section, founded in 1952 (see chapter 5). Before the advent of federal assistance, museums made up the second-largest institutional category of American planetaria. The longer and more robust pedigree of the AAM section enabled it to prosper through the 1960s, while the AAPO lay stillborn. Ten years elapsed before a second attempt was made to establish a continent-wide association of planetarium educators.

Directors from the nation's major planetaria wanted no part of the AAPO. They were not only reluctant to engage in formalized structures, but their professional interests differed considerably from those who directed smaller planetaria. More importantly, many of these individuals had attended the first international planetarium directors conference at New York's American Museum-Hayden Planetarium in 1959.

International Planetarium Directors Conference

At the American Museum-Hayden Planetarium, Chairman Joseph M. Chamberlain began laying plans for hosting the first international planetarium directors conference. Chamberlain acknowledged his desire to extend invitations to directors from the Soviet Union, Poland, and East Germany. Building on the collaborative spirit fostered by IGY scientific efforts, he confidently opined, "The question of inviting the planetarium people from behind the Iron Curtain arises. It is my judgment that we should do so. I can see no possibility of political embroilment." Chamberlain's invitation secured clearance from the Department of State in Washington, D.C., whose visa office envisaged no "major difficulty which might arise as a result of your invitation of these various gentlemen."[13]

At the meeting, Chamberlain wished to honor several planetarium benefactors along with Walther Bauersfeld, inventor of the Zeiss projector. Bauersfeld, however, declined the invitation because of poor health, saying, "I would have liked nothing better than to be present; this is sincerely meant."[14] Chamberlain submitted a request to the White House urging that "no gesture or gift of appreciation . . . would be more significant" to those attending than to receive a letter bearing the president's signature. Chamberlain intimated that Eisenhower's "personal interest in the [nation's] increasingly important scientific and cultural developments" would render this blessing "singularly appropriate."[15] A statement honoring Chamberlain's wishes was delivered as a telegram from Eisenhower:

> It is a pleasure to send greetings to those attending the first international meeting of planetarium executives. The support of private individuals is a basic element in the growth of many of our finest scientific and cultural developments. A good example is found in our country's planetariums which have been provided by generous benefactors. I am glad to join in gratitude to them and to express my hope that the study of astronomy be further advanced among the people of our land.[16]

This event marked the first occasion in which an American president formally recognized the significance of planetaria in fostering public science education.

From 11–16 May 1959, eleven overseas delegates, along with a representative from Carl Zeiss in Oberkochen, joined nineteen American planetarium personnel and three benefactors for a series of demonstrations and discussions. Three conference sessions addressed problems associated with "The Planetarium Lecture," "The Business of Operating a Planetarium," and "The Planetarium and the Future."[17] Afterward, delegates toured the Philadelphia and Boston planetaria and related astronomical facilities. A *New York Times* reporter noted how this gathering of "international astronomers" enjoyed far more convivial relations than a session of "foreign ministers" engaged in arms limitations talks at Geneva.[18] The conference demonstrated that professional disciplinary interests could circumvent Cold War political tensions affecting world affairs. It reaffirmed the elevated status that planetarium executives had acquired and largely severed those ties formed less than a year before with delegates at the Cranbrook symposium.

The space age delivered an important spur to planetarium executives, just as it had to operators of smaller planetaria. Both groups convened the largest and most diverse gatherings of colleagues ever assembled, and the acquired momentum dissipated slowly. In coming years, the urgency resulting from the IGY, *Sputnik,* and the creation of NASA was redirected into enlarged, yet ordinary activities dominated by rapid institutional growth at all levels. International planetarium directors did not reconvene until 1966, whereupon triennial meetings became the adopted norm.[19] Nonetheless, significant achievements were made in the areas of programming philosophy, curriculum development, and graduate-level education. Unlike school and collegiate planetaria, executives' institutions were largely unaffected by NDEA legislation (and its successors). New major planetaria were provided by private benefactors. With the exception of Canadian planetaria, few directors of major facilities played significant roles in the organization of regional associations.

Philosophies of Planetarium Programming

The sudden onset of space exploration triggered widespread introspection among education leaders. In the field of astronomy education, this moment of reflection prompted two of the planetarium community's leading spokesmen to commit their thoughts to print regarding the past, present, and future directions of their profession. Armand N. Spitz presented his views on planetarium education in the *Griffith Observer* in 1959. Spitz's essay remains the most complete statement of his educational philosophy and earned him a second nickname, "the philosopher."[20]

To Spitz, the word "planetarium" denoted more than the projection instrument, the room that housed it, or the building itself. It signified "an experience" that he argued "must be differently planned and executed than any single facet of operation." "Planetarium" was likened by Spitz to "Theatre." Spitz criticized any declaration that planetaria were "limited instrument[s]." Their ability to provide a spectrum of educational experiences, arising from the "appreciation, imagination, and inspiration" of operators, was no more restricted than the number of concertos that might be performed on the strings or keys of a musical instrument. "Few men so far have sensed the potentialities of Planetarium," he argued, out of hesitation to move beyond its astronomical confines. Instead, Spitz argued, this pedagogical device was capable of re-creating "major highlights of the history of the thinking of Man." Teachers and students of "mathematics, . . . history, literature, the fine arts, . . . the classics, and of course, all the sciences," stood to benefit from an appropriately fashioned presentation. The "intellectual gamut of Planetarium," Spitz suggested, "has never been run."[21]

Spitz described the pertinent characteristics of a successful planetarium director: an individual who need not be a professional astronomer, yet who combined "popular presentation with scientific accuracy." Few practicing scientists, in Spitz's experience, possessed the background skills and yet retained the ability to act as effective "interpreters." Besides imparting factual information, directors "must have a sense of the dramatic; . . . must remember the basic thesis that Planetarium is closely allied to theatre." From the myriad responsibilities incurred by directors, he argued, remuneration should be awarded at the level of a "full professor."[22] Spitz's essay placed his views on the planetarium's function in the vanguard of its professional community. Many, if not most, of Spitz's hopes for enriching educational curricula were adopted in coming years. Especially among smaller installations, Spitz's vision of the instrument's boundless capabilities "defined planetarium directorship thereafter."[23]

In New York City Joseph M. Chamberlain expressed his views in the pages of a new professional museum journal, *Curator,* launched by his parental institution in 1958. He noted that with some twenty-five million visitors to major American planetaria, and more than a hundred smaller school and collegiate units, "[u]nquestionably the planetarium is a significant force in the teaching of astronomy in this country. Or at least it should be!" His essay, "A Philosophy of Planetarium Lectures," provided the first in-depth statement of its kind and foreshadowed arguments later defended in his doctoral dissertation on the administration of a planetarium as an educational institution.[24]

Chamberlain's essay contained a well-informed discussion of the education-versus-entertainment debate over which directors had long struggled. His argument was fashioned upon the destruction of artificial strawmen, representing extremes of those twin stances. Chamberlain objected to a purely pedagogical approach on grounds that demonstrations of the planetarium's technical capabilities constituted little more than "projector-worship." In such cases, Chamberlain warned, it was the instrument, and not the lecturer, that had performed. No program composed of the "endless exposition of facts" could be construed as an effective use of the planetarium. But the opposite condition, labeled as "uninhibited soaring into the realm of science fiction" and often characterized by a "space trip replete with imaginary . . . rockets," left Chamberlain even more apprehensive. While admitting that these spectacles were justified in "economic terms," he argued that their lack of authenticity prevented them from gaining a permanent slot in the planetarium's repertoire. Attendance figures alone must not be the final arbiter of success, a conclusion seconded by the Morrison Planetarium's George W. Bunton.

After defending the planetarium as an "astronomical institution," Chamberlain urged that programmers "chart a mean course between the pedagogic and the fantastic." Once a subject or theme had been selected, a sequence or story line was developed to bring the conception to reality. Neither star instrument nor auxiliary effects, in his judgment, should become ends rather than means. Only by satisfying a "predetermined teaching objective" may such a program be judged as successful. Chamberlain's philosophy of moderation, while not particularly original, was adopted by many planetarium executives through the 1960s and beyond.[25]

Chamberlain's 1962 doctoral dissertation represented an outgrowth of his attempt to formulate a viable philosophy of planetarium operation. It presented the strongest argument yet for a planetarium's educational role in serving the needs of its community. His study focused almost exclusively on the ten major U.S. facilities then extant. While acknowledging the existence of some 250 smaller "classroom-size installations," he regarded their "purposes, potentialities, and problems" as quite "distinct" from larger institutions like his own.[26]

For Chamberlain, the planetarium was an educational institution dedicated to the teaching of astronomy that possessed a "unique potential and special pertinence" to citizens of the space age. A "composite of purposes," he argued, must be interwoven to formulate a mission statement for the planetarium, whose primary objective was "to provide educational services in astronomy and related fields to as many persons as practicable." In Chamberlain's judgment, a planetarium reached its greatest potential when "administered by professionally trained staff members" skilled in both astronomical and educational theories.[27]

The planetarium director assumed a host of duties, which Chamberlain identified as those of scientist, teacher, administrator, engineer, showman, and orator. His use of the term "scientist," however, requires clarification. It did not signify an astronomical researcher (as envisaged by George E. Hale in chapter 3), but rather an interpreter of astronomical facts, upon whose integrity the planetarium's reputation depended. Chamberlain was among the first to argue that these tasks constituted the responsibilities of a "planetarium astronomer." Pursuit of astronomical research, in his opinion, took valuable time away from the institution's more pertinent educational objectives. These facilities were better suited, he urged, for "encouraging an early interest in astronomy and identifying possible future astronomers" among younger audiences. When properly orchestrated, the planetarium experience enabled visitors to embark upon their own subsequent "voyage of discovery."[28]

In response to the existing shortage of "planetarium astronomers," Chamberlain recommended that a "concentrated effort" be undertaken to recruit and educate potential staff members. He urged cooperation with "degree-granting institution[s]" to establish graduate-level internship programs, whose certifications might be awarded in "science teaching" or "administration of education."[29] Chamberlain's characterization of American planetaria supported a viewpoint of these facilities as largely discrete, autonomously operating units. His portrayal reinforced the dominant attitude among planetarium executives and ignored the collaborative ventures arising within the many smaller institutions.

Recruitment and Training

One of the foremost challenges facing American planetaria after *Sputnik*'s launch was a perennial shortage of well-qualified instructors. Earth- and space-science curricula were being implemented rapidly in the nation's schools, with planetaria and astronomical observatories constructed to deliver those curricula. "Space science classrooms" were hailed by directors, manufacturers, and journalists alike as vital ingredients for imparting a space age education.[30] During years of peak activity in the late 1960s, roughly eighty new planetaria per year were added to the ranks of U.S. facilities. Planetarium directors became one of the most sought-after educational specialists of the time. District administrators gained a new awareness of these facilities and their roles in public education. Planetaria became "status symbols" for school boards, yet posed significant responsibilities for ensuring their fullest utilization. One essayist reminded officials that selection of a qualified instructor posed a far more important decision than the types of equipment to be purchased and installed.[31]

Transformation of the National Science Foundation's summer institute program enabled countless instructors to receive training in astronomy and planetarium education. NSF summer institutes originally were designed to provide advanced training for faculty members and to support fellowships for graduate-level education. But impending shortages of scientists and engineers redirected their missions toward in-service programs for training science and mathematics teachers. An early example is provided by Stanley J. Hruska of the Detroit Children's Museum. Before a Spitz planetarium was installed at his facility, Hruska attended the four-week NSF-sponsored Institute on Astronomy held in 1956 at the Eau Claire campus of the Wisconsin State Colleges system.[32] Hruska later chaired the AAM planetariums section in 1961. From the first summer institute created in fiscal year 1954, the number of similar offerings rose to 320 during fiscal years 1959 and 1960 and provided attendees with some 16,000 annual stipends.[33]

While the total number of astronomy instructors trained in this manner cannot be determined, evidently hundreds of teachers participated in one or more summer institutes, though not all were NSF-sponsored. Three particularly active centers were important in providing planetarium instruction. To meet job demands created by Pennsylvania's earth- and space-science curricula, workshops were organized by its Department of Public Instruction. State coordinator John E. Kosoloski reported that during the summer of 1960, seven Pennsylvania institutions trained some 40 earth- and space-science teachers apiece.[34] At the State University of New York's Oswego campus, a program for training novice planetarium instructors was initiated in 1962 by astronomer George E. Pitluga. These intensive courses delivered six credit hours of instruction in four weeks time.[35] Finally, under the direction of Michael A. Bennett, Spitz Laboratories developed its own summer institute program in 1969 that subsequently trained more than 800 participants.[36] These programs offered sufficient experience for large numbers of teachers to acquire positions in school and collegiate planetaria.

Graduate degree programs in astronomy education and planetarium education were founded by larger universities. These programs satisfied objectives framed by Joseph M. Chamberlain. In 1962 Chamberlain chaired the first U.S. conference on graduate education organized by the American Astronomical Society's Committee on Education in Astronomy (CEA). While not addressing the role of planetarium astronomers per se, the CEA heard discussion on "Renaissance of the M.A. Degree in Astronomy," wherein Harvard University astronomer William Liller urged that degree recipients could go primarily into teaching.[37] Restructuring of masters-level programs to include planetarium curricula allowed Chamberlain's vision to be realized.

Lastly, planetarium internships offered apprenticeships toward professional careers in the burgeoning planetarium community. Whether such training resulted in a diploma, certificate, or influential letters of recommendation, the internships proved invaluable to potential job seekers. Usually lasting a year's time, they gave candidates a range of teaching and show production experiences routinely found in major planetaria. Delivery of programs to school and public audiences was often complemented by topic research, scriptwriting, production design, photography, audio recording, special effects construction, supervision of artists and technicians, and interaction with mass media. Internships supplemented, but did not entirely supplant, the longer apprentice-style training offered by selected major planetaria (see chapter 3). A directory of institutions offering coursework in planetarium education was printed in the first number of the International Society of Planetarium Educators journal, *The Planetarian.*[38]

Regional Associations

A new stage of professionalization was reached with the formation of regional associations, whose constituents were chiefly personnel from school and district, and university and college planetaria.[39] During the mid–1960s the number of institutions and directors rose sharply from continued impact of federal assistance. Highest-growth areas included the "Mideast States," boasting 127 installations, and the "Great Lakes States," where 84 planetaria were becoming established. The next largest region of "Southeast States" supported 72 institutions, but these facilities were widely separated and their directors united more slowly. Other concentrations were found in California (36 planetaria, 26 of which belonged to institutions of higher education) and Texas (16 planetaria).[40] Among Canadian provinces, the first public planetarium was opened in Edmonton, Alberta, in 1960.[41] It was not joined by any major facilities in that country, however, until 1966. Mexico's first major planetarium was dedicated in 1967 and remained unique to that country's educational thrust before 1970.

A study of the regional associations reveals that attainment of a critical mass of institutions and personnel was a principal factor behind their creation. Other factors, however, were also responsible for explaining the appearance of regional associations. One was the failure of the American Association of Planetarium Operators (AAPO), created at the 1960 Cleveland symposium, to become a viable organization. In response, grassroots efforts to accommodate the needs of planetarium directors at regional levels were initiated. Successes garnered by the first associations spurred establishment of others in a bandwagon effect. While not intentionally competitive with the AAM planetariums section, regional associations likely contributed to the demise of that forum. Within five

TABLE 8.1 *Regional Planetarium Associations*

Name	Acronym	Year
Great Lakes Planetarium Association	GLPA	1965
Middle Atlantic Planetarium Society	MAPS	1965
Planetarium Association of Canada	PAC	1966
Southwestern Association of Planetariums	SWAP	1966
Pacific Planetarium Association	PPA	1967
Rocky Mountain Planetarium Association	RMPA	1969
Southeastern Planetarium Association	SEPA	1970
Great Plains Planetarium Association	GPPA	1973

Source: Author.

years, a successful, continent-wide association, stemming from unification of the regional associations, emerged and has prospered to the present time. Table 8.1 lists the names and years of creation for eight regional planetarium associations founded during this period.

No single professional association served as an exclusive model on which regional planetarium associations were fashioned. A host of national organizations catered to the regional interests of members through subdivisions. For example, the National Science Teachers Association (NSTA) sustained eight geographic "regions" in the continental United States. NSTA Region III elections committee member Margaret K. Noble founded the Middle Atlantic Planetarium Society (MAPS).[42] Other organizations composed of regional associations included the Astronomical League, whose first national convention was hosted by Philadelphia's Fels Planetarium in 1947.[43] Nothing particularly novel was required to conceptualize a regional planetarium association. Organizers merely adopted the structures and communication styles developed elsewhere. A detailed history of regional planetarium associations lies beyond the scope of this study.[44]

Steps toward the creation of a regional planetarium association in the Great Lakes region were taken on 21 November 1964 when thirty-eight instructors from Michigan, Ohio, and Illinois met to examine the new Spitz ISTP (Intermediate Space Transit Planetarium) projector installed at Michigan State University's Abrams Planetarium. Delegates were addressed by none other than James Stokley, professor of science journalism at the university since 1956. One year later, on 7 October 1965, the Great Lakes Planetarium Association (GLPA) was formally organized at the Grand Rapids (Michigan) Public Museum. Planetarium director David L. DeBruyn hosted the 1965 convention.[45] Within two years, GLPA had established the Armand N. Spitz Lecture as the keynote address of its annual conventions. Grace S. Spitz delivered the first such lecture at the association's 1967 Cleveland meeting. A publication serving the region's planetaria, the *GLPA Projector,* debuted the following year.

Personnel from nine planetaria clustered around the nation's capital met at Cordoza High School in Washington, D.C., in January 1965 and formed the nucleus of the Middle Atlantic Planetarium Society (MAPS). Chairperson of the first session was Margaret K. Noble, director of the District of Columbia schools planetarium. John M. Cavanaugh, director of Franklin and Marshall College's North Museum Planetarium in Lancaster, Pennsylvania, was elected first president. An annual lecture honoring Noble's founding role, along with a quarterly publication, the *MAPS Constellation*, were established. As reported by MAPS historian Claire J. Carr, the society adopted four main objectives:

1. To give assistance to newly formed planetariums and newly appointed planetarium teachers
2. To provide boards of education with recommendations on planetarium construction
3. To have its members become acquainted with new curriculum material at all grade levels, developed by planetarium educators across the country
4. To become familiar with any new projects, projectors, and audiovisual aids developed by members of MAPS[46]

These objectives were perhaps typical of guidelines adopted by other regional planetarium associations.

Across the border, celebration of the Dominion of Canada's centennial anniversary (1 July 1967) became a leading factor responsible for the wave of planetarium construction that swept its provinces. Major facilities appeared in five of the largest Canadian cities: Montreal, Calgary, Toronto, Winnipeg, and Vancouver.[47] Following this spurt of institutional growth, a Canadian planetarium association was organized. "With the opening of the space age and the consequent mushrooming of planetarium activity throughout the world," Canadian directors met at Edmonton's Queen Elizabeth Planetarium in 1966 with the object "of establishing a mutually beneficial association."[48] Two leading members of the Planetarium Association of Canada (PAC), Donald D. Davis and Dennis H. Gallagher, were recent arrivals from U.S. planetaria; Gallagher served as the PAC's founding president. The association's publication, *The Planetarium Journal,* was launched in 1969.

Within the United States, regional planetarium associations took their cues from school and district, and university and college directors, who sought to improve their educational strategies and professional communications. Regional associations complemented the AAM planetariums section that had represented planetaria in museums and science centers since 1952. All three institutional categories attained comparable levels of professionalism by 1970. Yet, the daunting challenge remained of uniting these factions, along with the Planetarium Association of Canada, into a viable continent-wide association.

A crucial test of the ability of regional planetarium associations to collaborate was demonstrated at the 1968 GLPA-MAPS Conference, hosted by the newly opened Strasenburgh Planetarium of Rochester, New York. From 24–27 October almost 200 delegates from the two associations assembled for the largest North American planetarium conference held to date. Ian McLennan, who left the Queen Elizabeth Planetarium to become Strasenburgh's first director, was one of the meeting's principal organizers.[49]

The success of this venture gave the green light to Abrams Planetarium director Von Del Chamberlain to lay plans for a continental meeting of planetarium personnel. Chamberlain had earned a bachelor's degree from the University of Utah (1958) and a master's degree from the University of Michigan (1960). After a four-year stint at Flint, Michigan's Longway Planetarium, he was appointed staff astronomer at the Abrams Planetarium (1964) and subsequently acting director (1968) and director (1969).[50] In 1967, Chamberlain held the presidency of GLPA and chairmanship of the AAM planetariums section.

Conference of American Planetarium Educators

Two years of planning went into Chamberlain's hosting of the continent-wide meeting that was called the Conference of American Planetarium Educators (CAPE). Two hundred and fifty-one registrants from seven regional associations and over thirty program support personnel convened at Michigan State University from 21–23 October 1970.[51] Delegates addressed the two most important matters of the conference: (a) whether to establish a continental association and (b) whether to subsidize a professional journal, whose main purposes would be "to disseminate information and coordinate activities . . . [of] conferences and . . . to support and strengthen the activities of existing and future regional groups."[52] By the meeting's conclusion, both measures were approved.

Seventeen members of the ad hoc Organization Committee met for two days preceding the CAPE conference to iron out differences regarding the proposed association's structure. Jack C. Howarth, director of the San Antonio College Planetarium, was elected chairman, and Norman Sperling, director of the Princeton Day School Planetarium, was elected secretary. Two competing proposals were (a) a federation of existing (and future) regional associations and (b) a more hierarchical organization consisting of regional subdivisions. Either type of association would depend upon member and institutional dues for its financial support.[53] One issue centered on inclusion and representation of non-U.S. associations, epitomized by the Planetarium Association of Canada (PAC). Sigfried Wieser, director of the Centennial Planetarium in Calgary and PAC representative to the CAPE meeting, had expressed these concerns to Chamberlain. Wieser affirmed that PAC's membership favored a federation of associations, "as national (Canadian) identity must be maintained." Potential U.S. domination of the planetarium discipline, he warned, could not be endorsed. Nonetheless, an "international organization," if adopted under certain conditions, would likely be supported.[54] Howarth's committee, although sensitive to Wieser's viewpoint, secured the Canadian's compromise for acceptance of the international organization.[55]

Establishment of a North American planetarium publication was the second major piece of business addressed by a pre-conference committee, the Publications Committee. Justification of this endeavor rested on multiple grounds: interregional communication could be improved; editorial efforts consolidated; unnecessary duplication eliminated; and most importantly, levels of professionalism increased.[56] Notions of creating such a journal had arisen in more than one regional association, including the PAC. Larry A. Gilchrist, a lecturer at Calgary's Centennial Planetarium and editor of the PAC's *Journal,* endorsed both the continental association and publication.[57] Gilchrist's analysis, submitted *in absentia* to the Publications Committee, revealed no insurmountable obstacles to the plan.[58] Previously, Frank C. Jettner, director of the Henry Hudson Planetarium Project at the State University of New York's Albany campus, was elected chairman, and Michael A. Bennett, education director at Spitz Laboratories, secretary of the Publications Committee.[59] Finding strong support for their objectives in Gilchrist's brief, Jettner's committee recommended a quarterly publication and future appointment of an editor-in-chief.[60]

Presentations from the Organization and Publications Committees were made to CAPE delegates on the conference's opening day, allowing time for the proposals to be debated. The next day, all regional planetarium associations met simultaneously to discuss the merits of a continental association and a publication. Votes of confidence were returned from each region and matters debated before the entire assembly. In the closing general session (October 23), both proposals were strongly endorsed, whereby the International Society of Planetarium Educators (ISPE) was established and plans for launching its journal were likewise adopted.

In a tribute to the individual chiefly responsible for establishing the North American planetarium community, host Von Del Chamberlain played an audio recording of greetings delivered by Armand N. Spitz, who was unable to attend the CAPE meeting because of poor health. The convention reportedly sat "hushed and deeply moved" by Spitz's statement, which affirmed that his "dreams of many years [were] beginning to come true; . . . the dream that Planetarium would [become] a profession."[61] Most delegates recognized that if it were not for Spitz's invention and marketing of the pinhole-style projector, their occupations as planetarium educators would not exist.

Spitz's recording served to introduce keynote speaker George O. Abell, chairman of the astronomy department of the University of California at Los Angeles and also chair of the American Astronomical Society's Committee on Education in Astronomy (CEA). At Chamberlain's request, Abell was invited to deliver this talk as the fourth Armand N. Spitz lecture. Chamberlain urged that Abell's address serve as "a message of inspiration for planetarium teachers" and summarize the viewpoints of professional astronomers "relating to planetarium instruction and programming." Abell's lecture conveyed, in the words of Spitz employee Jack Spoehr, appropriate sentiments regarding the "efficacy of the planetarium." Yet these proffered feelings were tinged with undisguised irony to Spitz, who acknowledged that years earlier, "[Abell] was mighty opposed to [the planetarium] and [q]uite chilly personally. Times do change," he mused. Six months later, Spitz passed away (on 14 April 1971) and a memorial fund in his honor was established by Chamberlain.[62]

In the decade after the 1960 Cleveland symposium, the number of North American planetaria had risen by a factor of five. But along with sheer numbers, the formation of regional associations and their status as affiliate "chapters" of ISPE, proved crucial to the latter's success. While not devoid of growing pains in the months to come, ISPE (which shortened its name to the International Planetarium Society, or IPS, in 1977) has remained the largest, most viable association of planetarium personnel to this day. It has held conferences in even-numbered years (following a pattern established before its existence by the Cranbrook and Cleveland symposia) at meeting sites in the United States, Canada, Mexico, and abroad. During the year 2000, IPS celebrated its thirtieth anniversary by hosting its fifteenth biennial conference in Montreal, Quebec. Its journal has been published without significant interruption since 1972. These traits are commonly regarded as signifying "crucial point[s] in the process of academic professionalization."[63] Hindsight reveals that the planetarium community's most pressing needs for disciplinary professionalization and communication, as understood by Von Del Chamberlain and addressed through appropriate committee actions, were fundamentally resolved at the CAPE conference. Forty years after the Adler Planetarium had opened its doors, the North American planetarium community had come of age.

ISPE Constitution Committee and Publications Committee

It was imperative that ISPE formally establish itself under a set of bylaws. At the CAPE meeting, nine members of the Constitution Committee were appointed, most of whom had served on the ad hoc Organization Committee. Jack C. Howarth was chosen executive secretary, while Paul R. Engle, director of the Pan American College observatory and planetarium in Edinburgh, Texas, was selected chairman (and first ISPE president). Howarth's and Engle's attainment of influential ISPE posts visibly demonstrated a significant attitude adjustment within the Southwestern Association of Planetariums (SWAP).[64] To affirm the organization's international character, Sigfried Wieser was appointed chairman-elect. This move underscored the Organization Committee's reconciliation of a potential divide along national lines and assuaged fears of marginalization among members of the Planetarium Association of Canada (PAC). Frank C. Jettner, who retained chairmanship of the Publications Committee, was appointed an *ex officio* member of the Constitution Committee.[65]

The ISPE Constitution Committee met at Baton Rouge, Louisiana, on 27–28 March 1971, prior to attending the 134th meeting of the American Astronomical Society. Jettner's absence from this meeting, and the displeasure he subsequently expressed over the committee's provisional draft, yielded the most contentious debates in the adoption of ISPE's bylaws. One of the articles to which he objected (and that was later discarded) was a stipulation requiring membership in one or more regional associations. Jettner's most vociferous arguments, however, were directed at proposals concerning the society's journal, actions that he regarded as affronts to his autonomy. From exchanges between Jettner, Wieser, and other committee members, it can be seen that the international issue remained volatile. Jettner's incautious remarks drew an adamant defense of the committee's actions from Wieser. By reminding Jettner that he was "in no position to demand anything," Wieser quenched the fires of discontent that threatened deliberations conducted, in his judgment, "as impartially and objectively as possible."[66] After these differences were settled, the ISPE Constitution was submitted for approval to the regional associations and was adopted by all.[67]

Concurrently, the nine-member ISPE Publications Committee, chaired by Jettner, began to design the society's quarterly journal, which remained unnamed. Assisting Jettner was Calgary planetarian Larry A. Gilchrist, chairman-elect, and representatives from each regional association, including Norman Sperling. Unlike its counterpart, the Publications Committee did not meet face-to-face, nor was it given an operating budget. Members were kept informed through correspondence issued by Jettner. One of the committee's first duties was to solicit publication support from commercial resources, including NASA. Responses to this venture, however, were uniformly negative.[68]

To determine "a realistic estimate of the [journal's] circulation base," Jettner sent a survey questionnaire to all North American planetaria and drafted an organizational chart defining eleven subcommittees. By overlooking the editorial format contained in Gilchrist's brief, Jettner reaffirmed Wieser's testimony of his dismissal of input from others, an action for which Jettner later apologized.[69] While declining any post on the editorial staff, Norman Sperling independently suggested the word "planetarian" as a label for any planetarium worker, regardless of rank. Sperling's term was chosen as the journal's official title, *The Planetarian.* In Jettner's estimate, this choice had precipitated the Publication Committee's "biggest disagreement."[70]

In June 1971 the Publications Committee issued its Special Report #1, *A Catalog of North American Planetariums (CATNAP),* containing entries on 743 institutions and personnel.[71] *CATNAP* was an outgrowth of Sperling's efforts to compile an updated mailing list for Chamberlain's CAPE meeting, a task that had been described as "impossible." With the addition of planetaria from outside North America and conversion into electronic format, this database has been continually updated and reissued periodically.

In late 1971 Larry A. Gilchrist was appointed executive editor of *The Planetarian,* being responsible to Jettner and the ISPE Council for its management and production. But Gilchrist's editorial role was short-lived; on 1 February 1972 he resigned, citing an increased "work load imposed upon me as the new Commanding Officer of my Regiment." Two weeks later, he and Jettner traded positions by mutual consent, with the latter appointing himself acting executive editor.[72] In fairness to Jettner, who called for nominations to fill the executive editor's role, his self-appointment allowed the journal's first issue to appear only three months behind schedule.

The inaugural issue of *The Planetarian,* dated 21 June 1972, bore a cover illustration by Strasenburgh Planetarium artist Victor Costanzo depicting the launch of a Saturn V rocket bearing the acronym ISPE. Its masthead displayed a hemispherical dome and urban skyline, resembling the kind employed by early Zeiss planetaria. This thirty-two-page publication brought to realization the wishes for an exchange of planetarium information across an international forum. With establishment of its quarterly journal and ratification of its bylaws, the North American planetarium community achieved professional status and lasting cooperation among its institutions and personnel.

Summary

Important signs of solidarity and collaboration emerged within American planetaria after 1957. Armand N. Spitz's influence as special consultant to the National Science Foundation paved the way for two nationwide symposia on the educational uses of planetaria. These gatherings affirmed the increasing visibility, integrity, and authority of the nation's smaller planetarium facilities, and the proceedings resulted in the community's first professional monographs. In similar fashion, the first conference of international planetarium directors invited personnel from Iron Curtain countries, thereby demonstrating how professional interests could circumvent political tensions affecting world affairs. Attempts to establish a national association of planetarium operators, however, proved surprisingly sterile. Another decade would pass before this agenda was successfully accomplished.

Dawn of the space age prompted two of the community's leading spokesmen to reflect on the foundations of planetarium programming. Armand N. Spitz endorsed the multidisciplinary aspects of planetarium usage; his goal remained the creation of a pluralistic yet educationally sound profession. Joseph M. Chamberlain urged that planetarium programs chart a mean course between the pedagogic and the spectacular, wherein the star instrument and auxiliary projectors served as means, rather than ends, to reaching a predetermined teaching objective. In his doctoral dissertation, Chamberlain defended planetaria as educational institutions to be administered by professionals ("planetarium astronomers") trained in scientific and educational theories. Both Chamberlain's and Spitz's philosophies strongly influenced programming strategies to the present time.

The recruitment and training of qualified personnel posed one of the discipline's most pressing needs. Creation of hundreds of new planetaria, especially school and district facilities, reflected the impact of federal assistance. A number of strategies were established to address this manpower shortage. Transformation of NSF-sponsored summer institutes, along with newly created graduate degree and internship programs, trained hundreds of new planetarium directors during the 1960s and beyond.

Regional associations of planetarium personnel, led by Midwestern and East Coast colleagues, sprang into existence to facilitate the exchange of professional skills and ideas through annual conferences and publications. These activities culminated in Von Del Chamberlain's 1970 CAPE conference, from which the International Society of Planetarium Educators (ISPE) was born. ISPE succeeded where a prior national association, lacking the regional affiliates model, had failed. Adoption of its bylaws, establishment of its quarterly journal, and continuation of its biennial conferences signified the planetarium community's transformation into a mature, stable professional association. ISPE's long-term success indicated that crucial needs for disciplinary professionalization and communication were clearly recognized and appropriately addressed at the CAPE conference. Armand N. Spitz's dream, that "Planetarium would [become] a profession," was finally brought to realization.

Epilogue

The number of American planetaria and their personnel continued to increase during the remaining years of the twentieth century. Because fundamental questions of how best to organize and professionalize its members had been answered, the discipline could turn its attention to more immediate concerns. Only a broad outline of the community's leading issues and developments can be described in this brief synopsis. A plethora of new media, technologies, and educational practices were initiated,[1] while the nature of its international character was expanded beyond the boundaries of the North American continent.

The American Planetarium Community after 1970

One of the principal changes concerned a reversal of the educational momentum that planetaria had acquired during the 1960s. Coinciding with the end of Apollo-era lunar exploration, the nation's planetaria dramatically slowed their rate of growth. In the wake of post-Viet Nam War recession, environmental activism, Watergate, and the rise of conservative political agendas, large segments of the population grew disenchanted with science, technology, and the federal policies responsible for creating and sustaining them.

School and district planetaria witnessed their first-ever decline in numbers. Rather than asking how a new planetarium could be staffed by an experienced instructor, some administrators confronted the task of phasing down, or phasing out, many of the curricula implemented during the 1960s. Planetaria were threatened with closure; their educational purposes seemed to reflect an era that had slipped away. University and college facilities were not immune to similar criticisms. Some were regarded as marginalized institutions that fell victim to downsizing and other cost-cutting measures. This phenomenon might have seemed unthinkable to the community's leaders only one or two decades before. Federal assistance, once allocated under the National Defense Education Act (and its successors), was reduced and finally eliminated. While new institutions continued to be built, greater reliance was again placed on private, rather than federal, funds for their construction.

This seemingly bleak outlook, however, did not affect a majority of the institutions created in the previous 40 years. Within the planetarium community, tremendous new opportunities and challenges emerged for expanding traditional audiences and programs. Digital technologies became one of the driving forces behind new planetarium programming. The rise of commercially marketed automation systems enabled directors to stage highly sophisticated programs, eclipsing those that were possible to run without automation by a single human operator. Increased use of slides and special effects reflected attempts to attract and retain more visually sophisticated audiences. While denying or disparaging direct competition with television and motion pictures, planetarium directors opened new rounds of discussion on the education versus entertainment issue. Automation brought about the virtual acceptance of taped programs within the planetarium community.

Planetarium architecture underwent considerable change, as the first tilted-dome theaters, led by San Diego's Reuben Fleet Space Theater, were opened in the early 1970s. This movement was a natural outgrowth of Spitz Laboratories's earlier success in creating the space science classroom. Under the inspiration of Donald M. Lunetta, the planetarium chamber's horizontal floor and dome were refashioned to accommodate tilted surfaces, completing the final stages of marriage between planetaria and motion picture theaters. Projection of scenery found below the traditional horizon provided a simulation resembling that of an airplane's or spacecraft's perspective. This projection geometry suited unification of the IMAX System Corporation's hemispherical OMNIMAX format with the planetarium chamber.

Digital technologies dramatically altered the way in which planetarium stars were projected. In 1983, the first-generation Digistar planetarium instrument was installed at the Universe Planetarium in Richmond, Virginia. Manufactured by the Evans and Sutherland Corporation of Salt Lake City, Utah, Digistar had no moving parts. Numerical data on some 6,800 stars was stored in the memory system of a VAX minicomputer. A graphics processor displayed the stars on a high-intensity cathode ray tube. From there, a fisheye lens system projected the stellar images onto the hemispherical dome. Unlike previous-generation planetarium instruments, Digistar allowed viewers to realistically explore the Sun's neighborhood within a radius of 650 light years. By the end of the twentieth century, dozens of Digistar projectors had been installed in planetaria throughout the United States. In response, these institutions exchanged software and original programs scripted exclusively for the new instrument.

On the other end of the financial scale, portable planetaria, complete with inflatable domes, were marketed by Learning Technologies, Inc., of Cambridge, Massachusetts. Called Starlab, these devices enabled astronomy educators to bring the stars to audiences that might never attend a permanent-type planetarium.

While the notion of traveling planetaria had been pioneered in the 1950s, the idea did not then catch on. But with the spread of these portable devices, Starlab users groups emerged to share the unique problems and prospects of this newer projection technology.

Perhaps the single greatest influence on the post–1970 planetarium community was the emergence of commercially distributed planetarium show packages. As early as the 1950s, planetarium executives had discussed the exchange of scripts, soundtracks, slides, and artwork, although the idea was not widely practiced. During the 1970s, two American planetaria succeeded in distributing show packages on a nationwide scale. Each employed significantly different means of doing so. At the Strasenburgh Planetarium of Rochester, New York, director Donald S. Hall sold mid-priced show production packages with the aim of turning a profit. Salt Lake City's Hansen Planetarium, under director Mark E. Littmann, began distribution of grant-funded program packages at considerably reduced prices. As a result, even the smallest planetarium could install and present a professional-quality program, featuring stereo soundtrack and abundant visual images. The availability of packaged shows acted as a great leveler among institutions of all sizes, erasing any significant differences in the quality of programs presented by large or small planetaria.

An important variant on the distribution of planetarium programs was the emergence of independent production companies. Foremost among these was Loch Ness Productions of Boulder, Colorado. The husband-and-wife team of Mark C. and Carolyn Collins Petersen offered competitive show packages that were used by hundreds of U.S. planetaria. It was unnecessary to be a planetarium director or a planetarium employee of any kind to make a significant contribution toward the community's development.

At Los Angeles's Griffith Observatory, the opening of Laserium in 1973 marked the advent of laser light shows in planetaria. The light show phenomenon sprang from a confluence of technical and cultural factors. Invented more than a decade earlier, the laser (Light Amplification by Stimulated Emission of Radiation) acquired a variety of scientific and commercial applications. Hardware grew smaller, less expensive, and increasingly diverse, as gas lasers were perfected that produced light in a range of coherent colors. When visual effects were teamed with sound (principally rock music), the laser light show was born. As a result, the nation's planetaria were thrust into a repeat of the situation first encountered in the late 1950s by Morrison Planetarium director George W. Bunton. Prospects of reaping significant short-term profits were weighed against the administrative costs of staging light shows and their attendant problems. The planetarium's identity and purpose were again called into question, as the education versus entertainment debate resurfaced with a vengeance.

Both competition and collaboration grew between planetaria and their specialized motion picture counterparts, chiefly the flat-screen (IMAX) and hemispherical (OMNIMAX) formats. Both types of theaters were constructed in conjunction with new and existing planetaria. The IMAX Theater at the National

Air and Space Museum in Washington, D.C., and its premiere motion picture, *To Fly,* drew audiences well in excess of those attending its adjoining Einstein Spacearium, a conventional Zeiss planetarium erected for the nation's bicentennial celebration. Essayists asked whether, in the face of such competition, planetaria might not become "cultural dinosaurs." For planetaria to survive, the argument ran, their objectives had to remain centered around the function at which they remained unexcelled—reproduction of the night skies.

Fisheye cinematography, in the original 35 mm format, underwent significant developments during the 1970s and after. Newer planetaria equipped to project these films were first constructed in Arizona, Mississippi, and Illinois. Their directors joined the University of Nevada-Reno's Fleischmann Atmospherium-Planetarium to form a consortium of fisheye motion picture producers, christened Cinema–360. This movement saw completion of the first feature-length films crafted on a variety of regional subjects. Using the proximity of director Richard H. Knapp's Davis Planetarium to the Florida launch site of NASA's space shuttles, a fisheye camera was flown in the payload bay aboard the mission on which the Solar Max satellite was repaired. Dramatic footage from this event was incorporated into the first truly out-of-this-world Cinema–360 film, *The Space Shuttle: An American Adventure.*

Planetarium programming experienced notable controversy when a host of new interpretations purporting to explain the origin and chronology of the Christmas Star emerged in the 1980s. Historical research published by biblical scholar Ernest L. Martin propounded a revised dating for the birth of Christ and placed the event much nearer the calendrical marker established by sixth-century monk Dionysius Exiguus. Martin's research was brought before the American planetarium community by the Griffith Observatory's John Mosley. Because the Christmas Star program remained a holiday offering at many planetaria, few institutions could fail to become involved in the debates that surrounded acceptance (or rejection) of the Martin-Mosley interpretations.

Research into planetarium education diversified, as individuals examined a variety of media applications to audience learning styles. One of this era's leading developments was an emphasis on participatory-oriented programs, pioneered by Allan J. Friedman and his staff at the Lawrence Hall of Science, Berkeley, California. Whether adopted as a form of scientific inquiry, or simply included as general audience-participation techniques, this approach demonstrated superior learning potential and retention of information. Thereafter, it became widely diffused among planetarium educators.

After the American planetarium community had marked its fiftieth anniversary in 1980, questions arose over the historic preservation of two original Zeiss installations erected during the formative period. The (renamed) Buhl Planetarium and Science Center was transferred from its location on Pittsburgh's North Side to the new Carnegie Science Center. Its Zeiss Model II projector, the last remaining instrument of its kind, was left behind and dismantled for storage; a Digistar projector took its place. The fate of the Model II instrument and the former Science Center building remained controversial, as alternatives were explored by the City of Pittsburgh.

In New York City the venerable American Museum-Hayden Planetarium was razed in the mid–1990s to allow for construction of the much larger and architecturally stunning Rose Center for Earth and Space (at a cost of some $210 million). That decision sparked a legal controversy among New Yorkers, who fought unsuccessfully to have the former Hayden Planetarium preserved, yet modernized, within its original structure. Replacement of this landmark signifies the continuing maturity of the American planetarium community, whose audiences have witnessed tremendous scientific and cultural changes in the decades since Chicago's Adler Planetarium opened its doors to the public in 1930. A fuller account of these and other social aspects awaits a future historian.

Planetaria in Cultural Contexts

Are there any reasons why the average person should be concerned with the historical developments, financial support structures, and social interactions that defined the American planetarium community across its first four decades of existence? Beyond their primary missions of communicating nontechnical information to children and adults alike, planetaria, as cultural institutions, have captivated and influenced the imaginations of artists, writers, filmmakers, and other individuals through their abilities to portray the movements of the heavens and their attempts to convey deeper meanings regarding humanity's place in the universe.

In this final section, a sampling is presented of ways in which planetaria have been incorporated into certain aspects of American culture. A full-scale account of these cultural influences, however, lies beyond the scope of this study.

As previous chapters have demonstrated, planetaria and their personnel have assumed active roles in World's Fair celebrations held at Chicago and New York; trained NASA's Mercury, Gemini, and Apollo astronauts; enabled researchers to uncover the stellar recognition abilities of nocturnally migrating birds; and received commendations from a sitting U.S. president. Their subject matters have explored religious symbols and traditions associated with the Star of Bethlehem, the Crucifixion, and Resurrection of Christ.

The design and construction of planetaria have given rise to numerous aesthetic considerations in their architecture and exhibit space. Both paintings and sculptures have depicted astronomical and mythological themes, often highlighting measures of progress achieved by the Western intellectual tradition. Nearly all major planetaria have commissioned works of art and sculpture to enhance their permanent exhibits. Oftentimes, these productions were contrasted with earlier, non-Western viewpoints of the heavens.

In formulating plans for a public observatory and Hall of Science, to be constructed on Mount Hollywood, Griffith J. Griffith responded to a suggestion from Throop Polytechnic Institute president James A. B. Scherer, who proposed the establishment of an "evolutionary museum," depicting the "formation of worlds" and tracing the "development of life on this planet from the most rudimentary organic forms to the highest civilization of the present time."[2] Such an exhibit philosophy was not far removed from contemporary plans envisioned by Henry Fairfield Osborn at the American Museum of Natural History (see chapter 2).

Instead, a different type of evolution came to be portrayed on the Griffith Observatory's walls. Funds from Griffith's estate were used to select a mural artist to execute a series of large panels for its central rotunda. Hugo Ballin (1879–1956), art director for Samuel Goldwyn's MGM studio in Hollywood, was chosen for his themes derived from classical mythology. Ballin's eight large murals painted in 1934 bear the collective title "Advancement of Science from Remote Periods to the Present Times." Thus, it was by artistic expression, rather than the display of natural history specimens, that an approximation to Griffith's original wishes was finally realized.[3]

At New York's Hayden Planetarium, a large triptych depicting astronomical lore of the Blackfeet Indians was commissioned from artist Charles R. Knight (1874–1953). While chiefly remembered for his paintings of extinct animals, especially mammals and dinosaurs described by Museum paleontologists, Knight composed the triptych's central panel for the opening of the Hayden Planetarium in 1935. On it was depicted the Sun God (in the guise of an Indian warrior) in pursuit of the Moon Goddess. Below them was illustrated a legend in which a muskrat successfully returns with a sample of mud collected from the bottom of the ocean, out of which the Earth will be formed. Above them were shown familiar stars of the Dipper and Pleiades. The triptych's wing panels, completed later by Knight, portrayed an auroral display as the procession of departed spirits, and the tribe's powerful Thunderbird.[4]

New York sculptor Sidney Waugh (1904–1963) received a commission to furnish six sculptures in raised relief that adorned the outside walls of Pittsburgh's Buhl Planetarium and Institute of Popular Science (now the site of the Children's Museum). Waugh's principal stone figures are "The Heavens," depicting a woman clasping the Sun, and "The Earth," a muscular, seated man grasping a sledge hammer. Behind this latter figure are shown leaves and plants of the Carboniferous period, a reminder of the importance of coal to the industrial history of Pittsburgh. Above the main entrance's paired revolving doorways are bronze-relief figures of a Native American of the Mingo tribe, representing "Primitive Science," and a laboratory worker, entitled "Modern Science." Two additional sculptures, located above the east and west emergency exit doorways, illustrate "Day" and "Night."[5]

Perhaps the most ambitious sculptural project associated with an early Zeiss planetarium facility is the six-sided Astronomers Monument located outside the main entrance to the Griffith Observatory in Los Angeles. The Monument is one of thousands of artworks created under the short-lived Public Works of Art Project (PWAP) initiated by President Franklin D. Roosevelt. Sculptor

Archibald Garner submitted the winning design, which positioned statues of six famous astronomers around a central pier, capped by an armillary sphere. The astronomers represented are Hipparchus, Copernicus, Galileo, Kepler, Newton, and Herschel.

To assist Garner in completing the nine-foot-high Art Deco statues, five other regional sculptors—Roger Noble Burnham, Djey el Djey, Arnold Foerster, Gordon Newell, and George Stanley—were employed. Each astronomer was first modeled in clay, then molded in plaster, from which its finished concrete cast was taken. The Astronomers Monument was dedicated 25 November 1934, six months before the Observatory's official opening. It was completely restored in 1994. Griffith's astronomical observer, Anthony Cook, has described the backgrounds and additional sculptures of these six Depression-era artists. The most famous (and coveted) piece of work executed by one of these artists is the Oscar statue awarded annually by the Academy of Motion Picture Arts and Sciences. It was sculpted in 1929 by George Stanley, after a design by MGM's art director Cedric Gibbons.[6]

Writers of poetry and fiction make up another group of individuals that has received inspiration through exposure to planetarium demonstrations or else used the planetarium (in a metaphorical sense) to fashion the outlines and meanings behind a fictional account. In verses contributed primarily by women poets during the formative period, distinctive reactions were expressed concerning the planetarium's depiction of the artificial heavens. These poetic themes frequently addressed the mechanical display of God's "handiwork" or "plan" as a recurrent emblem of twentieth-century natural theology (see chapter 4).

A metaphorical appropriation was undertaken in Nathalie Sarraute's novel *Le Planétarium* (1959). Sarraute's discarding of the usual narrative conventions and the ordering of reality invoked psychological turbulence among her characters that seemingly mirrored the behaviors of planets and satellites around their parent bodies. Sarraute's work became an important contribution to the twentieth-century movement known as *nouveau roman,* sometimes regarded as the literary counterpart of non-referential or abstract expressionist painting.[7]

Opening scenes from the 1955 Warner Brothers motion picture *Rebel Without a Cause,* starring James Dean and directed by Nicholas Ray, were filmed on location at Los Angeles's Griffith Observatory. In this film, the planetarium's lecturer addressed the theme of human insignificance in relation to the scale of the cosmos. Following a dramatic portrayal of the Big Bang, the lecturer continued: "In all the immensity of our universe, and the galaxies beyond, the Earth will not be missed. Through the infinite reaches of space, the problems of man seem trivial and naive indeed. And man, existing alone, seems himself an episode of little consequence."[8]

Dinsmore Alter, who remained Griffith's director until 1958, served as a role model for the film's lecturer. More importantly, the latter's words bear a striking resemblance to the social message that Alter himself imparted upon his school audiences, which numbered some 80,000 pupils per year: "There is much more to hold in their ambition than the mere accumulation of money or of physical comforts, . . . and that what happens even to a whole generation is comparatively unimportant when viewed from the totality of a universe."[9] Ray's classic film suggests that such startling perceptions, when unleashed upon disaffected youth of a postwar generation, could lead to social behaviors and consequences unanticipated by Alter.

A planetarium's ability to recreate the daytime or nighttime skies offers a valuable tool for solving a host of astronomical puzzles. Once the star instrument has been set to match an observer's location, an analog replica of celestial objects and their motions is at hand, without reliance upon more sophisticated calculations. This was especially true in the era before electronic calculators, personal computers, and planetarium-style software packages became readily available. On occasion, specific questions about the Sun's or Moon's placement in the sky at certain times of day or night have been brought before planetarium directors by law enforcement or judicial agencies. Answers supplied have been used as testimony in courtrooms, where civil lawsuits or the causes of accidents were being settled. Though seldom reported in the literature, these occurrences might be styled as cases of "forensic astronomy."

In one instance, Weldon D. Frankforter, director of the Sanford Museum Planetarium in Cherokee, Iowa, was asked to resolve a dispute between two neighbors over the proposed construction of a house. The plaintiff, who owned a solar home, was afraid that a nearby house, being built by the defendant, would cast a shadow on his living room window, and sought legal action to prevent its construction. In response, Frankforter set his planetarium Sun to December 21st (the winter solstice), when shadows cast by the real Sun would be the longest of the year. He then determined the Sun's altitude and azimuth for every hour of that day as seen from Cherokee. By applying this information to maps and architectural drawings of the proposed structure, Frankforter demonstrated that no shadow from the defendant's house would reach the plaintiff's window on that, or any other, day of the year and testified as such in court.[10]

The foregoing examples highlight a variety of ways in which the influence of American planetaria have extended well beyond their intended purposes of offering formal and informal astronomy instruction to audiences of all ages. As cultural institutions, the nation's planetaria have exerted a measured impact on the creative works of twentieth-century artists, sculptors, poets, novelists, and filmmakers, not to mention their occasional services to our legal system. More than merely sites at which the teachings and careers of their directors have been enacted, these institutions have provided a continuing locus for the creation and reflection of selected aspects of American culture almost from the time of their inception. A more complete account would reveal further aspects of these devices that arose from Oskar von Miller's pedagogical vision, and Walther Bauersfeld's realization, of the projection planetarium more than eighty years ago at the Deutsches Museum in Munich.

Appendix: North American Planetaria

Institutions

From data contained in ISPE's *A Catalog of North American Planetariums (CATNAP),* a well-defined, quantitative picture can be drawn of the community's institutional structure at the time its members united to form a continent-wide professional association. A total of 743 planetaria were either completed, under construction, or in planning stages at the date of publication (June 1971).[1] From this figure an institutional growth rate of roughly 80 new facilities per year in the decade's last four years (1967–1970) can be derived.[2] Outside of the United States 13 facilities opened in Canada and 1 major planetarium opened in Mexico City. Seven of the Canadian planetaria were associated with universities and colleges, while 6 were operated by museums and science centers. Among the latter category, 5 were major planetaria. Along with Mexico's lone institution, each major Canadian facility sported a late-model Zeiss projector. The Canadian pattern mimicked the rate of planetarium acquisition experienced in the United States during its formative decade (1930–1939).

Within the United States, 729 planetaria were located in every state but two (North and South Dakota) plus the District of Columbia. These were distributed among 102 museums and science centers (14.0%), 238 universities and colleges (32.7%), and 375 schools and districts (51.4%). Fourteen installations (1.9%) did not fit those institutional categories. Pennsylvania exceeded all states with 151 planetaria (20.7% of U.S. total), of which 125 were operated by schools and districts. This abundance reflected not only the Commonwealth's large number of unconsolidated city and borough school districts but also the dynamic approach to planetarium acquisition spearheaded by its Department of Public Instruction (see chapter 6).

CATNAP data corroborate the predominance of Spitz-manufactured equipment among U.S. planetarium facilities. At the same time, they allow analysis of the differential marketing success achieved by Spitz among the three institutional categories (table A.1). Altogether, 593 U.S. planetaria owned Spitz equipment (81.3%), of which 274 (37.6%) housed the A3P. Schools and districts had

TABLE A.1 *Distribution of 729 U.S. Planetaria by Institutional Category (S = Spitz, all models)*

School/district (375)	S = 299	79.7%
	A3P = 136	36.3%
University/college (238)	S = 215	90.3%
	A3P = 110	46.2%
Museum/science center (102)	S = 70	68.6%
	A3P = 26	25.5%
Other (14)	S = 9	64.3%
	A3P = 2	14.3%
Total (729)	S = 593	81.3%
	A3P = 274	37.6%

Source: *A Catalog of North American Planetariums (CATNAP).*

an intermediate percentage: 299 institutions (79.7%) made Spitz purchases, with 136 (36.3%) selecting the A3P. Universities and colleges acquired the highest percentages of Spitz apparatus: 215 institutions (90.3%) adopted Spitz equipment, with 110 (46.2%) investing in the A3P. Finally, museums and science centers displayed the lowest percentage of Spitz hardware: 70 institutions (68.6%) purchased Spitz equipment, with 26 choosing the A3P (25.5%).

To a large degree, these figures reflect the impact of federal assistance on post-*Sputnik* planetarium initiatives, especially the magnitude and duration of support among those categories. Schools and districts benefited from Title III NDEA and ESEA funds released after 1958, while universities and colleges received HEFA and HEA dollars for planetarium construction after 1963. Largely unable to muster any federal assistance, museums and science centers achieved the smallest relative gains (over the 41 pre-*Sputnik* installations) through private or foundational support.

Viewed on a percentage basis, however, Spitz's success among the categories is more readily explained in terms of a good fit between the institutions' average dome sizes and the optimum size for a Spitz projector (table A.2). Among 238 U.S. university and college planetaria, reliable dome diameters are available for 200 institutions, from which an average of 26.6 feet is determined. Dome diameters recommended for the Spitz A3P were either 24 or 30 feet, whose median value, 27 feet, closely matches the university and college tally.[3] Within this category, Spitz installations captured their highest market penetration (90.3%). Facilities of this size could accommodate either two or three sections of an undergraduate course per demonstration (50 to 75 students).

For U.S. school and district planetaria, comparable data are available for 294 institutions. Those facilities displayed an average dome diameter of 23.9

TABLE A.2 *Average Dome Sizes of U.S. Planetaria by Institutional Category and A3P Subcategory*

Institution	Number	Diameter (feet)
School/district	total = 294	23.9
	A3P = 125	26.9
University/college	total = 200	26.6
	A3P = 108	27.2
Museum/science center	total = 90	35.8
	A3P = 24	28.9

Source: *A Catalog of North American Planetariums (CATNAP).*

feet. Schools and districts constructed smaller planetaria that accommodated no more than one or two classes per demonstration (25 to 50 students). Spitz planetaria were located near the upper size and cost limit in which schools and districts were able to invest. Spitz's lower market penetration (79.7%) within this category is plausibly explained.

Among U.S. museum and science center planetaria, an even lower percentage occurred for the opposite reason. Figures for 90 institutions yield an average dome diameter of 35.8 feet, well in excess of the 30-foot dome size recommended for the A3P. This category accommodated the largest audiences (100 or more individuals). The A3P was rarely installed in domes exceeding 40 feet diameter, for which a more powerful star instrument was required. Spitz achieved its lowest market penetration (68.6%) among these institutions, which held the largest fraction of major planetaria. It was for this market venue that Spitz designed its Model B and ISTP projectors.[4]

CATNAP data likewise describe North America's major planetaria, 31 institutions making up 4.2% of the total number of planetaria (table A.3). Twenty-five were operated by museums and science centers (19 U.S. facilities), 5 by universities and colleges (all U.S.), and 1 by McGraw-Hill, Inc. (corporate owner of Spitz Laboratories). None was operated by schools and districts. Of the 31 projectors, 16 were built by Zeiss (Jena or Oberkochen), 11 by Spitz, and 4 by other manufacturers. After *Sputnik's* launch, 18 new major planetaria were constructed in the United States, while one new major planetarium was established in each of five Canadian provinces: Alberta, British Columbia, Manitoba, Ontario, and Quebec. The number of North America's major planetaria jumped from a pre-*Sputnik* level of 7 to 31 institutions in just over thirteen years.

Planetarium Directors

Unlike the rich assortment of *CATNAP* data on American planetarium institutions, demographic information is scarce concerning their personnel.

TABLE A.3 *Distribution of Thirty-One Major North American Planetaria by Institutional Category*

Museum/science center (25)	U.S.	=	19	76%
	Canada	=	5	20%
	Mexico	=	1	4%
University/college (5)	U.S.	=	5	100%
Corporate (1)	U.S.	=	1	100%
Total			31	

Source: *A Catalog of North American Planetariums (CATNAP).*

Members of the American planetarium community have remained historically obscure, even within their own specialty. Few individuals were awarded places in biographical encyclopedias. Little or nothing in the way of personal data is readily available on rank-and-file planetarians. Before comprehensive surveys of the planetarium community were undertaken, an irregular *Sky and Telescope* column called "Planetarium Notes" provided the fullest record of personnel and institutions after 1947.[5] But only those facilities that offered regular public programs received such listings. As a result, serious omissions exist regarding school and district, and university and college planetaria that did not satisfy this criterion.

In his analysis of the American astronomical community from 1859 to 1940, historian John Lankford divided 1,205 men and women into three "cohorts" that shared chronological experiences. Lankford's final two cohorts embraced forty-year periods each. Thus, it is reasonable to consider members of the American planetarium community between 1930 and 1970 as comprising a single cohort, which witnessed three developmental periods. For purposes of analysis, the remaining assessment is restricted to those individuals who, between 1930 and 1970, are known to have become planetarium directors (or an equivalent title), signifying managerial responsibility for the maintenance and operation of a permanent planetarium facility. Remuneration for the performance of duties, even at the highest levels of authority, cannot always be assumed.[6]

A total of 919 people constituted the known body of American planetarium directors between 1930 and 1970. One of its most evident characteristics was the preponderance of males over females: 841 (91.5%) were men and 78 (8.5%) were women.[7] Given the prevalence of cultural stereotypes and lack of appropriate role models for women,[8] it might appear surprising that so high a percentage was employed in directing American planetaria. However, this value is not at variance with women's constituency in the postwar American astronomical community. In 1960, the last year for which a distribution of American astronomers on the basis of gender was reported by the National Register of Scientific and Technical Personnel, 73 women made up 11.6% of the nation's

630 astronomers.[9] Such a finding is likewise congruent with the percentage of doctorates earned by women in astronomy (23 out of 208, or 11.1%) between 1947–48 and 1960–61, as reported by historian Margaret W. Rossiter.[10]

Adequate data are lacking on the highest degrees earned by the vast majority of American planetarium directors. Consequently, the conclusions advanced below must be regarded as merely suggestive rather than definitive. A comprehensive self-study was conducted during the 1970 CAPE conference, when Lansing Community College planetarium director Morton E. Mattson administered a forty-one-item survey questionnaire to 159 delegates (some 56% of registrants). Roughly 31.2% of CAPE respondents indicated a bachelor's degree as their highest educational level; 44.7% a master's; 2.8% an "Ed.S." or "sixth year" diploma; and 9.9% a doctorate. Some 11.3% reported "other" training (or a high school diploma) as their highest educational attainment. While not generalizable to remaining CAPE delegates, nor to the American planetarium community as a whole, these data nonetheless provide the most complete historical sketch of directors' educational backgrounds.[11]

Among planetaria of all sizes, the highest degree possessed by directors was strongly correlated with the institutional category in which employment was secured.[12] Directors of university and college planetaria were the most likely to possess advanced degrees; a master's degree can be fairly well assumed. A significant portion earned doctorates, especially if the planetarium was associated with the physics or astronomy department of a four-year college or research university. At smaller junior and community colleges, however, the planetarium director was more often an instructor whose primary duties involved teaching and little or no research. Directors who practiced their trades in museums and science centers possessed the widest range of educational experiences. At larger museums where research was conducted, there was a greater likelihood of directors having earned advanced degrees. The reverse was true among smaller, regional museums, whose planetarium directors were more closely attuned to methods of public education. Among directors employed by schools and districts, a bachelor's degree can be readily assumed. A master's degree was more likely to be obtained the longer one held this position, reflecting the need for career teachers to obtain permanent certification through completion of graduate-level coursework. School and district planetarium directors rarely acquired doctorates; Ann Arbor High School's John C. Rosemergy was a notable exception.

Eighty individuals out of 919 (8.7%) are identified as directors of major planetaria between 1930 and 1970. All but 8 worked inside the United States; the remainder served in Canada, Mexico, or South America.[13] Seventy-six were men, 4 were women.[14] It is impossible, however, to offer a reliable assessment of the importance of an advanced degree for the attainment of a major planetarium directorship.[15] Even among university and college facilities, a doctorate was not always required, as Morehead Planetarium's Anthony F. Jenzano demonstrated. Only five major planetaria were found at universities and colleges. Due to uncertainties in years of appointment, no precise determination of the average length of service can be calculated.[16] A handful of directors served for a year's time or less, while the longest tenure was held by Arthur L. Draper of Pittsburgh's Buhl Planetarium (30 years).[17] Other directors with lengthy records of service were Israel M. Levitt (22 years), Dinsmore Alter (19 years), Anthony F. Jenzano (19 years), F. Wagner Schlesinger (18 years), and Clarence H. Cleminshaw (15 years). The majority, however, served much shorter terms. Eight directors oversaw more than one major planetarium; none of the cohort directed more than two.[18]

Publication of journal articles (including conference proceedings) and book-length treatments on the nature and operation of planetaria, astronomy and science education, or astronomical history presents another means by which a director's standing within this cohort can be examined. Communication with peers and a desire to inform the public are the most obvious reasons behind an exchange of planetarium-related ideas. Both scientific and popular works enabled authors to secure credentials and enhance reputations during all three periods of planetarium development. Among directors as a whole, 129 (14.0%) became authors by 1970.[19] Of this group, 117 (90.7%) were men and 12 (9.3%) were women. Author distribution on the basis of gender closely matches the percentages of directors; neither men nor women authors were disproportionately represented in the literature of their field. This condition suggests that equal opportunities existed for publication by men and women directors.

Correlation of authorship with institutional category yields further insights into the nature of planetarium-related publications. Only 23 authors (18%) came from school and district planetaria. Despite the highest percentage (51.4%) of their facilities, publication was an activity seldom practiced among schoolteachers. Thirty-seven authors (29%) were employed in university and college positions, where scholarly publication was considered a professional norm. This figure is on a par with that category's institutional distribution (32.7%). Sixty-five authors (50%) held positions at museum and science center planetaria, often selecting popular magazines in which to publicize their facility's opening. While having the smallest categorical representation (14.0%), these directors proved themselves to be the most vocally active constituents of their peers. Four authors (3%) did not come from any category recognized in this study. Directors of major planetaria comprised 40 (31%) of the cohort's 129 authors. This proportion far exceeds the percentage of major facilities (4.2%) found within the American planetarium community.

Among the most prolific authors of this cohort were James Stokley, Roy K. Marshall, Henry C. King, and David A. Rodger. Sizeable numbers of publications were also contributed by William H. Barton, G. Clyde Fisher, and Charles F. Hagar. Five of the above (Barton, Fisher, King, Marshall, and Stokley) likewise coauthored textbooks on popular astronomy and its history. Other planetarium directors of this cohort, including Dinsmore Alter, Gordon A. Atwater, Maurice T. Brackbill, John M. Cavanaugh, Maribelle B. Cormack, Arthur L. Draper, Philip Fox, Israel M. Levitt, Marian Lockwood, O. Richard Norton, Norman Sperling, Julius D. W. Staal, and Philip D. Stern, either authored, co-authored, or edited one or more volumes on astronomical research, astronomy and science education, navigation, or planetarium development.

Before the late 1960s few internal reward systems were established within the American planetarium community; nor did external systems of peer recognition seem to hold any particular relevance to the advancement of directors' careers. Planetarians with research experience in astronomy might have belonged to the American Astronomical Society (AAS); those from museums and science centers possibly joined the American Association of Museums (AAM); and those employed by schools and districts might have acquired membership in the National Science Teachers Association (NSTA). A select few might have been admitted to more prestigious national or international scientific associations. Apart from their publication records, directors' peer recognition and rewards were bestowed for significant contributions made in the professional arena of planetarium education or administration. Some examples of these activities are collected below.

Directors whose institutions hosted either national or international conferences of astronomers or planetarium educators drew notable recognition from colleagues. These individuals included Philip Fox, James A. Fowler, Joseph M. Chamberlain, Dan Snow, and Von Del Chamberlain. On a reduced scale, the men and women who founded, hosted, and secured elected offices within regional planetarium associations performed related services in selected geographic areas. Similar honors were attached to the chairmanships of the AAM planetariums section, whose annual meetings were first hosted by Maxine (Begin) Haarstick. These conferences and associations spawned the discipline's first newsletters, monographs, journals, and related publications, whose production encompassed a growing number of directors. Graduate internship programs and summer institutes, which trained hundreds of new American planetarium directors, were coordinated by planetarians Charles F. Hagar, George E. Pitluga, and Michael A. Bennett.

Techniques for conveying the planetarium experience to wider audiences were pioneered by Anthony F. Jenzano (astronaut training), George W. Bunton (planetarium light shows), and O. Richard Norton (fisheye cinematography). Boston's Charles Hayden Planetarium received its twin-hemisphere star instrument (1958) via the technical ingenuity of Springfield, Massachusetts, director Frank D. Korkosz. Miss Charlie Noble had the planetarium of the Fort Worth, Texas, Museum of Science and History named after her. By 1967, the Great Lakes Planetarium Association (GLPA) established its annual lecture to recognize (and later memorialize) the contributions of Armand N. Spitz, without whom a majority of the nation's planetaria might never have been built.

Career Management Strategies

During the 1960s, an array of positions was opened to recently trained and veteran directors, assuming that mobility was not a hindrance to starting or advancing one's career. If one appointment proved unsuitable, another was certain to come along, potentially maximizing successful matchups between personnel and institutions. Informal networks fashioned among directors, institutions, and manufacturers channeled promising candidates into newly created openings. However, a more detailed analysis of the career management strategies applicable to the bulk of this cohort is precluded by the entry of hundreds of new directors to the profession during the final years of that decade. Instead, selected aspects of career management skills, as enacted within America's major planetaria and among women planetarium directors, are sketched below.

While nonexistent during the formative period (1930–46), and seldom accomplished during the postwar period (1947–57), a career pathway leading from smaller to larger institutions became a well-trodden route to the directorships of major planetaria during the space age (1958–70). Canadian-born Ian C. McLennan was appointed director at Rochester's Strasenburgh Planetarium after his five-year stint as head of the Queen Elizabeth Planetarium in Alberta. A reverse border crossing occurred with Dennis H. Gallagher's move from the Noble Planetarium of Fort Worth, Texas, to the Manitoba Planetarium in Winnipeg.

Persons immigrating from abroad likewise attained directorships of major American planetaria. Henry C. King, former director of the London Planetarium, took charge of the McLaughlin Planetarium in Toronto. Dennis P. Simopoulos and Julius D. W. Staal, who had gained experience in European facilities, directed new planetaria erected in Baton Rouge, Louisiana, and Atlanta, Georgia, respectively.

During the postwar period, chief technicians from two major planetaria became directors at other major installations. George W. Bunton departed Los Angeles's Griffith Observatory to head San Francisco's Morrison Planetarium, while Anthony F. Jenzano worked at Philadelphia's Fels and Chapel Hill's Morehead Planetarium before succeeding Roy K. Marshall at the latter facility.

On two occasions, planetaria that once operated as smaller installations enlarged their facilities. The Denver Museum of Natural History and the Miami Museum of Science constructed major planetaria under the leadership of extant directors Donald M. Lunetta and Vincent J. Gabianelli, respectively. This provided yet another, albeit uncommon, pathway to the directorship of a major planetarium.

Administrative turnover was especially high among personnel at the U.S. Air Force Academy Planetarium in Colorado Springs, Colorado. Rapid promotions characteristic of a military lifestyle yielded seven appointments across a span of twelve years. Significant cooperation with civilian planetaria was accomplished, however, in the career of Lt. Col. Edward R. Therkelsen, who founded the Rocky Mountain Planetarium Association (RMPA) and earned a Ph.D. from the University of Denver.

After heading the Morrison Planetarium for ten years, George W. Bunton stepped down to become director of the smaller Bishop Museum Planetarium in Honolulu, Hawaii. Bunton's reverse career move was adopted voluntarily, and no one could argue with his choice of location. While not the norm, such moves from major to smaller planetaria did not automatically signify career failure, but reflected a host of reasons for their adoption (perhaps the search for a less demanding role with fewer pressures).

Strictly speaking, the illustrations sketched above referred most accurately to the dominant gender of American planetarium directors. The cohort's seventy-eight women directors, by contrast, faced more restrictive opportunities and appear to have held different expectations than men for advancing their planetarium careers. Almost without exception, the career pathways opened to men, leading to the directorship of a major planetarium, remained closed to women during this period.

Women planetarium directors were rather evenly distributed among the institutional categories employed in this study. Thirty-one (40%) worked at museums and science centers, twenty-five (32%) at schools and districts, and twenty-one (27%) at universities and colleges. One woman (Sheila M. Duck) directed a planetarium at NASA's Goddard Spaceflight Center in Greenbelt, Maryland. At least five women directors earned doctorates; astronomer Katherine Bracher, who oversaw the Whitman College Planetarium in Walla Walla, Washington, provides an example.

If these figures are compared to those within each institutional category, however, several variations are readily noted. Women directors were significantly under-represented on the basis of gender among U.S. school and district planetaria (51.4% of the number of planetaria). By contrast, their numbers more nearly approximated the institutional representation of university and college facilities (32.7%). But an over-representation of women directors was apparent among museums and science centers, which made up the smallest institutional category (14.0%). That employment setting appears to have been more congenial to acquiring and retaining women planetarium directors, especially in smaller regional museums.

Only three women in the cohort directed major planetaria. Maude Bennot of Chicago's Adler Planetarium held the longest record of service (eight years) and earned the highest known degree (master's) within this select group. Marian Lockwood served nearly one year as acting director of New York's Hayden Planetarium. Both women were products of an apprentice-style training received during the formative period. An incomplete record of service is known for Elizabeth Hill, who directed the W. A. Gayle Planetarium at Troy State University in Montgomery, Alabama, in 1970.

Of the twelve women authors recognized in this cohort, nine were located at museums and science centers, two at schools and districts, and one at a collegiate facility. Reflecting the institutional distribution of author-directors as a whole, women from museums and science centers most often found expression in print. The most prolific writer was Marian Lockwood, who published numerous articles and coauthored three textbooks before acquiring the Hayden Planetarium directorship.

Women's record of service in the planetarium community was almost as lengthy as men's. Those with the longest appointments included Maxine B. Haarstick (19 years), Maribelle B. Cormack (17 years), Genevieve B. R. Woodbridge (14 years), and three women (Jacqueline Avent, Margaret K. Noble, and Mrs. Luther W. Kelly) with 12 years each. Five of those six were located in museums and science centers; Noble served in two school districts. Three women directed two institutions; Jeanne (Emmons) Bishop had charge of three. Eight women (10%) affixed the title of "Sister" to their names, revealing their devotions as nuns. For these women, the study and teaching of astronomy presumably offered strong affirmations of their spiritual values.

Few gender-specific differences appear to have operated among university and college or school and district planetaria. The largest differences instead resided within smaller, regional museums and science centers. Within such institutions, women achieved greater autonomy and professional advancement. Several women museum directors concurrently procured and managed Spitz planetaria, an experience not replicated at larger facilities. For those women who became both museum and planetarium directors, these career opportunities seemingly offered all of the professional responsibilities and rewards that they sought. But without added support from archival and primary source materials, the mobility patterns among women planetarium directors cannot be fully interpreted. The complex mixture of personal and professional motivations that guided their career management decisions has yet to be fully understood.

NOTES

List of Abbreviations

AMNH	American Museum of Natural History, New York, New York
AP	Abrams Planetarium Archives, Michigan State University, East Lansing, Michigan
APAM	Adler Planetarium and Astronomy Museum, Chicago, Illinois
CCNY	Carnegie Corporation of New York Archives, Columbia University, New York, New York
CPD	Chicago Park District Archives, Chicago, Illinois
FI	Franklin Institute Archives, Philadelphia, Pennsylvania
HL	Huntington Library, San Marino, California
HSP	Historical Society of Pennsylvania, Philadelphia, Pennsylvania
HSWP	Historical Society of Western Pennsylvania, Pittsburgh, Pennsylvania
IPS	International Planetarium Society Archives, Ash Enterprises, Bradenton, Florida
MP	Morehead Planetarium Archives, University of North Carolina, Chapel Hill, North Carolina
MSI	Museum of Science and Industry Archives, Chicago, Illinois
NU	Northwestern University Library, Evanston, Illinois
RSP	Richard S. Perkin Collection, Department of Astrophysics, American Museum of Natural History, New York, New York
UCLA	University of California at Los Angeles Library, Los Angeles, California
UP	University of Pennsylvania Archives, Philadelphia, Pennsylvania
YOA	Yerkes Observatory Archives, Williams Bay, Wisconsin

Introduction

1. See for instance Daniel J. Kevles, *The Physicists: The History of a Scientific Community in Modern America* (Cambridge, Mass.: Harvard University Press, 1987); John W. Servos, *Physical Chemistry from Ostwald to Pauling: The Making of a Science in America* (Princeton, N.J.: Princeton University Press, 1990); Lynn K. Nyhart, *Biology Takes Form: Animal Morphology and the German Universities, 1800–1900* (Chicago: University of Chicago Press, 1995); W. Conner Sorensen,

Brethren of the Net: American Entomology, 1840–1880 (Tuscaloosa: University of Alabama Press, 1995); Julie R. Newell, "American Geologists and Their Geology: The Formation of the American Geological Community, 1780–1865" (Ph.D. diss., University of Wisconsin-Madison, 1993); John Lankford, *American Astronomy: Community, Careers, and Power, 1859–1940* (Chicago: University of Chicago Press, 1997); and James H. Capshew, *Psychologists on the March: Science, Practice, and Professional Identity in America, 1929–1969* (New York: Cambridge University Press, 1999).

2. Sally Gregory Kohlstedt, *The Formation of the American Scientific Community: The American Association for the Advancement of Science, 1848–1860* (Urbana: University of Illinois Press, 1976); Jack Morrell and Arnold Thackray, *Gentlemen of Science: Early Years of the British Association for the Advancement of Science* (Oxford: Clarendon Press, 1981); Margaret W. Rossiter, *Women Scientists in America: Struggles and Strategies to 1940* (Baltimore: Johns Hopkins University Press, 1982); and Margaret W. Rossiter, *Women Scientists in America: Before Affirmative Action, 1940–1972* (Baltimore: Johns Hopkins University Press, 1995).
3. Lankford, *American Astronomy,* xvi.
4. Jordan D. Marché II, "Gender and the American Planetarium Community," *The Planetarian* 31, no. 2 (June 2002): 4–7, 36.
5. Lankford, *American Astronomy,* xviii.
6. Steven Shapin, "Science and the Public," in *Companion to the History of Modern Science,* eds. R. C. Olby, G. N. Cantor, J. R. R. Christie, and M. J. S. Hodge (London and New York: Routledge, 1990), 990.
7. Roger Cooter and Stephen Pumfrey, "Separate Spheres and Public Places: Reflections on the History of Science Popularization and Science in Popular Culture," *History of Science* 32 (1994): 254–255.
8. Literature on scientific instruments is vast; selected accounts include Albert Van Helden, "The Birth of the Modern Scientific Instrument, 1550–1750," in John G. Burke, ed., *The Uses of Science in the Age of Newton* (Berkeley and Los Angeles: University of California Press, 1983), 49–84; J. V. Field, "What Is Scientific About a Scientific Instrument?" *Nuncius* 3, no. 2 (1988): 3–26; Deborah Jean Warner, "What Is a Scientific Instrument, When Did It Become One, and Why?" *British Journal for the History of Science* 23 (1990): 83–93; and Albert Van Helden and Thomas L. Hankins, "Introduction: Instruments in the History of Science," *Osiris,* 2d ser., 9 (1994): 1–6. These last authors note that "*instruments can act as bridges between natural science and popular culture*" (emphasis in original, p. 5).

Chapter 1 ***Zeiss Planetaria in Europe, 1923–1929***

1. Charles F. Hagar, *Planetarium: Window to the Universe* (Oberkochen: Carl Zeiss, 1980), 7.
2. Hagar graphically depicts a conjunction of these streams, separately titled "Models of the Night Sky" and "Models of the Sun, Moon, and Planets," resulting in "Metamorphosis of the Zeiss Planetarium" (on pp. 100–101).
3. Besides Hagar's volume, five standard accounts showcase the technological history of celestial globes, orreries, and astronomical clocks. In reverse chronological order, these are Henry C. King, *Geared to the Stars: The Evolution of*

Planetariums, Orreries, and Astronomical Clocks (Toronto: University of Toronto Press, 1978); Harriet Pratt Lattin, *Star Performance* (Philadelphia: Whitmore Publishing, 1969); O. Richard Norton, *The Planetarium and Atmospherium: An Indoor Universe* (Healdsburg, Calif.: Naturegraph Publishers, 1968); Heinz Letsch, *Captured Stars,* trans. Harry Spitzbardt (Jena: Gustav Fischer Verlag, 1959); and Helmut Werner, *From the Aratus Globe to the Zeiss Planetarium*, trans. A. H. Degenhardt (Stuttgart: Gustav Fischer Verlag, 1957). See also Mark R. Chartrand III, "A Fifty Year Anniversary of a Two Thousand Year Dream," *The Planetarian* 2, no. 3 (Sept. 1973): 95–101.

4. Thomas L. Heath, *Greek Astronomy* (New York: Dover Publications, 1991), xxviii–xxix.
5. Werner, *Aratus Globe,* 14.
6. Wallace W. Atwood, "A New Way of Studying Astronomy," *Scientific American*, 21 June 1913, 557.
7. On Comstock's teaching, writing, and scientific illustrations, see Pamela M. Henson, "'Through Books to Nature': Anna Botsford Comstock and the Nature Study Movement," in *Natural Eloquence: Women Reinscribe Science,* ed. Barbara T. Gates and Ann B. Shteir (Madison: University of Wisconsin Press, 1997), 116–143.
8. Wallace W. Atwood, "Giant Celestial Sphere a Forerunner of Later Planetariums," *The Sky,* April 1938, 12–13.
9. See Derek J. de Solla Price, "An Ancient Greek Computer," *Scientific American*, June 1959, 60–67, and Price, *Gears from the Greeks: The Antikythera Mechanism, a Calendrical Computer from ca. 80 B.C.* (Philadelphia: American Philosophical Society, 1974).
10. King, *Geared to the Stars,* 10.
11. See F. A. B. Ward, ed., *The Planetarium of Giovanni de Dondi, Citizen of Padua* (London: Antiquarian Horological Society, 1974), a manuscript of 1397 translated by G. H. Baillie with additional material from another Dondi manuscript translated by H. Alan Lloyd.
12. Lattin, *Star Performance,* 132. Order of the two phrases has been reversed from the original.
13. Govert Schilling, "Eise Eisinga's Novel Planetarium," *Sky and Telescope,* February 1994, 28–30.
14. Werner, *Aratus Globe,* 26.
15. Thomas L. Hankins and Robert J. Silverman, *Instruments and the Imagination* (Princeton, N.J.: Princeton University Press, 1995), 37–71.
16. Edward P. Alexander, "Oskar von Miller and the Deutsches Museum: The Museum of Science and Technology," in *Museum Masters: Their Museums and Their Influence* (Nashville, Tenn.: American Association for State and Local History, 1983), 346–347, 343. See also Otto Mayr, *The Deutsches Museum: German Museum of Masterworks of Science and Technology, Munich* (London: Scala Books, 1990).
17. Oskar von Miller, "The German Museum in Munich," *Proceedings of the American Society of Civil Engineers* 52 (January 1926): 13.
18. Alexander, "Oskar von Miller," 353 (emphasis added).
19. Miller, "German Museum in Munich," 17.
20. King, *Geared to the Stars,* 341.
21. See Walther Bauersfeld, "The Great Planetarium of the German Museum in

Munich," *Proceedings of the United States Naval Institute* 51 (1925): 761–774. This is an English translation by G. N. Saegmuller of Bauersfeld's original description of the Zeiss planetarium, "Das Projektions-Planetarium des Deutschen Museums in München," *Zeitschrift des Vereines Deutscher Ingenieure* 68 (1924): 793–797.

22. Walther Bauersfeld, "Projection Planetarium and Shell Construction," part 1, *The Engineer,* 17 May 1957, 756.
23. Franz Fuchs, "Der Aufbau der Astronomie im Deutschen Museum," *Deutsches Museum Abhandlungen und Berichte* 23, no. 1 (1955): 1–68. Fuchs cites correspondence from von Miller to Zeiss, dated 20 March 1914, which reiterates the use of optical (i.e., projection) apparatus to depict the Sun, Moon, planets, and stars (p. 61). The 3 February 1917 press release featured artistic conceptions (reproduced as figs. 29 and 30, p. 62) of the Ptolemaic and Copernican planetaria, respectively.
24. See "Figure-Ground Perception" in Robert M. Goldenson, *The Encyclopedia of Human Behavior: Psychology, Psychiatry, and Mental Health,* vol. 1 (Garden City, N.Y.: Doubleday, 1970), 464–466; and "Perception" in Raymond J. Corsini, ed., *Encyclopedia of Psychology,* vol. 2 (New York: John Wiley and Sons, 1984), 498.
25. Thomas S. Kuhn, *The Structure of Scientific Revolutions,* 2d ed. (Chicago: University of Chicago Press, 1970), 111–135.
26. Bauersfeld, quoted in W. Villiger, *The Zeiss Planetarium* (London: W. and G. Foyle, 1926), on p. 11.
27. Bauersfeld likewise suggested the concrete-shell fabrication technique, which enabled "large-diameter domes without central supports to be erected with small cost in labour and materials" (Werner, *Aratus Globe,* 129). The Franklin Institute awarded its Edward Longstreth Medal to the Zeiss firm in recognition of the concrete-shell technique. See Bauersfeld, "Projection Planetarium and Shell Construction," part 2, *The Engineer,* 24 May 1957, 796–797.
28. See "Presentation in absentia of Dr. Walther Bauersfeld" and "Received by Franz Fieseler, Esq., on Behalf of Dr. Walther Bauersfeld," *Journal of the Franklin Institute* 216 (July-December 1933): 792–794.
29. I am indebted to Zeno G. Swijtink, formerly of the Department of History and Philosophy of Science, Indiana University, for emphasizing the significance of Miller's pedagogical objectives.
30. Author's translation from Franz Fuchs, "Die Planetarien des Deutschen Museums," *Berliner Tageblatt,* 7 Nov. 1923.
31. Science Service, "The 'Planetarium'," *Science — Supplement* 60 (5 September 1924): xii.
32. Bauersfeld, "Great Planetarium," 774.
33. G. H. Morison, "Heavens Built of Concrete," *Scientific American,* March 1925, 170–171. "Die Geheimnisse der Sterne" appeared in *Westermanns Monatshefte* 137 (February 1925): 576–580.
34. Morison, "Heavens Built of Concrete," 171.
35. David Todd, "A New Optical Projection Planetarium for Visualizing the Motions of the Celestial Bodies, as Seen by the Naked Eye, from the Earth," *Popular Astronomy,* August 1925, 448, 455, 456. Todd's final (unacknowledged) quotation is from William Blake's *Auguries from Innocence.*
36. "Astronomy de Luxe," *The Outlook,* 3 February 1926, 161.
37. Clyde Fisher, "The New Projection Planetarium," *Natural History,* July-August 1926, 403–405, 409. While Fisher's report was deliberately slanted to favor the pro-

jection planetarium, his assessments concerning its public reception remain historically significant.

38. Ibid., 409.
39. W. J. Luyten, "The New Projection Planetarium," *Natural History,* July-August 1927, 387; Albert G. Ingalls, "Canned Astronomy: What the New Planetariums for Chicago and Philadelphia Will Be Like," *Scientific American,* September 1929, 202.
40. O. D. Tolischus, "Seeing Stars: How an Intricate German Machine Reveals the Heavens," *The World's Work,* November 1927, 96–97; Ingalls, "Canned Astronomy," 202.
41. See Hagar, *Window to the Universe*, 96–98 and fig. 141 for Villiger's design proposal and reproduction of Bauersfeld's sketch.
42. Waldemar Kaempffert, "Now America Will Have a Planetarium," *New York Times Magazine,* 24 June 1928, 5; Ingalls, "Canned Astronomy," 202.
43. Illustrations are featured in Luyten, "New Projection Planetarium," 383–390; Tolischus, "Seeing Stars," 96–100; and "Cities of Germany Foster Enjoyment of Astronomy by Planetaria," *American City,* June 1928, 102–103. Many of these planetaria were damaged beyond repair during the Second World War.
44. For an exhaustive study of planetarium architecture, see Charles F. Gronauer, "The Planetarium, Its History, Functions, and Architecture: With Application for a Proposed Addition to the Florida State Museum" (M.A. thesis, University of Florida, 1978).
45. Tolischus, "Seeing Stars," 98, 96; Ingalls, "Canned Astronomy," 203, 204.
46. Tolischus, "Seeing Stars," 96, 97.
47. Jackson, quoted in Calvin Fraser, "The Zeiss Planetarium: A Fascinating and Instructive Theater of the Stars," *The Mentor,* January 1928, on p. 59.
48. Strömgren, quoted in Kaempffert, "Now America," on p. 4. Strömgren's label, the "Wonder of Jena," was widely applied to the planetarium. See Luyten, "New Projection Planetarium," 383.
49. Luyten, "New Projection Planetarium," 387.
50. Tolischus, "Seeing Stars," 100.
51. Ingalls, "Canned Astronomy," 201.
52. Kaempffert, "Now America," 5.
53. For statistics concerning planetarium locations, dome sizes, dates of opening, and occasional seating capacities, see Werner, *Aratus Globe,* 153. By general consensus, "major" planetaria are those with dome diameters of at least 16 meters (roughly 50 feet).
54. For two-year Munich attendance figures, see Luyten, "New Projection Planetarium," 390. Demonstration schedules and attendance records at the Munich and Zeiss factory planetaria were reported by Fisher, "New Projection Planetarium," 405, 409. First-year Jena planetarium attendance was gleaned from "Astronomy de Luxe," 161; Fraser, "Zeiss Planetarium," 58; Kaempffert, "Now America," 4; and Ingalls, "Canned Astronomy," 202. Berlin statistics were cited by Tolischus, "Seeing Stars," 100. Cumulative attendance figures were presented by Tolischus and Science Service, "The Planetarium," *Science — Supplement* 67 (22 June 1928): x.
55. The preliminary (1927) figure, obtained from Zeiss, was quoted by Tolischus, "Seeing Stars," 100. Intermediate (1928) figures appeared almost simultaneously in Science Service, "The Planetarium," x, and Kaempffert, "Now America," 4. Ingalls, "Canned Astronomy," 201, gave the final (1929) estimate.
56. Belt's (1925) projection is contained in his preface to Bauersfeld, "Great Planetarium," 762. Actual (1927) expenditures for Berlin's twenty-five-meter facility

were given by Tolischus, "Seeing Stars," 100. Further estimates (1928) are derived from Science Service, "The Planetarium," x, and Kaempffert, "Now America," 4.

Chapter 2 *Planetaria, Patrons, and Cultural Values*

1. In the United States, a marked decline in formal astronomical teaching affecting both secondary and collegiate instruction occurred in the years after 1900. For the reasons behind this transformation, see Jordan D. Marché II, "Mental Discipline, Curricular Reform, and the Decline of U.S. Astronomy Education, 1893–1920," *Astronomy Education Review* 1 (2002): 58–75.
2. Joel J. Orosz, *Curators and Culture: The Museum Movement in America, 1740–1870* (Tuscaloosa: University of Alabama Press, 1990), 8.
3. Steven Conn, *Museums and American Intellectual Life, 1876–1926* (Chicago: University of Chicago Press, 1998), 4, 5, 42.
4. Edward P. Alexander, "Carl Ethan Akeley Perfects the Habitat Group Exhibition," in *The Museum in America: Innovators and Pioneers* (Walnut Creek, Calif.: Altamira Press, 1997), 33–49.
5. Ronald Rainger, *An Agenda for Antiquity: Henry Fairfield Osborn and Vertebrate Paleontology at the American Museum of Natural History, 1890–1935* (Tuscaloosa: University of Alabama Press, 1991), 120.
6. Lengthier accounts of the origins of these planetaria may be found in Jordan D. Marché II, "Theaters of Time and Space: The American Planetarium Community, 1930–1970" (Ph.D. diss., Indiana University, 1999), chapter 3, 105–121. Their individual histories were likewise retold by planetarium director David H. Menke: "Phillip [sic] Fox and the Adler Planetarium," *The Planetarian* 16, no. 1 (January 1987): 46–48; "Clyde Fisher and the Hayden Planetarium," ibid. 16, no. 2 (April 1987): 54–58; "James Stokley and the Fels Planetarium," ibid. 16, no. 3 (July 1987): 39–43; "Dinsmore Alter and the Griffith Observatory," ibid. 16, no. 4 (October 1987): 68–78; and "Arthur L. Draper and the Buhl Planetarium," ibid. 17, no. 1 (March 1988): 58–61, 73. Menke, however, relied chiefly on oral histories from staff members and enlisted few primary sources in his narratives.
7. For a popular history of the catalogue merchandise store, see Lorin Sorensen, *Sears, Roebuck and Co. 100th Anniversary, 1886–1986* (St. Helena, Calif.: Silverado Publishing, 1985).
8. The Deutsches Museum's influence on Rosenwald's conception of the Museum of Science and Industry is recounted in Herman Kogan, *A Continuing Marvel: The Story of The Museum of Science and Industry* (Garden City, N.Y.: Doubleday, 1973). Manuscripts pertaining to the Museum's founding are contained in the Julius Rosenwald Correspondence, Museum of Science and Industry (MSI) Archives, Chicago, Illinois.
9. Unpublished manuscript, "The Adler Planetarium of Chicago (1930)," 3. Essays on the Adler Planetarium folder, History of Astronomy Research Center, Adler Planetarium and Astronomy Museum (APAM), Chicago, Illinois. Anecdotal evidence suggests that "[t]he gift was prompted by the impression made upon him by a performance seen in Munich." See "Star Chamber," *Time,* 19 May 1930, 72. His obituary, "Max Adler, Donor of Planetarium, 86," appeared in the *New York Times,* 6 November 1952.

10. Adler Planetarium director Philip Fox confirmed that Adler "offer[ed] first to build the Planetarium in connection with the Museum of Science and Industry, . . . but when the site nearer the center of the city . . . became available, he increased his gift to include construction of the building." Fox to George F. Kunz, 3 March 1931. Attendance folder, APAM.
11. See Official Proceedings, South Park Commissioners, vol. 36, 20 June 1928, 381, for the motion to accept Adler's proposal. Chicago Park District (CPD) Archives, Chicago, Illinois. "Agreement, Between South Park Commissioners and Max Adler, Relating to Adler Planetarium and Astronomical Museum, June 20, 1928," *Journal of the Proceedings of the South Park Commissioners,* 2 April 1930, 33–38.
12. Bruce Sinclair, *Philadelphia's Philosopher Mechanics: A History of the Franklin Institute, 1824–1865* (Baltimore: Johns Hopkins University Press, 1974), x.
13. Minutes, Board of Managers, 9 March 1927, Franklin Institute (FI) Archives, Philadelphia, Pennsylvania. Biographical data from McClenahan folder, University of Pennsylvania (UP) Archives, Philadelphia, Pennsylvania.
14. See Dale Phalen, *Samuel Fels of Philadelphia* (Philadelphia: Samuel S. Fels Fund, 1969). The principal repository of Fels materials is the Samuel Simeon Fels Papers, Historical Society of Pennsylvania (HSP), Philadelphia, Pennsylvania.
15. David Riesman, "The Zeiss Planetarium," *General Magazine and Historical Chronicle,* January 1929, 241.
16. Steven M. Spencer, "Meet the Donor of Planetarium," *Philadelphia Evening Bulletin,* 17 November 1933.
17. Quotation from a Mr. Brakeley, 3 March 1930, on donor card, Samuel Simeon Fels 1937 LL.D. Benefactor folder, UP Archives.
18. See Charles G. Clarke, "Griffith J. Griffith and His Park," *Griffith Observer,* May 1980, 2–10. Scholarship on Griffith has improved substantially with the release of Mike Eberts, *Griffith Park: A Centennial History* (Los Angeles: Historical Society of Southern California, 1996).
19. Griffith's principal statement on the preservation of public lands is contained in his *Parks, Boulevards and Playgrounds* (Los Angeles: Prison Reform League, 1910).
20. The first proposal to erect an observatory inside Griffith Park came from entrepreneur James W. Eddy in 1902. See Mike Eberts, "The Little Known Early History of the Griffith Observatory," *Griffith Observer,* May 1995, 2–18. Interest in Eddy's proposal was terminated by Griffith's shooting of his wife, Tina, in 1903, which left her an invalid. After serving a two-year sentence at San Quentin Prison, Griffith began a study of the American penal system and became secretary-treasurer of the Prison Reform League. Its volume, *Crime and Criminals* (Los Angeles: Prison Reform League, 1910), bears Griffith's editorial hand.
21. Griffith to Los Angeles City Council, 21 December 1912; reprinted in "The Story of Griffith," *Griffith Park Quarterly,* April/July 1978, 12, 13.
22. Excerpts from Griffith's will quoted in "The Story of Griffith," on pp. 20–21. Final disposition on Griffith's will was not ruled until 12 November 1924 by J. P. Wood, Judge in the Superior Court of the State of California.
23. Widney to California Institute of Technology, 16 April 1930, *Microfilm Edition, The George Ellery Hale Papers, 1882–1937* (Daniel J. Kevles, ed.), Carnegie Institution of Washington and California Institute of Technology, 1968, Reel 59; Adams to Widney, 18 April 1930, Folder 32.537, Walter Sydney Adams Papers,

Carnegie Observatories Collection, Huntington Library (HL), San Marino, California; Adams to Hale, 23 April 1930; Hale to Adams, 23 April 1930; Hale Papers, Reel 59.

24. Adams and Millikan to Widney, 11 July 1930, Folder 32.537, Adams Papers, HL.
25. John Michael Kennedy, "Philanthropy and Science in New York City: The American Museum of Natural History, 1868–1968" (Ph.D. diss., Yale University, 1968).
26. Rainger, *Agenda for Antiquity.*
27. Henry Fairfield Osborn, *Forty-Eighth Annual Report of the American Museum of Natural History for the Year 1916* (New York: N.p., 1 May 1917), 23.
28. Henry Fairfield Osborn, *The American Museum of Natural History: Its Origin, its History, the Growth of its Departments to December 31, 1909,* 2d ed. (New York: Irving Press, 1911).
29. An account of the Carnegie Corporation during its formative years has been provided by Robert E. Kohler, *Partners in Science: Foundations and Natural Scientists, 1900–1945* (Chicago: University of Chicago Press, 1991).
30. See Butler, "An Ideal Astronomic Hall," *Natural History,* July-August 1926, 393–398.
31. Documents relating to Osborn's campaign for the Astronomic Hall are housed at the Carnegie Corporation of New York (CCNY) Archives, Columbia University, New York, New York, Box 37, American Museum of Natural History folder. Related papers are found in Central Archives, American Museum of Natural History (AMNH), New York, New York, ser. 1178.5, 1925 folder.
32. See "Davison is Named to Head Museum," *New York Times,* 10 January 1933.
33. "Summary Report to Directors of the Reconstruction Finance Corporation on Self-Liquidating Project," 23 June 1933, AMNH ser. 1178.5, June 20–23, 1933 folder; "Planetarium Here Assured by Loan," *New York Times,* 27 June 1933.
34. The financier's donation was spelled out in Hayden to Davison, 20 November 1933, AMNH ser. 1178.5, November 20–31, 1933 folder. Previously, Hayden had donated $2,500 towards establishment of the Astronomic Hall. Hayden to Fisher, 27 April 1927, Richard S. Perkin (RSP) Collection, Department of Astrophysics, American Museum of Natural History, ser. 4:5, Box 11, Folder 27. The J. W. Fecker Company of Pittsburgh received a contract to produce the Museum's Copernican Planetarium.
35. Marian Lockwood, "The Hayden Planetarium," *Natural History,* October 1935, 189.
36. Sources include "Henry Buhl, Jr.," in Charles F. Lewis, *The Buhl Foundation: A Report by the Director upon its Work to June 30, 1942* (Pittsburgh: Farmers Bank Building, n.d.); and unpublished manuscript, "Some Chronological Notes on the Boggs and Buhl Department Store," 6 September 1941, Buhl Foundation Records, Box 1, Folder 8, Library and Archives Division, Historical Society of Western Pennsylvania (HSWP), Pittsburgh, Pennsylvania.
37. *Will of Henry Buhl, Jr., Deceased* (Pittsburgh: Buhl Foundation, 1929), sec. eighteenth, 9–10. Buhl Foundation Records, Box 1, Folder 1, HSWP.
38. Buhl to Hon. Robert S. Frazer et al., 1 November 1926, Buhl Foundation Records, Box 2, Folder 1, HSWP.
39. See "Charles F. Lewis, Led Conservancy," *New York Times,* 20 March 1971.
40. Fox related their plans to Allegheny Observatory director Heber D. Curtis as "a campaign for a planetarium for your city." Fox to Curtis, 28 May 1930, 1930–1960

folder, APAM. Scanlon's Valley View Observatory was described in "Amateur Astronomy in Pittsburgh," *Scientific American,* July 1931, 30–32.

41. A tribute to Evans is found in the "Acceptance Address" delivered by Pittsburgh Mayor Cornelius D. Scully, in *Dedication of the Buhl Planetarium and Institute of Popular Science* (N.p., n.d.), 25.
42. [Lewis], "To the Board of Managers," 30 November 1936, Buhl Foundation Records, Box 33, Folder 7, HSWP.
43. Robertson to Lewis, 12 April 1937, Buhl Foundation Records, Box 33, Folder 7, HSWP.
44. Minutes of the Board of Managers were not entrusted to the HSWP. Consequently, only an incomplete picture of the board's plans for the Institute of Popular Science has been obtained.
45. A sum of $750,000 was appropriated for the project. "Appropriation Resolution No. 55," 24 June 1937, Buhl Foundation Records, Box 33, Folder 7, HSWP. See also "Buhl Foundation Gives Pittsburgh Planetarium," *Greater Pittsburgh,* May 1937, 34–36.
46. Three additional appropriation resolutions were passed 18 March and 17 August 1938 and 19 October 1939. Buhl Foundation Records, Box 33, Folder 7, HSWP.
47. Helen Wright, *Explorer of the Universe: A Biography of George Ellery Hale* (New York: E. P. Dutton, 1966), 156, and note 38, 444.
48. Edwin [sic] Widney, "Bank, as Trustee, Builds Greek Theatre and Astronomical Observatory," *Trust Companies* 53 (1931): 336.
49. Griffith, quoted in Eberts, "Little Known Early History," on p. 6.
50. Samuel S. Fels, *This Changing World: As I See Its Trend and Purpose* (Boston: Houghton Mifflin, 1933), 66, 68.
51. Adler, quoted in Kaempffert, "Now America," on p. 21.
52. Fels, *This Changing World,* 96. Fels added, "Our prejudices are prejudices no matter how we have received them, and being ours, we can change them if we want to. The important thing is that people shall *want* to change them" (on p. 96).
53. For an account of natural theology's lengthy impact, see John Hedley Brooke, *Science and Religion: Some Historical Perspectives* (Cambridge: Cambridge University Press, 1991).
54. "Banker to Religion via Stars," *Time,* 15 January 1934, 36–37.
55. "$150,000 by Hayden for Planetarium," *New York Times,* 5 January 1934, 23.
56. Rockefeller Jr., to Hayden, 6 January 1934, AMNH ser. 1178.5, January 5–7, 1934 folder.
57. Fels, *This Changing World,* 158.
58. Ibid., 277, 25.
59. Ibid., 279, 294.
60. Fisher to Osborn, 26 October 1931, AMNH ser. 1178.5, 1931 folder.
61. Hale's articles comprised three sub-chapters titled "Solar Research for Amateurs" in Albert G. Ingalls's *Amateur Telescope Making, Book One,* 4th ed. (New York: Scientific American, 1967), 180–214, and featured cutaway drawings by Russell W. Porter.
62. See correspondence between Hale and the Franklin Institute, Hale Papers, Reel 59.
63. McClenahan to Hale, 14 April 1926; 17 February 1927; Hale to McClenahan, 24 February 1927; McClenahan to Hale, 24 May 1927; Hale Papers, Reel 59.

Astronomer Harlow Shapley accepted the Franklin Medal in absentia and delivered Hale's paper, "The Sun as a Research Laboratory."

64. Hale to McClenahan, 27 January 1931, Hale Papers, Reel 59. Hale evidently thought well enough of McClenahan's selection of James Stokley as planetarium director to report, "I am sure you have chosen wisely in selecting Mr. Stokley. . . . His experience should enable him to organize a very attractive and useful exhibit."
65. Hale to Stokley, 27 March 1931. Stokley replied, "Yes, we shall surely have a spectroheliosope and will be glad to take part in your international scheme of cooperative observations." Stokley to Hale, 2 April 1931, Hale Papers, Reel 59.
66. Hale to Adams, 23 April 1930, Hale Papers, Reel 59. Citing ownership of Chicago's planetarium by the South Park Commissioners, Hale queried "whether the same arrangement could be made here." Hale to Widney, 25 April 1930, Hale Papers, Reel 59.
67. Hale to Widney, 3 June 1930, Hale Papers, Reel 59. For Hale's views on evolutionary theory, see Wright, *Explorer of the Universe,* 230–242.
68. Adams and Millikan to Widney, 11 July 1930.
69. Hale to Widney, 3 June 1930, Hale Papers, Reel 59.
70. Adams and Millikan to Widney, 11 July 1930.
71. See Berton C. Willard, *Russell W. Porter: Arctic Explorer, Artist, Telescope Maker* (Freeport, Maine: Bond Wheelwright, 1976). For an account of Ingalls's association with Porter, see Thomas R. Williams, "Albert Ingalls and the ATM Movement," *Sky and Telescope,* February 1991, 140–143.
72. Anthony Cook, "The Secret History of Griffith Observatory: A Belated Acknowledgment to the Patron Saint of the Telescope Makers," *Griffith Observer,* May 1994, 2–18.
73. Hale to Davison, 21 June 1933; Hale to George H. Sherwood, 24 January 1934, Hale Papers, Reel 56. Clyde Fisher, "The Hayden Planetarium," *Natural History,* May 1934, 258.
74. See Stokley to Fisher, 17 December 1934, RSP ser. 5:4, Box 26, Folder 3, for notice of Cook's fifteen-inch siderostat telescope.
75. Leslie Shepard, ed., *Encyclopedia of Occultism and Parapsychology,* 3d ed., 2 vols. (Detroit: Gale Research, 1991), vol. 1, 959.
76. "Planetarium to be Erected Here by Rosicrucians," *San Jose Mercury Herald,* 19 February 1936, 13.
77. This description is drawn from the vintage-era poster, "The Theater of the Sky," preserved by the Rosicrucian Egyptian Museum and Planetarium, San Jose, California. Lewis's projector was replaced in 1950 with a Spitz A–1.
78. "Rite Dedicates Rosicrucians' Planetarium," *San Jose Mercury Herald,* 14 July 1936, 16. Quotation from inscription above planetarium doorway (visited by the author in 1996).
79. Laurence Vail Coleman, *The Museum in America: A Critical Study,* 3 vols. (Washington, D.C.: American Association of Museums, 1939), vol. 3, 486.
80. Thomas F. Gieryn, "Boundaries of Science," in *Handbook of Science and Technology Studies,* ed. Sheila Jasanoff, Gerald E. Markle, James C. Petersen, and Trevor Pinch (Thousand Oaks, Calif.: Sage Publications, 1995), 405, 424, 434; Gieryn, *Cultural Boundaries of Science: Credibility on the Line* (Chicago: University of Chicago Press, 1999).
81. For a sketch of the Springfield, Mass., planetarium and its designer, see Richard

Sanderson, "Frank Korkosz and the First American Planetarium," *Griffith Observer,* May 1986, 14–17.

82. "New American Planetarium for Springfield, Massachusetts," *Popular Astronomy,* November 1937, 518; "Fifth American Planetarium Opened to Public," *School Science and Mathematics* 37 (1937): 1139.
83. "Home-Made Planetarium Reveals Sky-Glories to Springfield, Mass.," *New York Times,* Sunday, 7 November 1937, sec. 2, p. 7.
84. James Stokley, letter to the editor, *Popular Astronomy,* March 1938, 175 (emphasis in original).
85. Wm. H. Barton, letter to the editor, *The Sky,* July 1938, 20.
86. James Stokley, "America's Fifth Planetarium," *The Sky,* October 1939, 3–6, 25.
87. For a description of the Korkosz instrument at Boston's Museum of Science, see John Patterson, "Boston's Planetarium Opens," *Sky and Telescope,* November 1958, 12–15.
88. Aitken and Adams quoted in Clyde Fisher, "The Drama of the Skies," *Natural History,* March 1931, on p. 152–153. Struve to George A. Hoadley (Franklin Institute), 28 June 1932, Box 139, Folder 2. Papers of the Director, Yerkes Observatory Archives (YOA), Williams Bay, Wisconsin.
89. [Lewis], "To the Board of Managers," 30 November 1936.
90. Waldemar Kaempffert, "New Planetarium is the First in America," *New York Times,* Sunday, 11 May 1930, sec. 10, p. 8.
91. Mabel V. Socha, "Address at the Formal Opening of the Griffith Observatory and Planetarium, May 14, 1935," *Publications of the Astronomical Society of the Pacific* 47 (1935): 157. A synopsis of the ceremony is given by W[illiam] F. M[eyer], "Dedication of the Griffith Observatory," ibid., 170–171.
92. Adams quotations derived from an audio recording of the radio broadcast made over the Columbia-Donnelley Network, 14 May 1935. A full transcription of this recording has been furnished by Anthony Cook, Griffith Observatory.
93. Hale's statement, "The Griffith Observatory," delivered in absentia 19 June 1933, appears in Hale Papers, Reel 59.
94. Transcription of audio recording, 14 May 1935 (cit. no. 92).
95. Charles F. Lewis, "Presentation Address," in *Dedication of the Buhl Planetarium and Institute of Popular Science* (N.p., n.d.), 15, 19.
96. Samuel S. Fels, "Donation of the Planetarium to the Franklin Institute," *Journal of the Franklin Institute* 216 (July-December 1933): 791.
97. See John Lankford, ed., *History of Astronomy: An Encyclopedia* (New York: Garland Publishing, 1997), 224, and Robert W. Smith, *The Expanding Universe: Astronomy's 'Great Debate' 1900–1931* (Cambridge: Cambridge University Press, 1982).
98. Heber D. Curtis, "The Importance of the Planetarium to Science, to Education, and to Recreation," *Journal of the Franklin Institute* 216 (July-December 1933): 795–797. Although Curtis remained agnostic on the matter of "unanswered questions," he included himself in the category of "religious believers" (on p. 797).
99. Lewis, "Presentation Address," 17, 18.
100. Finley, quoted in William L. Laurence, "'Tour of Sky' Opens Planetarium; 800 Get a New Vision of Universe," *New York Times,* 3 October 1935, on p. 21.
101. Horace J. Bridges, "Speech at the Dedication of the Adler Planetarium, Chicago, May 10th, 1930." Invocation folder, APAM.

102. Transcription of audio recording, 14 May 1935 (note 92).
103. Davison, quoted in Laurence, "'Tour of Sky' Opens Planetarium," on p. 21.
104. Charles F. Lewis, *The Buhl Foundation: A Report by the Director upon its Work to June 30, 1942* (Pittsburgh: Farmers Bank Building, n.d.), 65–67.
105. Lewis, "Presentation Address," 14, 15.
106. Scully, "Acceptance Address," 27.
107. "Noted Scientist to Speak at Telescope Dedication," *Pittsburgh Press,* 19 November 1941. Astronomer Harlow Shapley was the invited speaker.
108. See [Charles A. Federer, Jr.], "The Editor's Note," *Sky and Telescope,* February 1944, 2, concerning planetaria and the wartime education of servicemen.

Chapter 3 ***Personnel, Training, and Careers***

1. "Director" is the term most widely recognized within the planetarium community to signify an individual on whom chief responsibility for the operation of the planetarium instrument, its programs, and institution resides. Directors have been assigned various titles, such as "curator," "chairman," or "president" by their institutions (and themselves).
2. Years of employment expended before an institution's public opening are not included in this tally. Such a decision excludes Caltech physicists Edward H. Kurth and Rudolph M. Langer, both of whom served as temporary directors of the Griffith Observatory before 1935.
3. Sec. eleventh, "Agreement, Between South Park Commissioners and Max Adler," 36.
4. Various celestial and meteorological phenomena, including solar eclipses, meteor showers, and the aurora borealis, could not be demonstrated by the Zeiss projector alone. Accordingly, Philip Fox designed apparatus for these effects that was constructed by Adler Planetarium technician Curt Richter. See Fox, "Auxiliary Apparatus for Planetarium Demonstrations," *Popular Astronomy,* April 1932, 188–194. Plans and/or duplicates of equipment were offered to the Zeiss firm. Fox to K. A. Bauer, 23 February 1931, Zeiss Planetarium Information folder, APAM.
5. Fox's part-time assistants ranged from demonstrator F. Wagner Schlesinger, son of astronomer Frank Schlesinger, to a team of fourteen lecturers who were paid $5 for each performance. Several of these lecturers (e.g., Dinsmore Alter, Roy K. Marshall, Harry E. Crull, G. Clyde Fisher, and James Stokley) achieved distinguished careers as planetarium directors. Adler Planetarium Staff 1930 folder, APAM.
6. The Adler Planetarium hosted the forty-fourth meeting of the American Astronomical Society in 1930. This gathering of 125 members was estimated by Fox to be "the largest attendance the Society has ever had." Fox, "September 1930 Report" to George T. Donoghue, on p. 2. Monthly reports folder, APAM.
7. Oliver Justin Lee, "Philip Fox, 1878–1944," *Popular Astronomy,* October 1944, 367; Joel Stebbins, "Philip Fox," *Science* 100 (1 September 1944): 185; and Robert G. Aitken, "Philip Fox, 1878–1944," *Publications of the Astronomical Society of the Pacific* 56 (1944): 180.
8. Data from Alumni Biographical Records folder, Northwestern University (NU) Library, Evanston, Illinois. The most complete sketch of Bennot appears in Winfield Scott Downs, ed., *Encyclopedia of American Biography,* new ser., vol. 13 (New York: American Historical Company, 1941), 24–26.

9. Bennot's thesis, bearing Fox's signature, was accepted 18 August 1927. Her research appeared as "Proper Motions of Forty Stars," *Astronomical Journal* 36 (8 June 1926): 177–181.
10. Papers relating to the Bennot controversy are found in Box 13, Folder 10, College of Liberal Arts-Dept. of Astronomy, 17 May 1921–15 December 1930 Records, University Archives, NU Library. In April 1929 Raymond A. Kent, dean of the College of Liberal Arts, secured the consent of university president Walter Dill Scott to "den[y] the privileges of Dearborn Observatory" to Bennot. On Scott's decision and Fox's resignation, see Scott to Fox, 1 May 1929; Fox to Scott, 6 May 1929.
11. Fox to Adler, 7 February 1929, Philip Fox folder, APAM. Fox declared that his work "shall be in the nature of conferences and consultations on plans for the building itself and for the organization of the museum, library, and other technical equipment."
12. Fox to Adler, 16 February 1929. Bennot's salary was fixed at $2,700 per year, considerably more than the $2,000 she earned while a "Computer" or "Research Assistant" at Dearborn Observatory. Adler Planetarium Staff 1930 folder, APAM.
13. On Alter's appointment, see Hale to Fisher, 30 July 1935, Hale Papers, Reel 56. Historian John Lankford has styled Alter's research program, in the questionable application of statistical methods to analysis of periodicities in climatological and solar phenomena, as a "failed career." See Lankford, *American Astronomy,* 167–168.
14. Menke, "Dinsmore Alter and the Griffith Observatory," 73–74.
15. Shapley, quoted in Menke, "James Stokley and the Fels Planetarium," on p. 41.
16. Among the best sketches of Fisher's life are D. R. Barton, "He Brought the Stars to America," *Natural History,* June 1940, 59–63, and Marian Lockwood, "Clyde Fisher: Naturalist and Teacher," *Sky and Telescope,* March 1949, 111–113. Andrews, quoted in Lockwood, on p. 112.
17. A section of the Hale Papers titled "Science News Service, 1924–1937" contains letters exchanged between Stokley, Hale, and other American astronomers, including Edwin Hubble. These contacts were likely influential in allowing Stokley to land the Fels planetarium directorship. Hale Papers, Reel 65.
18. Unpublished manuscripts from the James Stokley folder, UP Archives, include a biographical memorandum, dated 4 May 1928. Published sketches of Stokley's life include [Armand N. Spitz], "Who's Who in The Franklin Institute: James Stokley," *The Institute News,* December 1936, 4. Stokley's European planetarium visit was reported in "Where the Stars Stand Still," *Publications of the Astronomical Society of the Pacific* 40 (1928): 24–30.
19. For a sketch of Barton's life, see Wayne M. Faunce, "Interpreter of the Heavens," *Sky and Telescope,* September 1944, 5–7. Short sketches of Draper, Bennett, and Lockwood appear in the inaugural issue of *The Drama of the Skies,* November 1935, 8. Annual salaries for the Hayden staff were as follows: curator Fisher, $6,750; associate curator Barton, $4,000; assistant curators Draper, Bennett, and Lockwood, $2,400 each. A planetarium technician was paid $2,400, while freelance lecturers received $6 per performance. RSP ser. 5:2, Box 12, Folder 1.
20. On Fisher's administrative problems with the Amateur Astronomers Association, see Faunce to Roy Chapman Andrews, 9 January 1936, AMNH ser. 1191.1, 1934–1952 folder. On the recommendation supporting Barton's replacement of Fisher, see Faunce to Andrews, 8 November 1937, AMNH ser. 1186.1, 1937 folder (and motion passed by the Museum's board of directors, 20 December 1937). I am

indebted to Thomas R. Williams, whose interpretations of Fisher superseded my initial reconnaissance. See Williams, "Getting Organized: A History of Amateur Astronomy in the United States" (Ph.D. diss., Rice University, 2000).

21. Menke, "Dinsmore Alter and the Griffith Observatory," 68. Besides himself, Menke lists some thirty-three former Griffith employees who became astronomers, planetarium directors, or observatory directors (p. 68).
22. Menke, "Clyde Fisher and the Hayden Planetarium," 57.
23. Fox to Adams, 3 May 1937. Adams offered "heartiest congratulations" and reminded Fox that "you are still primarily an astronomer, and we shall welcome you for some astronomical observing whenever you feel you can spare the time." [Adams] to Fox, 11 May 1937, Folder 24.434, Adams Papers, HL.
24. Aitken, "Philip Fox," 181. Aitken added, "This action was deeply resented, not only by Fox's personal friends . . . but by scientific men throughout the country" (on p. 181).
25. Kogan, *Continuing Marvel,* 56.
26. "'Playing' the Planetarium," *The Literary Digest,* 12 October 1935, 18; Arthur L. Draper, "How Man Controls the Universe," *The Sky,* December 1936, 22–23, 28.
27. Fisher to Stokley, 29 May 1933; Stokley to Fisher, 27 June 1933; Fisher to Stokley, 1 July 1933; Stokley to Fisher, 5 July 1933; RSP ser. 5:4, Box 26, Folder 2.
28. Rossiter, *Struggles and Strategies,* 167, 268.
29. Kay Hall, "Woman Plays Lead in Celestial Show," *New York Times,* Sunday, 16 January 1938, sec. 6. Upon Bennot's earlier appointment as assistant director, she was elected treasurer (1930) and second vice president (1931) of Sigma Delta Epsilon, the national graduate women's scientific fraternity. See Durward Howes, ed., *American Women: The Standard Biographical Dictionary of Notable Women,* vol. 3 (Los Angeles: American Publications, 1939), 73.
30. Before Bennot's appointment was confirmed, Fox was questioned about her chances of succeeding him. "That is a matter for the Park District to determine. Miss Bennot will do much of the work until the board selects a new director," he replied. See "Woman Likely to Succeed Fox as Adler Planetarium Head," *Chicago Herald and Examiner,* 27 April 1937.
31. Sydney J. Harris, "Here is Chicago," *Chicago Daily News,* 24 March 1944,16.
32. Examination of monthly reports filed by Bennot for the period from January to November 1944 shows that, with minor exceptions, the titles and order of programs given were identical to those of 1932. Monthly reports folders, APAM.
33. Rossiter, *Before Affirmative Action,* 28, 27.
34. Three newspaper articles, published 27 December 1944, establish the principal means for interpreting this event. They are: "Maude Bennot Will be Ousted at Planetarium," *Chicago Tribune;* "Parks to Oust Maude Bennot at Planetarium," *Chicago Sun;* and "Fight Woman's Ouster as Planetarium Chief," *Chicago Times.* Clippings preserved, Alumni Biographical Records folder, NU Library. Subsequent quotations are extracted from these articles.

    Three further accounts, published 28 December 1944, released Schlesinger's name as the Park District's new appointee. See "New Director Named for Planetarium," *Chicago Sun;* "Philadelphian Picked as Head of Planetarium," *Chicago Tribune*; and "Philadelphian Named Planetarium Director," *Chicago Daily News.* Clippings in YOA, Box 215, Folder 3.
35. Schlesinger had earned only a bachelor's degree in mathematics and physics from

Yale University in 1923. He accomplished no astronomical research and published little or nothing of importance. In both education and experience, Schlesinger was clearly inferior to Bennot.

36. Adler admitted that he and the park board had attempted to "find the right man for the job" for two years, and was "delighted with the result." Adler, quoted in "New Director Named for Planetarium," 28 December 1944, on p. 13.
37. On Lockwood, see Howes, *American Women,* vol. 3, 534.
38. See "Atwater, Navy Man, to Head Planetarium," *New York Times,* 28 August 1945; "Starry-eyed," *American Magazine,* January 1946, 138.
39. For hints on her post-planetarium career, see Lockwood to Robert R. Coles, 30 October 1951, RSP ser. 5:2, Box 14, Folder 1.
40. Fox to Adler, 7 February 1929; Adler to Fox, 11 February 1929; Fox to Adler, 16 February 1929; Philip Fox folder, APAM.
41. Before accepting the Dearborn Observatory's directorship, Fox began a study under Hale's guidance on the rotation of the Sun from measurements of 285 plates exposed between 1903 and 1908. This work was published as Fox's "Rotation of the Sun" in 1921. See Lee, "Philip Fox," 367–368.
42. Fox to Hale, 1 March 1929, Hale Papers, Reel 55; Fox to Adams, 6 April 1929, Folder 24.434, Adams Papers, HL; Hale to Fox, 27 June 1929, Hale Papers, Reel 55.
43. A photograph (fig. 69) of the coelostat and vertical telescope is reproduced in Philip Fox, *Adler Planetarium and Astronomical Museum: An Account of the Optical Planetarium and a Brief Guide to the Museum* (Chicago: Lakeside Press, 1933), 57.
44. Fox to Hale, 11 October 1929; Fox to Hale, 26 December 1929; Hale Papers, Reel 55. According to Kaempffert, "New Planetarium is the First in America," the Mensing Collection was purchased by Adler for $55,000.
45. Historian Liba Taub has argued that Adler's purchase of the Mensing Collection was undertaken in response to a popular article prepared by Albert G. Ingalls ("'Canned Astronomy' or Cultural Credibility: The Acquisition of the Mensing Collection by the Adler Planetarium," *Journal of the History of Collections* 7 (1995): 243–250). The collection of instruments that Adler purchased did enhance the prestige of his planetarium, but as Fox's own correspondence attests, it was not obtained under the speculative pretenses that Taub suggests.
46. For a definitive account of the investigations made upon this structure, see William Graves Hoyt, *Coon Mountain Controversies: Meteor Crater and the Development of Impact Theory* (Tucson: University of Arizona Press, 1987).
47. On Fisher's association with Nininger, see Kathleen Mark, *Meteorite Craters* (Tucson: University of Arizona Press, 1987), 81–83. Fisher's thoughts on Meteor Crater were delivered as the Barnwell Address in Philadelphia (1934) and published as *Meteor Crater, Arizona* [Guide leaflet ser., no. 92] (New York: American Museum of Natural History, 1936). His examination of the Estonian craters was reported at the American Astronomical Society's fifty-ninth meeting and published as *The Meteor Craters in Estonia. 'Footprints' left by visitors from outer space—Evidence of an astronomical collision that occurred perhaps two thousand years ago* (New York: American Museum of Natural History, 1936).
48. Fox to Curtis, 28 May 1930. The six cities mentioned were Pittsburgh, Philadelphia, New York, Cleveland, St. Louis, and Montreal. Fox admitted, however, that if German planetaria were operated "as part of the school system and are budgeted by the school boards, then there is justification for almost any number."

49. Fox to Raymond Hood, 14 January 1932, Attendance folder, APAM. Fox repeated his previous argument almost verbatim: "I can see the value of having at least half a dozen in the country but we certainly should avoid an excess of them; I believe it is very decidedly overdone in Germany."
50. James Stokley, "Planetarium Operation," *Scientific Monthly,* 16 October 1937, 307 and Table II. Stokley's original paper, delivered 11 May 1936, is preserved in the Buhl Foundation Records, Box 33, Folder 8, HSWP.
51. On the RAS nominations, see Stokley to Fisher, 19 October 1931, Stokley to Fisher, n.d., 1932, and Barton to Stokley, 28 January 1938. On lectures exchanged before the Amateur Astronomers Association and Rittenhouse Astronomical Society, see Fisher to Stokley, 4 September 1929, Stokley to Fisher, 4 September 1929 (telegrams), Stokley to Fisher, 3 January 1934, Stokley to Fisher, 11 January 1934, Stokley to Fisher, 17 January 1934, Fisher to Stokley, 24 January 1934; Fisher to Stokley, 23 February 1934, Fisher to Stokley, 23 December 1935, Stokley to Fisher, 19 August 1936, Fisher to Stokley, 10 September 1936. On the exchange of public and school lectures, see Stokley to Fisher, 27 February 1934, Fisher to Stokley, 1 February 1935, Stokley to Fisher, 18 October 1935, Fisher to Stokley, 22 October 1935. On the attendance "race," see Stokley to Fisher, 25 October 1935. First-year Hayden figures were supplied in Fisher to Stokley, 5 October 1936 and 14 October 1936. On publication and review of Stokley's *Stars and Telescopes,* see Stokley to Fisher, 1 September 1935, Stokley to Fisher, 17 December 1935, and Fisher to Stokley, 17 January 1936 (twin docs.). On Fisher's *Exploring the Heavens,* see Stokley to Fisher, 13 January 1938, Fisher to Stokley, 18 January 1938. All correspondence found in RSP ser. 5:4, Box 26, Folders 2–4.
52. In his 1964 address to the AAM planetariums section, Armand N. Spitz asserted the lack of any "spirit of fraternity" among planetarium directors in the formative period. Unpublished manuscript, 28 May 1964. Armand N. Spitz Memorial Fund folder, International Planetarium Society (IPS) archives, Ash Enterprises, Bradenton, Florida, on p. 9. Excerpts were published as Spitz, "The Philosopher's Point of View," *Museum News,* December 1964, 22.
53. Before the R.F.C. loan was secured, Stokley assured Fisher that "if I can do anything to assist you in getting your planetarium, please call on me." Stokley to Fisher, 31 May 1933. Fisher's expressions of gratitude appear numerous times, e.g., following "another pleasant and profitable visit to the Fels Planetarium," Fisher to Stokley, 5 November 1934. Shortly before his own planetarium was opened, Fisher opined, "You [Stokley] have been extremely generous and helpful all along. We are most grateful," Fisher to Stokley, 27 September 1935. Finally, "your generous statement . . . is another evidence of your fine spirit. We appreciate the high compliment especially since it comes from you," Fisher to Stokley, 12 July 1937. RSP ser. 5:4, Box 26, Folders 2–4.
54. Apart from those reasons advocated in the text, other factors that might have contributed to the brevity of the meetings were (1) Barton's replacement of Fisher as planetarium curator, effective December 1937; and (2) Stokley's growing involvement as a consultant to the Buhl Foundation.
55. Schlesinger, quoted in Stokley to Fisher, 6 November 1936. RSP ser. 5:4, Box 26, Folder 3. Lankford, *American Astronomy,* p. 335, has argued that "much of the business of the American astronomical community" was transacted at meetings of the "Neighbors."
56. Stokley to "Dear Sir," 7 January 1937, RSP ser. 5:4, Box 26 Folder 4. This invita-

tion almost certainly excluded the women curators on Fisher's staff. Bartkey's visit was sponsored by the Rittenhouse Astronomical Society.

57. Stokley to Barton, 14 April 1937; Barton to Stokley, 16 April 1937; Stokley to Barton, 20 April 1937. RSP ser. 5:4, Box 26, Folder 4.
58. Correspondence between planetaria and observatories regarding astronomical photographs is too abundant to document individually. On the portraits supplied by Fisher, see YOA, Box 149, Folder 1.
59. Fox participated on the Yerkes Observatory expedition to witness the 31 August 1932 solar eclipse from New England. Fox to Clifford C. Crump, 27 July 1932. YOA, Box 139, Folder 2. The eclipse experiences of Fisher, Barton, and Stokley are recounted in Lockwood, "Clyde Fisher," 113; Faunce, "Interpreter of the Heavens," 6; and Stokley, "America's Fifth Planetarium," 25.
60. On the invitation to Corning, see Stokley to Fisher, 20 March 1934; Fisher to Stokley, 14 April 1934. On Einstein's visit to the Fels Planetarium, see Fisher to Stokley, 20 May 1935. RSP ser. 5:4, Box 26, Folder 3. On the 1943 anniversary, see Bennot to Struve, 30 April 1943. YOA, Box 207, Folder 7.
61. On Fisher's review, see YOA, Box 149, Folder 1. On Alter's review, see Box 167, Folder 1. On Fox's two reviews, see Box 160, Folder 4; Box 187, Folder 2; Box 195, Folder 3. Fox also contributed to Frost's obituary. Box 154, Folder 6. On Stokley's textbook, see Box 150, Folder 5.
62. On Sherman's plight, see Struve to Fox, 17 January 1938, 12 April 1938, 16 May 1938. YOA, Box 173, Folder 5. On Sawtell's placement, see Fox to Struve, 10 September 1940. YOA, Box 187, Folder 2.
63. Struve to Fox, 14 November 1938. Fox harbored doubts that Freundlich "would fit well with American audiences." Fox to Struve, 17 November 1938. YOA, Box 173, Folder 5. Alter to Struve, 3 March 1939; Struve to Alter, 7 March 1939. YOA, Box 178, Folder 8.

Chapter 4 ***Planetaria and Popular Audiences***

1. Philip Fox, "The Adler Planetarium and Astronomical Museum of Chicago," *Popular Astronomy,* March 1932, 152. Fox's account was completed in parts 2–4: June–July 1932, 321–351; November 1932, 532–549; and December 1932, 613–622.
2. Stokley, "Planetarium Operation," 313.
3. Fox, "Adler Planetarium and Astronomical Museum," 155. In 1943 Bennot described "Architecture of the Heavens" as "the most profoundly scientific of the various popular lectures presented in the different months of the year; a splendid lecture for the Christmas season." December 1943 monthly report, APAM.
4. Fox, "Adler Planetarium and Astronomical Museum," 155.
5. Fox to Curtis, 28 May 1930.
6. Fox added, "The experience in all the German cities . . . is that the attendance is large in the initial months and thereafter falls off very sharply." Fox to George Bramwell Baker, 23 November 1935. Attendance folder, APAM.
7. Fox to Hood, 14 January 1932.
8. Undated visitor survey form and responses, Adler Planetarium Visitor Responses folder, APAM. That this survey was conducted prior to the Century of Progress Exposition is indicated by one tabulated reaction: ". . . will be a treat for World's Fair Visitors."

9. Mildred M. Lutz, "The Fels Planetarium: Philadelphia, Pennsylvania," 6–7, unpublished manuscript, 28 December 1936, Buhl Foundation Records, Box 33, Folder 8, HSWP.
10. "Report by Mr. Starrett on Visit to Hayden Planetarium," unpublished manuscript, 9 April 1937, Buhl Foundation Records, Box 33, Folder 8, HSWP.
11. Poems referenced (but not reproduced) in this section are: Elizabeth A. Stein, "The Planetarium," *The Sky,* December 1936, 28; Lisa Odland, "At the Planetarium," *Popular Astronomy,* January 1938, 14; Grace Maddock Miller, "Planetarium," *The Sky,* May 1939, 10; Ruth Ingalls, "The Hayden Planetarium," *The Sky,* July 1939, 24; and Florence S. Edsall, "Planetarium," *The Institute News,* May 1944, 4 (reprinted from *The New Yorker*).
12. Horace Shipp, "In the Planetarium," *Palimpsest* (London: Sampson Low, Marston, 1930), 39. In the Adler reproduction, Shipp's final adjective, "shilling," was replaced by the less-derogatory term, "man-made." Adler Planetarium General Brochures folder, APAM.
13. James Stokley, "The Fels Planetarium of the Franklin Institute," *The Telescope,* April 1934, 29–32; Stokley, "Benjamin Franklin Memorial and the Franklin Institute: Opening of the Fels Planetarium," *Popular Astronomy,* December 1933, 535–538.
14. "Planetarian," *Time,* 24 April 1939, 61. This descriptor (for one who works in association with a planetarium) marks its earliest known appearance in the English language. It was later adopted as the title of the International Society of Planetarium Educators's quarterly journal (1972).
15. In 1978, biblical historian Ernest L. Martin presented a revised chronology for the time of Christ's birth, while Griffith Observatory planetarium supervisor John Mosley proposed a cadre of celestial events fitting the Star's description. See Ernest L. Martin, *The Birth of Christ Recalculated* (Pasadena, Calif.: Foundation for Biblical Research Publications, 1978), and John Mosley and Ernest L. Martin, "The Star of Bethlehem Reconsidered: An Historical Approach," *The Planetarian* 9, no. 2 (summer 1980): 6–9.
16. "Planetarium to Reproduce Ancient Skies," *Science—Supplement* 78 (22 December 1933): 12. As early as 1925, planetaria were deemed capable of depicting "the conjunction of [planets] which formed the so-called Star of Bethlehem." Morison, "Heavens Built of Concrete," 171. Whether any European operator performed a related demonstration before Stokley remains unknown.
17. An authoritative study of the astronomical and historical evidence supporting the 7–6 B.C. planetary conjunction theory is David Hughes, *The Star of Bethlehem: An Astronomer's Confirmation* (New York: Walker, 1979). For an annotated bibliography on the subject, see Ruth S. Freitag, *The Star of Bethlehem: A List of References* (Washington, D.C.: Library of Congress, 1979).
18. The inspiration for Kepler's proposal came from the appearance of a supernova in the year 1604. That star's sudden appearance was preceded by a triple conjunction of Jupiter and Saturn in the constellation Pisces, along with the addition of Mars. Kepler causally associated this planetary grouping with the supernova, and calculated backwards to determine that a similar triple conjunction had occurred in Pisces during 7–6 B.C. These events, he reasoned by analogy, had triggered, but were not themselves, the Magi's Star. See R[oy] K. M[arshall], "Star of Bethlehem?" *Sky and Telescope,* December 1943, p. 15, for discussion and reference to Kepler's original writings.

19. Stokley's five-page script is preserved in Buhl Foundation Records, Box 33, Folder 8, HSWP. It was likely collected by Mildred M. Lutz on her visit to the Fels Planetarium in December 1936. It not only represents the oldest Christmas Star script, but also the earliest script known to be preserved from any American planetarium program. Its distribution may have arisen from audience requests for copies of Stokley's lecture.
20. "The Christmas Star," *Museum News,* December 1964, 11.
21. Ian C. Barbour, *Issues in Science and Religion* (New York: Prentice-Hall, 1966), esp. 44–50.
22. David C. Lindberg and Ronald L. Numbers, eds., *God and Nature: Historical Essays on the Encounter between Christianity and Science* (Berkeley: University of California Press, 1986).
23. Brooke, *Science and Religion,* 268. For a treatment of the American crisis of faith, see James Turner, *Without God, Without Creed: The Origins of Unbelief in America* (Baltimore: Johns Hopkins University Press, 1985).
24. Michael R. Molnar, *The Star of Bethlehem: The Legacy of the Magi* (New Brunswick, N.J.: Rutgers University Press, 1999), suggests an occultation of Jupiter as the astrologically auspicious event, while Mark Kidger, *The Star of Bethlehem: An Astronomer's View* (Princeton, N.J.: Princeton University Press, 1999), supports the appearance of a nova in the constellation Aquila.
25. While no script of Stokley's Easter program has been located, his account of it appears in "Easter: The Awakening," *The Sky,* March 1940, 5–6, and describes its "scientific portion." This remark suggests that a dramatic retelling of the Crucifixion and Resurrection was conducted in the program's second half.
26. "Planetarian," *Time,* 61. By this account, "[s]tuffy astronomers were shocked by this fiction." Stokley, however, declared the program's contents as a product of imagination "guided by a knowledge of exact facts."
27. Stokley, "Planetarium Operation," 314.
28. For two popular articles on Adonis, see Henry Norris Russell, "A Catastrophe That Did Not Occur," *Scientific American,* May 1936, 248–249, and Gaylord Johnson, "How Will the World End?" *Popular Science Monthly,* October 1936, 52–53, 123.
29. "How Will the World End?" *The Institute News,* September 1938, 4.
30. See "Trip to the Moon," *The Institute News,* April 1939, 1–2; and "Moon Trip Continued," *The Institute News,* June 1939, 2, 1.
31. For a sketch of Schlesinger, see [Armand N. Spitz], "Who's Who in The Franklin Institute: Wagner Schlesinger," *The Institute News,* September 1937, 4.
32. Descriptions of these programs appear in "What is Astrology?" and "Spring Stars a la Disney," *The Institute News,* April 1941, 1.
33. Fisher's monthly program topics for January-December 1936 were: "The Winter Constellations," "Eclipses," "The Seasons," "The Southern Cross," "The Planets and Their Motions," "The Midnight Sun," "The Summer Constellations," "Comets and Meteors," "Timekeeping and Navigation," "Five Thousand Years," "Stars and Nebulae," and "The First Christmas." See *The Drama of the Skies*, *The Sky*.
34. "Hayden Planetarium Shows Four Ways in Which the World May End," *Life,* 1 November 1937, 54–58; Wm. H. Barton Jr., "The End of the World," *The Sky,* October 1937, 3–4, 14, 19. The program was revived after the Second World War (see "The End of the World!" *Science Illustrated,* October 1946, 6–7). Audience

reactions depicted mixtures of fascination and fright at the evident realism. In a post-Hiroshima world, few were said to laugh upon seeing the cometary collision.

35. For a classic study in the behavior of mass hysteria, see Hadley Cantril, *The Invasion from Mars: A Study in the Psychology of Panic* (New York: Harper and Row, 1966).
36. Dorothy A. Bennett, "The Mysterious Moon," *Natural History,* October 1935, 262–264; G. Edward Pendray, "To the Moon via Rocket?" *The Sky,* November 1936, 6–7, 21.
37. Wm. H. Barton Jr., "Exploring the Moon," *The Sky,* April 1938, 3–5, 23; Emile C. Schnurmacher, "Explorers of Space," *Popular Mechanics,* November 1939, 680–682.
38. Faunce to Alan R. Ferguson, 13 June 1938, AMNH ser. 1276, New York World's Fair, 1939 inc. 1938 folder.
39. For an account of the *Observer*'s founding and golden anniversary, see Susan Arminé Injejikian, "The Universe in Your Mailbox: Observing the *Observer*'s Anniversary," *Griffith Observer,* February 1987, 2.
40. Otto Struve, "Giant Mystery Star . . . The Great Ghost Companion of Epsilon Aurigae," *The Sky,* March 1938, 3–5, 27.
41. Struve to Fisher, 8 March 1938; Fisher to Struve, 18 March 1938. YOA, Box 173, Folder 5.
42. From its inaugural issue in July 1933, *The Telescope* had been edited by Harvard astronomer Loring B. Andrews, who was succeeded in the May-June 1937 issue by Donald H. Menzel.
43. Thomas R. Williams, personal communication, 19 May 1999; see also Williams, "Getting Organized," 175–177, 186–189, 234–236.
44. Robert W. Rydell, *World of Fairs: The Century of Progress Expositions* (Chicago: University of Chicago Press, 1993), 5, 9, and 10.
45. Fox to Committee on Astronomy of the 1933 World's Fair, 18 December 1929, Hale Papers, Reel 55. Fox's colleague, Northwestern University physicist Henry Crew, headed the Exposition's Division of Basic Sciences. Rydell, *World of Fairs,* 97.
46. Fox to Frost, 14 October 1931, YOA, Box 134, Folder 8; Rydell, *World of Fairs,* 99. An account of this feat is given by Charles D. Stewart, "A Beam from Arcturus," *Atlantic Monthly* 151 (1933): 331–336. The four cooperating observatories were Harvard, Allegheny, Illinois, and Yerkes.
47. Fox to Baker, 23 November 1935.
48. Fox to Struve, 21 December 1934 (Box 149, Folder 1); Struve to Fox, 17 January 1938 (Box 173, Folder 5). YOA.
49. Rydell, *World of Fairs,* 106–114.
50. Gerald Wendt, "A Proposal for a Daily Illumination Ceremony Utilizing Cosmic Rays," 29 December 1938; Faunce to Wendt, 6 September 1938, AMNH ser. 1276, New York World's Fair 1939 inc. 1938 folder; "Cosmic Rays Start Brilliant Display," *New York Times,* 1 May 1939.
51. Correspondence between Ferguson, Faunce, and other organizers is found in AMNH ser. 1276 (New York World's Fair).
52. Mike Wallace, "Mickey Mouse History: Portraying the Past at Disney World," *Radical History Review* 32 (1985): 33–57.
53. George Buchanan Fyfe, "Time and Space Dramatized at the World's Fair," *The Sky,* May 1939, 13.

54. H. B. Rumrill, "The Revival of Popular Interest in Astronomy," *Popular Astronomy,* June-July 1937, 316, 311.
55. "The Committee on Science and its Social Relations," *Popular Astronomy,* June-July 1938, 351.
56. Bart J. Bok, "Report on Astronomy," *Popular Astronomy,* August-September 1939, 359–360.

Chapter 5 *Armand N. Spitz and Pinhole-Style Planetaria*

1. Curtis, "Importance of the Planetarium," 799.
2. Lars Broman, "A Spitz-Type Planetarium from 1935," *The Planetarian* 15, no. 4 (fourth quarter 1986): 10–11.
3. King, *Geared to the Stars,* 356–357.
4. "A High School Planetarium," *Popular Astronomy,* October 1946, 438–439; "Central High School Dedicates Planetarium," *Sky and Telescope,* January 1947, 13.
5. See full-page advertisement, "The Universe in Your Classroom," *Sky and Telescope,* January 1947, 27; King, *Geared to the Stars,* 357.
6. On his role as an educational consultant, see Armand N. Spitz, "Hall of Earth, Air, and Sky of the Polytechnic Institute of Puerto Rico," *Sky and Telescope,* April 1947, 3–5 (with a photograph of the Peerless projector), and [Spitz], "Puerto Rican Visit," *The Institute News,* August 1945, 2.
7. Brent P. Abbatantuono, "Armand Neustadter Spitz and His Planetaria, with Historical Notes of the Model A at the University of Florida" (M.A. thesis, University of Florida, 1994); Abbatantuono, "Armand Spitz—Seller of Stars," *The Planetarian* 24, no. 1 (March 1995): 14–22. Sketches of Spitz's life are also found in *The National Cyclopedia of American Biography,* vol. 56 (Clifton, N.J.: James T. White, 1975), 421–422; and C[harles] A. F[ederer], "Armand N. Spitz—Planetarium Inventor," *Sky and Telescope,* June 1971, 354–355. For a partial list of Spitz's publications, see Jordan D. Marché II, "A Bibliography of Armand N. Spitz (1940–1972)," *The Planetarian* 31, no. 1 (March 2002): 10.
8. "Fac-totum," *The Institute News,* December 1943, 3.
9. Spitz was active in both the Lansdowne and Newtown Square, Pennsylvania, Friends Meetings. According to Spitz's second wife, Grace (Scholz) Spitz, one of Armand's favorite recreations was to repaint Quaker meetinghouses. The 1950s sale of Spitz Laboratories to the Worth brothers was predicated in part on their Quaker faith. Grace S. Spitz, telephone conversation with the author, 29 June 1996.
10. Frederick B. Tolles, *Quakers and the Atlantic Culture* (New York: Macmillan, 1960), 60–63; Brooke Hindle, "The Quaker Background and Science in Colonial Philadelphia," in *Early American Science* (New York: Science History Publications, 1976), 173.
11. See Armand N. Spitz, "Meteorology in the Franklin Institute," *Journal of the Franklin Institute* 237 (January-June 1944): 271–287, 331–357.
12. "Spitz and the Planetarium," original manuscript transcribed and annotated by Grace S. Spitz, from Armand's recollections dictated on tape and delivered before the Great Lakes Planetarium Association (GLPA) conference in 1967. A much-edited version of this talk was published as Mrs. Armand Spitz, "First Armand Spitz Lecture," *Great Lakes Planetarium Association (GLPA) Projector* 1 (1968): 6–8. I am indebted to Grace S. Spitz for providing a copy of the original.

13. "Spitz and the Planetarium" (cit. no. 12).
14. Robert Fleischer, "Make Your Own Planetarium," *The Sky,* November 1936, 22–23.
15. Armand N. Spitz, *The Pinpoint Planetarium* (New York: Henry Holt, 1940), 1, 3, 78. The text's illustrations were supplied by G. Carter Morningstar.
16. Marian Lockwood, review of *The Pinpoint Planetarium* by Armand N. Spitz, *The Sky,* June 1940, 11.
17. During a 1954 interview with journalist Steven M. Spencer, Spitz explained: "I had at first no idea of making one to sell. I simply thought it would be nice to have a little projector to shine stars on [her] bedroom ceiling" (on p. 97). Spencer, "The Stars are His Playthings," *The Saturday Evening Post,* 24 April 1954, 42–43, 97–98, 100, 102–103. Twenty years earlier, Spencer had interviewed Samuel S. Fels regarding his donation of a Zeiss planetarium to the Franklin Institute.
18. "Man-Made Moon," *Science Illustrated,* August 1947, 8–9.
19. "Spitz and the Planetarium" (cit. no. 12).
20. See James R. Benford, "New Projector for Navigation Stars," *Sky and Telescope,* November 1942, 3–4.
21. "Spitz and the Planetarium" (cit. no. 12). The Wolff family reportedly inherited a substantial fortune from a grandfather's invention of metal-edge cardboard boxes. The family's complete identification remains unknown.
22. An icosahedron-shaped projector was independently devised by William Schultz Jr. of the Cranbrook School, Bloomfield Hills, Michigan, and exhibited at the institution's 1958 planetarium symposium. See Schultz, "A 45-Cent Planetarium Projector," in *Planetaria and Their Use for Education,* ed. Miriam Jagger (Bloomfield Hills, Mich.: Cranbrook Institute of Science, 1959), 167–169.
23. The only correspondence preserved between Spitz and Einstein is a reply by the latter to an inquiry from Spitz that no longer exists. Einstein's answer consisted of a four-line German verse that quipped how it was impossible to answer Spitz's query because of its vague and confused expression. [Einstein] to Arnold [sic] B. [sic] Spitz, 3 March 1938. Einstein Papers Project, Boston University, Boston, Massachusetts. This note establishes a prewar correspondence between the two men.
24. M. T. Brackbill, letter to the editor, *Sky and Telescope,* December 1947, 35. Brackbill's institution later fabricated a mechanized ceiling (Copernican) orrery, patterned after those devices constructed for the Hayden and Morehead planetaria. See "Harrisonburg Observatory Expands Facilities," *Sky and Telescope,* July 1956, 396–397.
25. A brief institutional history of Spitz Laboratories, Inc., has been furnished by Abbatantuono, "Armand Neustadter Spitz and His Planetaria," 49–54.
26. "Planetarium, Junior Version," *Business Week,* 20 February 1954, 112–114.
27. Wolff, quoted in Abbatantuono, "Armand Neustadter Spitz and His Planetaria," 46.
28. Informal announcement of Spitz's projector appeared in "Mr. Spitz Brings Science Down to Earth," *Science Illustrated,* September 1947, 26–29.
29. "Spitz and the Planetarium" (cit. no. 12). No copy of Marshall's purported criticism has been located.
30. [Spitz], "Profile of Prodigy," *The Institute News,* August 1943, 3.
31. Roy K. Marshall, "Franklin Lecturer Perfects Poor Man's Planetarium," *Philadelphia Evening Bulletin,* 25 November 1947. Courtesy of Temple University Library, Philadelphia, Pennsylvania. Marshall's expression, the "poor man's planetarium,"

is reminiscent of the phrase "poor man's telescope" chosen by Russell W. Porter and Albert G. Ingalls to promote the amateur telescope-making movement in America. See Williams, "Getting Organized," 140–154.

32. See full-page Science Associates advertisement, "KORKOSZ Planetarium," *Sky and Telescope,* December 1947, 55.
33. "Spitz and the Planetarium" (cit. no. 12).
34. Spitz, unpublished address before the AAM planetariums section, 28 May 1964, 9.
35. Spitz, "Kite, Apple, and Tea-Kettle, 1946," *The General Magazine and Historical Chronicle* 48 (spring 1946): 178.
36. "Spitz and the Planetarium" (cit. no. 12).
37. Cyrus F. Fernald, "Thirty-Sixth Annual Meeting of the AAVSO Held at Harvard College Observatory, Cambridge, Mass., October 10–11, 1947," *Variable Comments* 4, no. 18 (n.d.): 83–86. Fernald reported that the "quoted prices of this instrument should enable it to be bought by all colleges and larger high schools which teach astronomy" (on p. 83). I am indebted to Thomas R. Williams for supplying a copy of Fernald's report.
38. Spitz, "First Armand Spitz Lecture," 8. Details were embellished by Federer, "Armand N. Spitz—Planetarium Inventor," 354.
39. These events were retold by Charles F. Hagar, "Through the Eyes of Zeiss," *Griffith Observer,* June 1961, 62–70, from materials supplied by the Zeiss firm and Hagar's personal interview with Walther Bauersfeld, then Zeiss's director, only months before the latter's death in 1959.
40. This claim appears in "Spitz and the Planetarium" (cit. no. 12) as having been announced by the Oberkochen division of Zeiss.
41. Werner, *Aratus Globe,* 153.
42. Abbatantuono, "Armand Neustadter Spitz and His Planetaria," 2.
43. See David H. Menke, "Tony Jenzano and the Morehead Planetarium," *The Planetarian* 17, no. 3 (September 1988): 51–55; Roy K. Marshall, "The Morehead Planetarium," *Sky and Telescope,* August 1949, 243–245.
44. "Astronomy Professor May Devote Full Time to Delivering Commercials on Television," *Printer's Ink* 234 (2 March 1951): 74; "Morehead Director Resigns," *Sky and Telescope,* May 1951, 161.
45. See Robert Cunningham Miller, "Galaxy by the Golden Gate," *Pacific Discovery,* November-December 1952, 11–17. Porter's sketch is reproduced in Leon E. Salanave, "Russell W. Porter [1871–1949]," ibid., 35. Salanave, "San Francisco Planetarium," *Sky and Telescope,* December 1952, 31–34, offers an abridged account of the *Pacific Discovery* narratives.
46. David H. Menke, "George Bunton and the Morrison Planetarium," *The Planetarian* 17, no. 2 (June 1988): 50–55.
47. Richard H. Emmons, "A Report on a School Planetarium: Its Design; Its Development as a Group Project; Its Utility as an Instructional Aid; and Its Program in School Community Relations" (M.A. thesis, Kent State University, 1950).
48. Ibid., 8.
49. Ibid., 66, 64.
50. "The North Canton Planetarium in Ohio," *Sky and Telescope,* March 1956, 214.
51. David DeVorkin, "Organizing for Space Research: The V–2 Rocket Panel," *Historical Studies in the Physical and Biological Sciences* 18, pt. 1 (1987): 1–24, and

DeVorkin, *Science With a Vengeance: How the Military Created the US Space Sciences after World War II* (New York: Springer-Verlag, 1992).

52. DeVorkin, "Organizing for Space Research," 22.
53. Howard E. McCurdy, *Space and the American Imagination* (Washington, D.C.: Smithsonian Institution Press, 1997), 35–41. *Collier*'s had a circulation of some 3.1 million subscribers, making it one of the nation's top ten weekly magazines.
54. McCurdy, *Space and the American Imagination*, 43.
55. Walter A. McDougall, . . . *The Heavens and the Earth: A Political History of the Space Age* (New York: Basic Books, 1985), 100.
56. McCurdy, *Space and the American Imagination*, 140, 145.
57. Neil de Grasse Tyson, "Onward to the Edge," *Natural History,* July 1996, 60.
58. McCurdy, *Space and the American Imagination*, 227–229. *Discover* magazine editor-in-chief Paul Hoffman seconded McCurdy's argument in a May 1997 editorial, "Marsstruck." Hoffman noted how views of Earth from space offered "a startling new perspective on just how fragile and small [our] home was within the immensity of the universe, a perspective that helped [to] fuel the modern environmental movement" (on p. 8).
59. The two most complete accounts of the I.G.Y.'s multifarious activities are those by Walter Sullivan, *Assault on the Unknown: The International Geophysical Year* (New York: McGraw-Hill, 1961), 4, and the participant-history recounted by Canadian geophysicist J. Tuzo Wilson, *I.G.Y., The Year of the New Moons* (New York: Alfred A. Knopf, 1961). An important analysis of the IGY satellite proposals is furnished by Rip Bulkeley, *The Sputniks Crisis and Early United States Space Policy* (Bloomington/Indianapolis: Indiana University Press, 1991). Bulkeley is completing a study of scientific cooperation during the IGY.
60. See James H. Capshew and Karen A. Rader, "Big Science: Price to the Present," *Osiris,* 2d ser., 7 (1992): 3–25.
61. Singer's idea was not without precedent. In 1948, Van Allen had proposed that a satellite be orbited at an altitude of 1,000 kilometers. Sullivan, *Assault on the Unknown,* 31.
62. Bulkeley, *Sputniks Crisis,* 100, 120.
63. For an official history of Project Vanguard, see Constance McLaughlin Green and Milton Lomask, *Vanguard: A History* (Washington, D.C.: Smithsonian Institution Press, 1971). Green and Lomask argued that the Stewart Committee received instructions that "noninterference with ballistic missile development was imperative," and that the satellite program was to be regarded as "purely scientific rather than politically significant" (on p. 36).
64. For contrasting interpretations on the overflight issue, see Dwayne A. Day, "Cover Stories and Hidden Agendas: Early American Space and National Security Policy," in *Reconsidering Sputnik: Forty Years Since the Soviet Satellite,* eds. Roger D. Launius, John M. Logsdon, and Robert W. Smith (Amsterdam: Harwood Academic Publishers, 2000), 161–195; and Michael J. Neufeld, "Orbiter, Overflight, and the First Satellite: New Light on the Vanguard Decision," ibid., 231–257. In turn, Kenneth A. Osgood, "Before Sputnik: National Security and the Formation of U.S. Outer Space Policy," ibid., 197–229, presents evidence that issues of national prestige, technological progress, and peaceful intentions colored the decision to support America's Vanguard proposal.

65. Green and Lomask, *Vanguard,* 99.
66. For a history of the SAO prior to its relocation, see Bessie Zaban Jones, *Lighthouse of the Skies: The Smithsonian Astrophysical Observatory, Background and History, 1846–1955* (Washington, D.C.: Smithsonian Institution, 1965). Jones notes that its transfer to Cambridge marked the "beginning of a new and more publicly active role for the Observatory" (on p. ix).
67. For an account of the SAO's satellite tracking program, see E. Nelson Hayes, *Trackers of the Skies* (Washington, D.C.: Smithsonian Institution, 1967). See also Shirley Thomas, *Satellite Tracking Facilities: Their History and Operation* (New York: Holt, Rinehart and Winston, 1963), 43–51.
68. Records of Project Moonwatch are contained within the Moonwatch Division, Smithsonian Astrophysical Observatory, 1956–1975, Collection 255, Smithsonian Institution Archives. See *Guide to the Smithsonian Archives* (Washington, D.C.: Smithsonian Institution Press, 1996), 252–253. A history of Project Moonwatch's significance has yet to be written.
69. See Rennie Taylor, "Special Instruments Will Trace Satellite," *Columbus Dispatch,* 12 June 1956.
70. Spitz, quoted in Abbatantuono, "Armand Neustadter Spitz and His Planetaria," 39.
71. An important record of Moonwatch activities is contained in the nine issues of the SAO's *Bulletin for Visual Observers of Satellites,* published from July 1956 to July 1958. Copies were inserted into issues of *Sky and Telescope* by editor Federer, whose promotion of the campaign aided Spitz's personal contacts.
72. Whipple, quoted in Green and Lomask, *Vanguard,* on p. 149.
73. For copies of the dedication program and clippings pertaining to the Weitkamp Observatory and Planetarium, I am indebted to the Otterbein College Archives, Westerville, Ohio.
74. On the formation of the Astronomical League, and committee debates over whether scientific or recreational observing was to be promoted, see Williams, "Getting Organized," 305–350.
75. See the entry on Grace Scholz Spitz in *Who's Who of American Women: A Biographical Dictionary of Notable Living American Women,* vol. 1 (Chicago: A. N. Marquis, 1958), 1210.
76. "National MOONWATCH Committee Meets at Smithsonian Observatory Headquarters," *Sky and Telescope,* April 1957, 277.
77. Armand N. Spitz, "MOONWATCH Preparations Swing Into High Gear," *Bulletin for Visual Observers of Satellites* no. 3 (November 1956): 1.
78. See Bulkeley, *Sputniks Crisis,* 112–115, and also Bulkeley, "The Sputniks and the IGY," in Launius, et al., *Reconsidering Sputnik,* 125–159, esp. 129–130.
79. Hayes, *Trackers of the Skies,* 55.
80. John C. Rosemergy, "Roots and Routes: Ptolemy, Copernicus, Sputnik, and G.L.P.A." Unpublished manuscript, delivered as the Twelfth Annual Armand Spitz Lecture, Great Lakes Planetarium Association, 20 October 1978, on p. 16.
81. Spitz's address, delivered 20 May 1949, was titled, "Use of a Portable Planetarium as Part of a Museum Program." The demonstration was performed by David M. Ludlum. See *The Museum News,* 15 May 1949, 5. At the same conference, F. Wagner Schlesinger described "The [Adler] Planetarium's Instrument and its Possibilities" to delegates.

82. This panel was conducted by the AAM's education section during its 1950 conference in Colorado Springs. See *The Museum News,* 15 May 1950, 6.
83. "AAM Philadelphia Meeting Program Announced," *The Museum News,* 15 May 1951, 1. For Spitz's earlier announcement, see "Planetarium Conference Announced for May 30," *The Museum News,* 1 April 1951, 1. Spitz, unpublished address delivered before the AAM planetariums section, 28 May 1964.
84. For details of this first AAM planetariums section meeting, see the conference program in *The Museum News,* 1 May 1952, 6.
85. Andrew Fraknoi and Donat Wentzel, "Astronomy Education and the American Astronomical Society," in *The American Astronomical Society's First Century*, ed. David H. DeVorkin (Washington, D.C.: American Astronomical Society, 1999), 194.
86. Thornton Page, "Motion Pictures as an Aid in Teaching Astronomy," *Popular Astronomy,* March 1951, 117–128. Pages's device was modeled after that of Fletcher G. Watson, "A Tin Can Planetarium," *The Science Teacher,* November 1950, 180–183.
87. Joseph Miles Chamberlain, "The Development of the Planetarium in the United States." *Annual Report of the Board of Regents of the Smithsonian Institution . . . For the Year Ended June 30, 1957* (Washington, D.C.: Government Printing Office, 1958), 274.
88. See Spencer, "The Stars are His Playthings," 98; Ernest T. Ludhe [sic], et al., "Trail Blazing with Spitz Planetariums," *Sky and Telescope,* January 1949, 66–69. This article heralded the adoption of Spitz planetaria as opening "a new and significant chapter in the general story of astronomical education" (on p. 66).
89. See Coles to Spitz, 27 August 1951, RSP ser. 5:4, Box 25, Folder 4. Coles was also appointed as "Special Consultant" to Spitz Laboratories, as announced in the August 1953 *Sky and Telescope,* 269. Only the second issue, titled *Spitz Planetarium Pointer* and bearing the subtitle "Being a Bulletin for Planetarium People Projecting News and Ideas," winter 1953, has been located. RSP ser. 5:4, Box 25, Folder 4.
90. Menke, "Clyde Fisher and the Hayden Planetarium," 57.
91. Margaret Noble, "A Survey of Planetarium Programs," in *Planetariums and Their Use for Education,* vol. 2, ed. Richard H. Roche (Cleveland, Ohio: Cleveland Museum of Natural History, n.d.), 26.
92. Bunton, "What Price Change?" *Curator* 2, no. 1 (1959): 25. Bunton added, "[T]here is a stimulus to the major planetaria that arises from the existence of the lesser ones. Executives . . . feel they must maintain the superiority of the major installation by exceeding the capabilities of the classroom instrument as far as possible" (on p. 25).
93. No information regarding the first two executives' meetings has been located; possibly, no written records were kept. However, documentation of executives' meetings, starting with the so-called "third" conference hosted at Pittsburgh's Buhl Planetarium in 1954, is preserved in the papers of Joseph M. Chamberlain, Department of Astrophysics, AMNH. All subsequent citations are extracted from RSP ser. 1:5, Box 5, Folders 4–12.
94. John Patterson acted as informal "reporter," afterward distributing typed summaries of proceedings to executives.
95. Alter, "The Planetarium as a University Classroom," *Griffith Observer,* December 1947, 148. Alter's essay was symptomatic of contemporary fears raised by atomic

weapons and the accelerating Cold War. He argued that without significant improvement in the population's understanding of science, there was little "hope of successful conduct of our country as a democracy" and avoidance of a "dictatorship" (on pp. 146, 154).

96. Undated letter (ca. 1954) from Patterson to planetarium executives regarding the AAM section meeting at Santa Barbara, California. RSP ser. 1:5, Box 5, Folder 4.
97. Quotations from executives' conference, Fels Planetarium, 13–14 May 1957, "General Report and Topics Discussed," 7–8. RSP ser. 1:5, Box 5, Folder 7. Some twenty-three delegates were present, including representatives from Flint (Mich.) Junior College's Longway Planetarium and the U.S. Air Force Academy Planetarium in Colorado Springs (both of which awaited delivery of Spitz Model B projectors).
98. On Griffith Observatory's "spectacle," see Alter, "A Trip to the Moon," *Griffith Observer,* July 1948, 81–82; August 1948, 86–88, 94. On Jenzano's twin programs, see Billy Arthur, "Please Fasten Seat Belt, Next Stop The Moon," *Tarheel Wheels* 15, no. 2 (February 1958): 3.
99. See "Report of Planetarium Officials' Conferance [sic], Morehead Planetarium, June 7 and 8, 1955." RSP ser. 1:5, Box 5, Folder 5.
100. The social convention of listing a married woman's identity by her husband's name has made it impossible to attribute this accomplishment under Sullivan's own name.
101. Details concerning Noble's career and awards are drawn from Spencer, "The Stars Are His Playthings," 102, and "Notes on a Convention in Miami," *Sky and Telescope,* September 1956, 485–488.
102. For a sketch of Haarstick's life, see Rodney M. Nerdahl and Gary Tomlinson, "In Memoriam . . . Maxine B. Haarstick, 1922–1985," *The Planetarian* 14, no. 3 (third quarter 1985): 15–16. Haarstick, "How to Succeed in the Planetarium," *Museum News,* December 1964, 17–23.
103. Haarstick, "How to Succeed," 20.
104. This list was compiled from data gathered by Ruth Anne Korey, "Contributions of Planetariums to Elementary Education" (Ph.D. diss., Fordham University, 1963), along with Korey's article, "Planetariums in the United States," *Sky and Telescope,* September 1963, 147–148. Additional data and corrections have been applied from Norman Sperling, ed., *A Catalog of North American Planetariums (CATNAP),* Special Report #1 (N.p.: International Society of Planetarium Educators, June 1971), monthly advertisements placed in *Sky and Telescope* by Science Associates and Spitz Laboratories, and from that periodical's irregular column, "Planetarium Notes." The tally deliberately excludes institutions judged to be of only temporary, portable, or questionable status, along with those privately owned.
105. Of the 151 planetaria cited in a 1966 survey conducted by the National Aeronautics and Space Administration as having received at least partial federal support, only two, the U.S. Merchant Marine Academy, Kings Point, Long Island, and the U.S. Army Artillery School, Fort Sill, Oklahoma, predated 1958. See unpublished typescript on 421 permanent planetarium installations compiled by Myrl H. Ahrendt, "Planetarium Installations in the United States: Report of a Survey Made in November-December 1966." Copy preserved in Smithsonian Institution Libraries. Ahrendt was Instructional Resources Officer, Educational Programs Division, Office of Public Affairs.
106. For evidence that these changes were recognized early in the postwar period, see

"Children's Museums Grow in Number and Quarters," *The Museum News,* 15 December 1947, 1.

107. Spencer, "The Stars are His Playthings," 98, 100.
108. Arthur Zilversmit, *Changing Schools: Progressive Education Theory and Practice, 1930–1960* (Chicago: University of Chicago Press, 1993), 91.
109. Harry E. Crull, "Astronomy in the Junior High School Curriculum," *School Science and Mathematics* 49 (1949), 371. Crull, a professor of mathematics and astronomy, went on to direct the school's Holcomb Observatory and Planetarium.

Chapter 6 **Sputnik *and Federal Aid to Education***

1. American physicist Edward Teller seemingly originated this analogy, declaring that the United States had lost a battle "more important and greater than Pearl Harbor." Teller, quoted in Robert A. Divine, *The Sputnik Challenge* (New York: Oxford University Press, 1993), on p. xvi. For an in-depth account of the satellite's political and social ramifications, see Paul Dickson, *Sputnik: The Shock of the Century* (New York: Walker Publishing, 2001).
2. McDougall, *Heavens and the Earth,* 132. McDougall has retracted an earlier claim that *Sputnik* and the dawning space age represented a metaphorical "saltation." Pearl Harbor itself, he affirms, marked the "real watershed" in federal involvement with science and technology. See McDougall, "Introduction: Was Sputnik Really a Saltation?" in Launius, et al., *Reconsidering Sputnik,* on p. xviii.
3. Jordan D. Marché II, "Sputnik, Planetaria, and the Rebirth of U.S. Astronomy Education," *The Planetarian* 30, no. 1 (March 2001): 4–9.
4. James R. Killian Jr., *Sputnik, Scientists, and Eisenhower: A Memoir of the First Special Assistant to the President for Science and Technology* (Cambridge, Mass.: MIT Press, 1977), 7. Virtually the same phrase, a "crisis in self-confidence," was applied by historian Stephen E. Ambrose in his biography *Eisenhower: The President,* vol. 2 (New York: Simon and Schuster, 1984), on p. 424.
5. Sullivan, *Assault on the Unknown*, 2, 70. Historian Rip Bulkeley contends that the Eisenhower administration held a strong interest in "portraying Soviet actions in the first space race as cheating, and their own as irreproachable." Bulkeley, *Sputniks Crisis,* 120.
6. Bulkeley, *Sputniks Crisis,* 121.
7. J. Allen Hynek, quoted in Hayes, *Trackers of the Skies,* on p. 58. Newer historiography argues that the causes of surprise were largely "self-generated within the United States." Glenn P. Hastedt, "Epilogue: Sputnik as Technological Surprise," in Launius, et al., *Reconsidering Sputnik,* on p. 417.
8. Robert A. Divine, *Since 1945* (New York: John Wiley, 1975), 89; Bulkeley, *Sputniks Crisis,* 5.
9. McDougall, *Heavens and the Earth,* 148 (emphasis in original). McDougall asserted that public knowledge of *Sputnik* was "no more or less than what the news media imparted" (on p. 142).
10. Jack Lule, "Roots of the Space Race: Sputnik and the Language of U.S. News in 1957," *Journalism Quarterly* 68, nos. 1/2 (spring/summer 1991): 80, 85 (emphasis in original).
11. Wilson, *Year of the New Moons,* 64. Wilson served as president of the International Union of Geodesy and Geophysics (IUGG) during the IGY.

12. McDougall, *Heavens and the Earth,* 7.
13. Hayes, *Trackers of the Skies,* 58. The fate of those records remains unknown. While the bulk of SAO documents was transferred to the Smithsonian Institution Archives, correspondence of the sort described by Hayes apparently was not preserved.
14. Eisenhower, quoted in Divine, *Sputnik Challenge,* on pp. 7, 17. Signs of Eisenhower's perplexity emerged during a 15 October 1957 PSAC meeting, wherein he remarked, "I can't understand why the American people have got so worked up over this thing" (quoted on p. 12).
15. Gretchen J. Van Dyke, "Sputnik: A Political Symbol and Tool in 1960 Campaign Politics," in Launius, et al., *Reconsidering Sputnik,* 365–400.
16. Physicist Edward Teller declared that Americans had suffered "a very serious defeat—in the classroom." Teller, quoted in Divine, *Sputnik Challenge,* on p. 15.
17. Lawrence A. Cremin, *The Transformation of the School: Progressivism in American Education, 1876–1957* (New York: Random House, 1964), 347.
18. Life-adjustment education was intended to equip American youths "to live democratically with satisfaction to themselves and profit to society as home members, workers, and citizens." USOE, quoted in Diane Ravitch, *The Troubled Crusade: American Education, 1945–1980* (New York: Basic Books, 1983), on p. 66.
19. Bestor, quoted in Cremin, *Transformation of the School,* on p. 344. Cremin regarded Bestor's writings as "by far the most serious, searching, and influential criticisms of progressive education" to appear before the launch of *Sputnik* (on p. 344).
20. Scott L. Montgomery, *Minds for the Making: The Role of Science in American Education, 1750–1990* (New York: Guilford, 1994), 179, 178, 169.
21. See Ronald Lora, "Education: Schools as Crucible in Cold War America," in *Reshaping America: Society and Institutions, 1945–1960,* eds. Robert H. Bremner and Gary W. Reichard (Columbus: Ohio State University Press, 1982), 223–260.
22. Zilversmit, *Changing Schools,* 179.
23. Joel Spring, *The Sorting Machine: National Educational Policy Since 1945* (New York: David McKay, 1976). Spring argues that federal support of new science, mathematics, and foreign language curricula "was a direct result of national concern about meeting manpower needs in the cold war and the demands of the civil rights movement" (on p. 2).
24. PSSC Statement of Purpose, quoted in Spring, *Sorting Machine,* on p. 115. For a detailed account of these curricular initiatives, see John L. Rudolph, *Scientists in the Classroom: The Cold War Reconstruction of American Science Education* (New York: Palgrave, 2002).
25. Divine, *Sputnik Challenge.* Killian quoted on p. 49; Eisenhower quoted on p. 55.
26. Divine, *Sputnik Challenge,* 70. Eisenhower's personal memoirs record his dislike of "the use of federal funds to pay for the normal operation of schools" (quoted on p. 91).
27. Barbara Barksdale Clowse, *Brainpower for the Cold War: The Sputnik Crisis and the National Defense Education Act of 1958* (Westport, Conn.: Greenwood Press, 1981), 144.
28. Eisenhower, quoted in Divine, *Sputnik Challenge,* on pp. 164–165.
29. Ravitch, *Troubled Crusade,* 5, 8. Ravitch's interpretation of the rise of federal aid is styled a "crusade for equal educational opportunity" (on p. xi). In her account, a threefold process opened successive windows of opportunity in American education. The nineteenth century established universal primary schooling. During the

first half of the twentieth century, secondary-level education became available to all students. After the Second World War, expectations were raised that a college education might be attained by every aspiring adult.

30. Spring, *Sorting Machine,* 96, 108.
31. Concerning the ESEA, Clowse has written, "In 1958 the politically [correct] concept was 'defense'; in 1965 it was 'poverty.' The mode of obtaining federal aid in 1965 was the technique of 1958 writ large to suit the spirit of . . . [Lyndon Johnson's] Great Society at its zenith. The order of the day was now to abolish poverty rather than to beat the Russians" (on pp. 157–158).
32. Montgomery, *Minds for the Making,* 207. Conant's proposal was borrowed from Admiral Hyman G. Rickover, a noted critic of progressive education. Spring, *Sorting Machine,* 36; Ravitch, *Troubled Crusade,* 228.
33. John A. Douglass, "A Certain Future: Sputnik, American Higher Education, and the Survival of a Nation," in Launius, et al., *Reconsidering Sputnik,* 327–363.
34. For the complete text of the National Defense Education Act of 1958 (Public Law 85–864), see "Chapter 17—National Defense Education Program," in *United States Code, 1958 Edition, Containing the General and Permanent Laws of the United States, in Force on January 6, 1959,* vol. 4, Title 16—Conservation to Title 21—Food and Drugs (Washington, D.C.: Government Printing Office, 1959), 3702–3714. All subsequent citations refer to this edition.
35. Ibid., 3706. The nearest that legislators came to specifying what types of equipment were eligible was given in sec. 443, (a), (1), under "State plans," wherein "laboratory and other special equipment" might include "audio-visual materials and equipment and printed materials (other than textbooks)." Likewise, "minor remodeling" constituted the appropriate redesign of "laboratory or other space" for meeting intended instructional purposes (on p. 3707).
36. "'. . . The First Work of These Times . . .': A Description and Analysis of the Elementary and Secondary Act of 1965," *American Education,* April 1965, 17, 18. On the consolidation of assistance awarded under NDEA and ESEA legislation, see Clowse, *Brainpower for the Cold War,*152–161.
37. *Notes and Working Papers Concerning the Administration of Programs Authorized Under Title III of National Defense Education Act.* Prepared for the Subcommittee on Education of the Committee on Labor and Public Welfare, U.S. Senate, 90th Congress, 1st Session (Washington, D.C.: Government Printing Office, 1967),130, 150.
38. A school planetarium created through NDEA assistance was that installed at the Newburgh (New York) Free Academy, as illustrated in Joyce Rothschild, "The NDEA Decade," *American Education,* September 1968, 9. The planetarium housed a Spitz A3P projector whose significance is described below.
39. See "Chapter 21.—Higher Education Facilities," in *United States Code, 1958 Edition, Cumulative Supplement V, Containing the General and Permanent Laws of the United States Enacted During the 86th and 87th Congresses and 88th Congress, First Session,* vol. 2, Title 20—Education to Title 39—The Postal Service (Washington, D.C.: Government Printing Office, 1964), 1118. All subsequent citations of the Higher Education Facilities Act of 1963 (Public Law 88–204), enacted 16 December 1963, refer to this edition.
40. Ibid., sec. 701, "Congressional findings and declaration of policy," 1115–1116.
41. Ibid., 1122.

42. Clowse, *Brainpower for the Cold War,* 158.
43. See "Federal Money for Education," *American Education,* April 1965, 27. While not specified therein, the planetarium projector obtained by Clarke College was a Spitz A3P.
44. See Jack Spoehr, "Herbert N. Williams: A Remembrance," *The Planetarian* 14, no. 1 (first quarter 1985): 16, along with the unpublished manuscript, "A Brief History of Spitz Space Systems," n.d., courtesy of Spitz, Inc. Williams became sales manager and eventually company vice president.
45. Herbert N. Williams, "Recent Developments in Planetarium Projectors and Accessories," in *Planetariums and Their Use for Education,* vol. 2, ed. Richard H. Roche (Cleveland, Ohio: Cleveland Museum of Natural History, n.d.), 86–88.
46. Jagger, Preface to *Planetaria and Their Use for Education* (Bloomfield Hills, Mich.: Cranbrook Institute of Science, 1959), 12.
47. Williams, "Recent Developments in Planetarium Projectors," 86.
48. This argument does not consider Spitz's Model B projector, of which only three units were ever sold. For a description of the remarkable Model B, see Norton, *Planetarium and Atmospherium,* 101–103.
49. For an account of the A3P's planetary analogs, see Norton, *Planetarium and Atmospherium,* 73–82.
50. The A3P had a third axis of motion for the demonstration of precession, while a xenon arc lamp offered superior star images of greater brilliance, truer color, and smaller size. This "prime-sky" version of the A3P was first installed at the University of Nevada-Reno's Fleischmann Atmospherium-Planetarium in 1963 (see chapter 7).
51. Wallace E. Frank, telephone conversation with the author, 5 March 1997. At the time of Frank's appointment, Spitz Laboratories was owned by brothers Richard, William, and Edward Worth, who had relocated the company from Philadelphia to Elkton, Maryland (ca. 1953) and then to Yorklyn, Delaware (ca. 1954).
52. *A Space Science Classroom* (Yorklyn, Del.: Spitz Laboratories, n.d.) states that Spitz became the "leading producer of planetariums in the free world . . . during the last 14 years" (on p. 9). The company's next most recent booklet (identical in size and format, ca. 1959), titled *Spitz Planetariums: A Fundamental of Modern Education* (Yorklyn, Del.: Spitz Laboratories, n.d.), contains the preface "What is a Planetarium?" authored by Spitz, and features the Model A2 projector and its accessories.
53. *Space Science Classroom,* 1. An adjoining photograph of the Moon was captioned, "Reaching this goal is only a matter of time." The remark is perhaps an allusion to President John F. Kennedy's 25 May 1961 proclamation of achieving that goal. See John Noble Wilford, *We Reach the Moon* (New York: Bantam Books, 1969).
54. *Space Science Classroom*, 3. While opposed by other planetarium directors at the time, the notion of introducing children of preschool age to the planetarium was advocated by Armand N. Spitz. His essay, "Pre-School Children in the Planetarium," appeared in *Planetariums and Their Use for Education,* vol. 2, ed. Richard H. Roche (Cleveland, Ohio: Cleveland Museum of Natural History, n.d.), 180–185.
55. For an assessment of Bruner's impact on post-*Sputnik* curricular reforms, see Montgomery, *Minds for the Making,* 208–215.
56. *Space Science Classroom,* 3.
57. See Richard A. Duschl, *Restructuring Science Education: The Importance of*

Theories and Their Development (New York: Teachers College Press, 1990), 21–24. Both the American Geological Institute's Earth Science Curriculum Project (ESCP) text, *Investigating the Earth* (1967) and *The [Harvard] Project Physics Course* (1970) offered significant units on astronomy, while an Elementary-School Science (ESS) curriculum introduced the subject in grades 1–6. See Stanley P. Wyatt Jr., "Astronomy at the Lower School Levels," in *Education in and History of Modern Astronomy,* ed. Richard Berendzen. *Annals of the New York Academy of Sciences* 198 (25 August 1972): 178–191.

58. *Space Science Classroom,* 3.
59. Jack Spoehr, personal communication, 8 February 1997, on p. 2. In Spoehr's recollection, this aspect "may very well have been introduced" by a small company, Janson Industries, which "served [Spitz Laboratories] for a number of years as a marketing agency" in Ohio and Pennsylvania. Spoehr cautions, however, that in addition to this pair, "a good many people developed those changes" associated with the space science classroom (on p. 5).
60. Roger Neil Early, *The Use of the Planetarium in the Teaching of Earth and Space Sciences* (Yorklyn, Del.: Spitz Laboratories, 1960).
61. Ibid., 1–2, 4, 5.
62. For a complete listing of the thirty-three lessons contained within the *Handbook,* issued between 1971 and 1975, see Jordan D. Marché II, "A Bibliography of the Spitz *Planetarium Director's Handbook,*" *The Planetarian* 18, no. 3 (September 1989): 36–37.
63. Earl W. Newton and Phillip D. Stern, "New Planetarium Opens at Bridgeport," *Sky and Telescope,* March 1962, 132–134.
64. Congressional debates over NASA's creation are recounted in McDougall, *Heavens and the Earth,* 157–176, which highlights the agency's prescribed ambiguities toward military and entrepreneurial limitations.
65. In conjunction with the University of Bridgeport, NASA sponsored the publication of a fifty-two-page booklet, *The Planetarium: An Elementary-School Teaching Resource,* ed. Bartlett A. Wagner (Bridgeport, Conn.: University of Bridgeport, 1966).
66. Quotations from the introduction to Ahrendt, "Planetarium Installations in the United States," on first four unnumbered pages. As explained elsewhere by Ahrendt, a "policy decision, made after the report had been prepared for the press" ruled out publication at NASA expense and urged that it be financed by another non-profit agency. None were willing to underwrite the report, however, and the survey's findings became obsolete. See Myrl H. Ahrendt, "Services of NASA to Astronomy Education," in *Education in and History of Modern Astronomy,* ed. Richard Berendzen. *Annals of the New York Academy of Sciences* 198 (25 August 1972): 197–201, on p. 198.
67. Questionnaire, "Survey of Planetarium Installations . . . ," on p. 2, appended to Ahrendt, "Planetarium Installations in the United States."
68. Besides the institutional data contained in Ruth Anne Korey's doctoral dissertation (1962), the American Geological Institute's Earth Science Curriculum Project (ESCP) documented eighty-four additional institutions by 1964. This catalogue, however, was restricted to the institutional name, city, and state of each facility. See William H. Matthews III, *Sources of Earth Science Information* (Englewood Cliffs, N.J.: Prentice Hall, 1964), 6–11. Sperling, *Catalog of North American Plan-*

etariums (CATNAP), collected data on instrument type, year of opening, and personnel through mid–1971, but nothing on funding sources.

69. Only two installations, the U.S. Air Force Academy Planetarium, Colorado Springs, Colorado, and the Airborne Operator School, Cherry Point, North Carolina, were devoted to military aviation training.
70. Frederick Robert Mayer, "The Pennsylvania Department of Public Instruction's Earth and Space Science Curriculum Promotional Program and the Status of Earth and Space Science Education in the Public Secondary Schools of Pennsylvania, 1963–1964" (Ed.D. diss., Temple University, 1965). Mayer reported that 545 of the State's 657 districts had offered an earth- and space-science course prior to the conduct of his survey.
71. Charles H. Boehm, "Development of the Earth and Space Science Program in Pennsylvania," *The Pennsylvania Geographer* 1, no. 1 (March 1963): 2.
72. Quoted by Evan Evans, executive director, National Aviation Education Council, in the preface to *Earth and Space Guide for Elementary Teachers* (Harrisburg, Pa.: Commonwealth of Pennsylvania, Department of Public Instruction, 1961; reprint, Washington, D.C.: National Aviation Education Council, November 1961).
73. John E. Kosoloski, "The Pennsylvania Earth and Space Science Program," *Geotimes* 7, no. 3 (October 1962): 15–17, and especially table 1, "Number of Pennsylvania high schools teaching earth and space science, 1958–1962," on p. 15.
74. John E. Kosoloski, "Establishing a Space Science Curriculum," in *Planetariums and Their Use for Education,* vol. 2, ed. Richard H. Roche (Cleveland, Ohio: Cleveland Museum of Natural History, n.d.), 10, 11. By contrast, superintendent Boehm conveyed a much stronger (if erroneous) impression of the DPI's implementation. Boehm wrote, "In 1959, . . . Pennsylvania became the first state to *require* a planetarium, or an observatory, or an Earth and Space Laboratory, in new high schools." Boehm, "Earth and Space Science Program," on p. 2 (emphasis added). This statement is perhaps the source of misconceptions that a legislative mandate had been passed.
75. While no copy of the original manual has been located, a later edition, *Recommended Design Criteria for School Facilities* (Harrisburg, Pa.: Pennsylvania Department of Education, March 1977), lists "optional" plats of "500–1200 sq. ft." for a planetarium and "100–250 sq. ft." for an observatory (sec. 3–321, "Instructional Use," on p. 6).
76. "A Planetarium Can Enrich Curriculum," *Pennsylvania Education,* November-December 1969, 4–5.
77. Boehm, "Earth and Space Science Program," 3, 2.
78. Kosoloski, "Establishing a Space Science Curriculum," 13, 16.
79. *A Suggested Teaching Guide for the Earth and Space Science Course* (Harrisburg, Pa.: Commonwealth of Pennsylvania, Department of Public Instruction, 1959), ii, iii, 1, 3. Except where noted, all citations refer to this (first) edition of the *Teaching Guide.*
80. Ibid., 3, 6, 27.
81. Topics examined in the eight subunits were: (a) Aspects of the Sky; (b) The Earth in Motion; (c) The Moon; (d) The Solar System; (e) The Sun; (f) The Stars; (g) The Milky Way; and (h) The Universe. *Teaching Guide,* 27–52.
82. Ibid., 52–55.
83. *Earth and Space Science: A Guide for Secondary Teachers* (Harrisburg, Pa.: Commonwealth of Pennsylvania, Department of Public Instruction, 1963; reprint 1965),

71, 72, 75, 76, 80. NDEA Title III funds were used to publish this revision, whose distribution reached more than 40,000 copies. An edition of the *Guide* was distributed as late as 1973.

84. *Earth and Space Guide for Elementary Teachers*, 2, 3.
85. Ibid., 5, 7.
86. Ibid., 43, 54.
87. While the earth- and space-science curricula were implemented rapidly, planetarium and observatory construction proved a slower task. Mayer, "Earth and Space Science Curriculum Promotional Program," reported that through 1964, only 13 out of 197 Pennsylvania school districts had installed professional planetaria and only 4 districts a permanent telescope (p. 108).

CHAPTER 7 ***New Horizons in Planetarium Utilization***

1. No copy of the Morehead proposal has been located. Correspondence between Jenzano and Henry A. Pearson, head of the Flight Mechanics Branch, Aero-Space Mechanics Division, nonetheless provides clues as to how the agreement between Morehead and NASA was secured. See Preliminary Negotiations File, Morehead Planetarium (MP) Archives, University of North Carolina, Chapel Hill, North Carolina.
2. Jenzano to H. A. Pierson [sic], 27 October 1959. Pearson responded by noting that "[t]he objectives of this part of the training would be to so familiarize the astronauts with the appearance and position of the constellations that they could use them as a back up for the yaw orientation system." Pearson to Jenzano, 13 November 1959, Preliminary Negotiations File, MP Archives.
3. John Glenn, quoted in Paul Goerz Langfeld, "Astronauts' Star Recognition Training Intensifies," *Missiles and Rockets,* 19 November 1962, on p. 28. Additional details of the training program's early history are provided by Rose Bennett Gilbert, "Space Age Education," *Museum News,* November 1965, 25–30. For an account of the technical upgrade, see "Two Eastern Planetariums Re-Equipped," *Sky and Telescope,* March 1960, 271–273.
4. On this skepticism, see Charles Heatherly, "Spotlight: Morehead Planetarium," in *North Carolina Report* (N.p.: Department of Natural and Economic Resources, n.d.), Mercury-Gemini File, MP Archives.
5. Henry Howard, "East Carolina Professor Helped Train Astronauts," *High Point Enterprise,* 11 May 1961. Project Mercury/Pre-Orbital Information File, MP Archives.
6. Donald S. Hall, "Astronaut Training," in *What You Should Know About Astronaut Training at Morehead Planetarium* (Chapel Hill, N.C.: Morehead Planetarium, 1966), and Hall, "What's 'Up' in Space," *Review of Popular Astronomy,* June 1966, 6–8.
7. Howard, "East Carolina Professor."
8. Zedekar, quoted in George Myrover, "A Little of This . . . A Little of That," *Asheville Citizen,* 31 January 1963. Newspaper clipping file, MP Archives.
9. Walter Cunningham, quoted in Jim Clotfelter, "All 14 Astronauts at UNC Scheduled for Space Trips," *Durham Herald,* 28 March 1964. Newspaper clipping file, MP Archives.
10. Roy Thompson, "Chapel Hill Helped Cooper 'Find Himself' in Space," *Charlotte News,* 18 May 1963. Project Mercury/MA7/ Gordon Cooper File, MP Archives.

11. Cooper, quoted in "Astronauts End Study Time in NC," *Gastonia Gazette,* 3 March 1960. Project Mercury/Pre-Orbital Information File, MP Archives.
12. Borman, quoted in Bryan Haislip, "Astronauts Say Training at Planetarium Valuable," *Greensboro Record,* 30 January 1963. Newspaper clipping file, MP Archives.
13. Armstrong, quoted in Lawrence Maddry, "Astronauts Praise Course at Chapel Hill Planetarium," *Raleigh Times,* 27 March 1964. Newspaper clipping file, MP Archives.
14. Knapp, quoted in Joan Hill, "UNC 'Alumni' Reach Moon," *Greensboro Daily News,* 18 July 1976. Apollo/(G) Mission 11 File, MP Archives. Virtually all of NASA's astronauts were named honorary "alumni," or Tarheels, for their having passed through "astronaut college" at UNC's Morehead Planetarium.
15. Descriptions of the sighting devices and Kollsman simulator are found in Sidney Stapleton, "NASA and NC," *Wachovia,* July-August 1971, 11.
16. Ibid., 13.
17. Undated manuscript, "12-hour Mini Course in Celestial Studies Proposed for Skylab and Apollo/Soyuz Test Project Astronauts," Evaluations, Proposals, Schedules File, MP Archives.
18. Alphabetical and chronological listings of the sixty-two astronauts, training dates, and missions flown have been compiled by Morehead director Lee T. Shapiro, "Astronauts Who Trained at Morehead Planetarium and the missions they flew," Morehead Planetarium History/Astronaut Training Program File, MP Archives.
19. Jenzano, quoted in Heatherly, "Spotlight: Morehead Planetarium," on p. 18. I am indebted to Lee T. Shapiro for these insights.
20. This narrative is derived from Bunton's confidential 1958 manuscript, "A Special Report to the Planetarium Executives' Conference Meeting at the Robert T. Longway Planetarium, Flint, Michigan," Morrison Planetarium Archives, California Academy of Sciences, San Francisco, California.
21. Louis M. Brill, "Planetarium Lightshows—Past, Present and Future," *The Planetarian* 13, no. 1 (first quarter 1984): 4; Donna M. Stein, *Thomas Wilfred: Lumia, A Retrospective Exhibition* (Washington, D.C.: Corcoran Gallery of Art, 1971).
22. While an unproven conjecture, it is suggested that ideas concerning Wilfred's art form were communicated to the Bay area by noted art critic Alfred Frankenstein, who not only reviewed Wilfred's efforts in 1938 for the *San Francisco Chronicle,* but attended the 1957 Vortex premiere and reported on it in the same newspaper. For references to Frankenstein, see the bibliography in Stein, *Thomas Wilfred,* 97. Excerpts from Frankenstein's review (29 May 1957 *Chronicle*) were reprinted in the third Vortex program. Morrison Planetarium Archives.
23. "Gross" ticket receipts from the last four concerts were reported as follows: Vortex 2 (October 1957): $744; Vortex 3 (January 1958): $1,552.85; Vortex 4 (May 1958): $3,422.05; Vortex 5 (January 1959): $6,900 (est). Jacobs and Belson to Robert C. Miller, n.d., Morrison Planetarium Archives.
24. Jacobs and Belson to Bunton, 28 July 1961, Morrison Planetarium Archives.
25. Bunton, "What Price Change?" 21, 24.
26. John Hare, "(In-House) Laser Shows . . . A Long Term Proposition," *The Planetarian* 13, no. 1 (first quarter 1984): 8–9.
27. Norton, *Planetarium and Atmospherium,* 129.
28. O. Richard Norton, "Techniques of Extreme Wide-Angle Motion-Picture

Photography and Projection," *Journal of the Society of Motion Picture and Television Engineers* 78, no. 2 (February 1969): 81–85.

29. See "Wrap-Around Film Projection System Helps Skygazers to Learn Wonders of Atmosphere," *Business Screen* 28 (October 1967): 54.
30. See O. Richard Norton, "Nevada's Atmospherium-Planetarium . . . The Sky's the Limit!" *Review of Popular Astronomy,* July 1964, 4–7; Roy K. Marshall, "University of Nevada's Atmospherium-Planetarium," *Sky and Telescope,* December 1963, 318–321.
31. O. Richard Norton, "New Advances in Planetaria," *Astronomical Society of the Pacific Leaflet No. 444* (June 1966), 4 (emphasis in original).
32. Norton, "Techniques of Extreme Wide-Angle Motion-Picture Photography," 85.
33. O. Richard Norton, "A Major Planetarium for Tucson, Arizona," *Sky and Telescope,* March 1975, 143–146.
34. Norton, "Techniques of Extreme Wide-Angle Motion Picture Photography," 85.
35. Arthur W. Johnson Jr., "Recent Advances in 35-mm Hemispheric Cinematography for the Planetarium," *The Planetarian* 12, no. 2 (second quarter 1983): 5–6.
36. Pamela D. Crooks, "San Diego's Adventure in Space," *Sky and Telescope,* February 1983, 127–129.
37. Victor J. Danilov, "IMAX/OMNIMAX: Fad or Trend?" *Museum News,* August 1987, 32–39.
38. German ornithologist Gustav Kramer noted that blackcap warblers and red-backed shrikes exhibit seasonal tendencies to orient themselves in the direction of their migration routes (so-called *Zugunruhe* behavior). Other species utilizing celestial cues are the bobolink (*Dolichonyx oryzivorus*), European mallard (*Anas platyrhynchos*), and white-throated sparrow (*Zonotrichia albicollis*).
39. Franz Sauer and Eleonore Sauer, "Zur Frage der nächtlichen Zugorientierung von Grasmücken," *Revue Suisse de Zoologie* 62 (1955): 250–259.
40. Franz Sauer, "Zugorientierung einer Mönchsgrasmücke (*Sylvia a. atricapilla,* L.) unter künstlichem Sternhimmel," *Die Naturwissenschaften* 43 (1956): 231–232; Sauer, "Astronavigatorische Orientierung einer unter künstlichem Sternhimmel verfrachteten Klappergrasmücke, *Sylvia c. curruca* (L.)," *Die Naturwissenschaften* 44 (1957): 71; Sauer, "Die Sternorientierung nächtlich ziehender Grasmücken (*Sylvia atricapilla, borin* und *curruca*)," *Zeitschrift für Tierpsychologie* 14 (1957): 29–70; E. G. F. Sauer, "Celestial Navigation by Birds," *Scientific American,* August 1958, 42–47. In related experiments, warblers transported to southwestern Africa and exposed to natural starry skies at spring and autumn migration periods displayed in-phase orientation abilities. Franz Sauer and Eleonore Sauer, "Nächtliche Zugorientierung europäischer Vögel in Südwestafrika," *Die Vogelwarte* 20 (1959): 4–31.
41. E. G. Franz Sauer and Eleonore M. Sauer, "Star Navigation of Nocturnally Migrating Birds: The 1958 Planetarium Experiments," *Cold Spring Harbor Symposia on Quantitative Biology* 25 (1960): 463–473; E. G. Franz Sauer, "Further Studies on the Stellar Orientation of Nocturnally Migrating Birds," *Psychologische Forschung* 26 (1961): 224–244.
42. R. Robin Baker, *Bird Navigation: The Solution of a Mystery?* (New York: Holmes and Meier, 1984), 88.
43. Stephen Thompson Emlen, "Experimental Analysis of Celestial Orientation in a

Nocturnally Migrating Bird" (Ph.D. diss., University of Michigan, 1967); Stephen T. Emlen, "Migratory Orientation in the Indigo Bunting, *Passerina cyanea,*" parts 1 and 2, *The Auk* 84 (1967): 309–342, 463–489; Stephen T. Emlen, "The Celestial Guidance System of a Migrating Bird," *Sky and Telescope,* July 1969, 4–6.

44. Stephen T. Emlen, "Celestial Rotation: Its Importance in the Development of Migratory Orientation," *Science* 170 (11 December 1970): 1198–1201; Emlen, "The Stellar-Orientation System of a Migratory Bird," *Scientific American,* August 1975, 102–111.
45. Korey, "Contributions of Planetariums to Elementary Education," 6. Korey was no stranger to planetaria, having taught units on astronomy for many years. See Korey, "Making a Planetarium: A Unit on the Sky," *The Grade Teacher* 70 (December 1952): 43, 91–92.
46. John Thomas Curtin, "An Analysis of Planetarium Program Content and the Classification of Demonstrators' Questions" (Ed.D. diss., Wayne State University, 1967).
47. Dale E. McDonald, "The Utilization of Planetaria and Observatories in Secondary Schools" (Ed.D. diss., University of Pittsburgh, 1966).
48. Maurice Gene Moore, "An Analysis and Evaluation of Planetarium Programming as it Relates to the Science Education of Adults in the Community" (Ph.D. diss., Michigan State University, 1965).
49. On this well-known criticism of top-down communication models, see Terry Shinn and Richard Whitley, eds., *Expository Science: Forms and Functions of Popularisation,* Sociology of the Sciences Yearbook, vol. 9 (Dordrecht, Netherlands: D. Reidel, 1985), and Stephen Hilgartner, "The Dominant View of Popularization: Conceptual Problems, Political Uses," *Social Studies of Science* 20 (1990), 519–539.
50. George Reed, "Is the Planetarium a More Effective Teaching Device than the Combination of the Classroom Chalkboard and Celestial Globe?" *School Science and Mathematics* 70 (1970): 491.
51. Billy Arthur Smith, "An Experimental Comparison of Two Techniques (Planetarium Lecture-Demonstration and Classroom Lecture-Demonstration) of Teaching Selected Astronomical Concepts to Sixth-Grade Students" (Ed.D. diss., Arizona State University, 1966); John Charles Rosemergy, "An Experimental Study of the Effectiveness of a Planetarium in Teaching Selected Astronomical Phenomena to Sixth-Grade Children" (Ph.D. diss., University of Michigan, 1967); Delivee Loraine Cramer Wright, "Effectiveness of the Planetarium and Different Methods of Its Utilization in Teaching Astronomy" (Ed.D. diss., University of Nebraska, 1968); George Francis Reed, "A Comparison of the Effectiveness of the Planetarium and the Classroom Chalkboard and Celestial Globe in the Teaching of Specific Astronomical Concepts" (Ed.D. diss., University of Pennsylvania, 1970).
52. Glenn E. Warneking, "Planetarium Education in the 1970s—Time for Assessment," *The Science Teacher* 37 (October 1970): 14–15.
53. Reed, "Is the Planetarium a More Effective Teaching Device?" 492, 491.

Chapter 8 ***New Routes to Professionalization***

1. Telephone contact in 1995–96 with NSF officials revealed that no archival records of Spitz's role as special consultant were preserved.

2. Robert T. Hatt, "Symposium on Planetaria and Their Use for Education," in *Planetaria and Their Use for Education,* ed. Miriam Jagger (Bloomfield Hills, Mich.: Cranbrook Institute of Science, 1959), 7.
3. Ray J. Howe, "The Planetarium Program and the School Curriculum," in *Planetaria and Their Use for Education,* ed. Miriam Jagger (Bloomfield Hills, Mich.: Cranbrook Institute of Science, 1959), 104.
4. John C. Rosemergy, "The Use of the Planetarium in Secondary Schools," in *Planetaria and Their Use for Education,* ed. Miriam Jagger (Bloomfield Hills, Mich.: Cranbrook Institute of Science, 1959), 60.
5. Jagger, preface to *Planetaria and Their Use for Education,* 12, 13.
6. See "Planetarium Operators' Symposium at Cranbrook Institute," *Sky and Telescope,* December 1958, 80–81, which reproduced its group photograph.
7. Cleveland's Mueller planetarium had opened in 1952 with a Spitz A–2 instrument. Mention should be made, however, of the institution's earlier, non-projection device, the Hanna Star Dome, which was installed in 1938. This 16-foot diameter copper dome was wired with some 3,000 radio panel lamps and an equal number of electric-filament bulbs to depict forty-eight constellations seen during the year. See Harold L. Madison Jr., "The Hanna Star Dome," *The Sky,* June 1938, 8–9, and Don H. Johnston, "The Hanna Star Dome," *Popular Astronomy,* January 1940, 53–54.
8. Richard H. Roche, preface to *Planetariums and Their Use for Education,* vol. 2 (Cleveland, Ohio: Cleveland Museum of Natural History, n.d.), 1.
9. Ibid.
10. James A. Fowler, "Qualifications for Planetarium Personnel," in *Planetariums and Their Use for Education,* vol. 2, ed. Richard H. Roche (Cleveland, Ohio: Cleveland Museum of Natural History, n.d.), 134–145.
11. A biographical sketch of John M. Cavanaugh, then director of the North Museum Planetarium, Franklin and Marshall College, Lancaster, Pennsylvania, reports the "Am. Assn. of Planetarium Operators" as one of several professional associations to which he belonged. See Robert C. Cook, ed., *Who's Who in American Education: An Illustrated Biographical Directory of Eminent Living Educators of the United States and Canada,* 20th ed. (Nashville, Tenn.: Who's Who in American Education, 1961–62), on p. 266. Data on "Planetarium Association Organizational Meeting" taken from pre-conference brochure, "Symposium on Planetariums and Their Use in Education —II," RSP ser. 1:5, Box 5, Folder 8.
12. Howard Pegram, "Choosing and Training Planetarium Personnel," in *Planetariums and Their Use for Education,* vol. 2, ed. Richard H. Roche (Cleveland, Ohio: Cleveland Museum of Natural History, n.d.), 165–171. Pegram remarked, "I was wondering if once the association begins functioning well, if a list of such people could not be kept and sent to any planetarium desiring to hire a full-time operator or demonstrator" (on p. 165).
13. Chamberlain to directors of nine major U.S. planetaria, 27 September 1957; Joseph S. Henderson to Chamberlain, 24 February 1958. RSP ser. 1:5, Box 6, Folder 9.
14. Bauersfeld to Chamberlain, 9 February 1959. RSP ser. 1:5, Box 6, Folder 10. Later that same year, Bauersfeld passed away.
15. Chamberlain to Major General Wilton B. Persons, 18 February 1959; Persons to Chamberlain, 7 March 1959; RSP ser. 1:5, Box 6, Folder 9. The three benefactors

in attendance were Robert S. Adler (son of the late Max Adler), John M. Morehead, and Joseph Rulon, advocate of a national planetarium for Washington, D.C.

16. Eisenhower to Chamberlain, 7 May 1959. RSP ser. 1:5, Box 6, Folder 10.
17. See "Planetarium Representatives Attend International Conference," *Sky and Telescope,* July 1959, 503. For delegate attendance, see unpublished typescripts in RSP ser. 1:5, Box 6, Folder 10.
18. "Star-Level Parley Has No Geneva Woe," *New York Times,* 14 May 1959, 18.
19. The next two international conferences were held in Bochum/Munich, West Germany and Vienna, Austria. See Charles F. Hagar, "Second International Planetarium Executives Conference," *Sky and Telescope,* October 1966, 205, and George Lovi, "Vienna Planetarium Conference," *Sky and Telescope,* October 1969, 236–239. By the third meeting, the label "Executives" had been replaced with "Directors" (on p. 236).
20. Armand N. Spitz, "Planetarium: An Analysis of Opportunities and Obligations," *Griffith Observer,* June 1959, 78–87. As an idiosyncrasy, Spitz used the singular noun, "planetarium," without a definite article preceding it, and preferred "planetariums" to "planetaria," because of correspondence with the plural noun, "museums." Previously, Spitz was known as the "Henry Ford of planetaria."
21. Ibid., 78, 79, 80.
22. Ibid., 82, 81, 82.
23. Abbatantuono, "Armand Neustadter Spitz and His Planetaria," 57.
24. Joseph Miles Chamberlain, "A Philosophy for Planetarium Lectures," *Curator* 1, no. 1 (1958): 52–60. *Curator*'s founding editor was the American Museum of Natural History's department chairman of vertebrate paleontology, Edwin H. Colbert.
25. Ibid., 57, 53, 59, 60. Guidelines on the preparation and delivery of lectures were subsequently furnished by Hayden Planetarium astronomer Thomas D. Nicholson, "The Planetarium Lecture," *Curator* 2, no. 3 (1959): 269–274.
26. Joseph Miles Chamberlain, "The Administration of a Planetarium as an Educational Institution" (Ed.D. diss., Columbia University, 1962), 7.
27. Ibid., in unpaginated abstract; 128–129 (emphasis in original).
28. Ibid.,105–106, 93, 96, 124. Chamberlain noted three exceptions in the research of Dinsmore Alter (selenography), Israel M. Levitt (astronautics), and Kenneth L. Franklin (radio astronomy). Few planetarium directors, he added, were able to conduct research, while others "do not particularly wish to do so" (on p. 93).
29. Ibid., 100–101.
30. See John C. Rosemergy, "The Planetarium—Stars and Space in School," *School Executive* 79 (October 1959): 70–71; Herbert N. Williams, "The Planetarium in Modern Education," *American School Board Journal* 141 (September 1960): 40; and Theodore Berland, "Classroom for the Space Age," *Science Digest,* January 1961, 42–47.
31. Martin J. Steinbaum, "The Chosen," *The Science Teacher* 35 (January 1968): 49–50. Steinbaum, scientific assistant at the American Museum-Hayden Planetarium, reiterated that a planetarium "is something that a teacher uses and not something that teaches astronomy" (on p. 50).
32. "Personals," *The Museum News,* 1 November 1956, 3.
33. Duschl, *Restructuring Science Education,* table 2.1 on p. 18. Duschl attributed the "tenfold increase" in sponsorship of NSF programs between 1956 and 1957 to

acceptance of the "implied threat of Soviet superiority" argued by director Alan Waterman during congressional testimony in 1955 (on pp. 20–21).

34. Kosoloski, "Establishing a Space Science Curriculum," 16.
35. "Training Planetarium Operators," *Sky and Telescope,* April 1967, 207.
36. George F. Reed, personal communication; Polly H. Vanek, "The Role of Astronomers in Elementary and Secondary Education. I," in Proceedings of the Conference on Education in Astronomy, held at the 130th Meeting of the American Astronomical Society, 11–14 August 1969, at the State University of New York at Albany, ed. Frank C. Jettner. *Bulletin of the American Astronomical Society* 2 (1970): 272–275.
37. Joseph M. Chamberlain, "Proceedings of the First U.S. Conference on Graduate Education in Astronomy," *The Astronomical Journal* 68, no. 3 (April 1963): 215–219. Princeton University astronomer Martin Schwarzschild remarked, "It is a calamity that we astronomers are suddenly popular; we are not used to it. Suddenly the world wants astronomers. What a happy calamity!" (on pp. 218–219).
38. "A Directory of Institutions Offering Coursework in Planetarium Education," *The Planetarian* 1, no. 1 (June 1972): 24, 32.
39. School and district (171) and university and college (162) planetaria comprised the largest combined percentage (79%) of the 421 U.S. installations surveyed in late 1966. Ahrendt, "Planetarium Installations in the United States," unpaginated table titled, "Distribution by States of American Planetariums."
40. Ibid.
41. Ian C. McLennan, "Canada's First Public Planetarium," *Sky and Telescope,* August 1962, 86–88. McLennan's institution hosted the foundational meeting of the Planetarium Association of Canada (PAC) in 1966.
42. A map depicting eight continental "NSTA Regions" was printed in *The Science Teacher,* October 1964, 5. Noble is listed as an NSTA elections committee member.
43. Charles E. Allen III, "A Golden Celebration," *Sky and Telescope,* May 1997, 102–105.
44. Partial histories of most regional associations not featured in this study may be gleaned from the following sources: "Transcript of the First Annual Meeting of the Southwestern Association of Planetariums," SWAP folder, IPS Archives; "Pacific Planetarium Association—A Brief History," PPA folder, IPS Archives; Bob Risch to IPS historian Paul Engle, 19 April 1982, RMPA folder, IPS Archives; "Planetarium Associations," *Sky and Telescope,* May 1969, 299; and Jane P. Geohegan, "The Origin of SEPA: An Intermittent Family Outing," SEPA folder, IPS Archives. The Great Plains Planetarium Association (GPPA) was not organized until 1973.
45. "Planetarium Conference," *Sky and Telescope,* March 1965, 140; Von Del Chamberlain, "Education Dawns in the Planetarium," *Great Lakes Planetarium Association (GLPA) Projector* 1 (1968): 52–53. For an account of Michigan State University's Abrams Planetarium, see Victor H. Hogg, "The Abrams Planetarium in Michigan," *Sky and Telescope,* June 1964, 336–338.
46. Claire J. Carr, "MAPS History," MAPS folder, IPS Archives.
47. Donald D. Davis, "New Skies for a New City," *Sky and Telescope,* April 1966, 196–198; David A. Rodger, "Current Planetarium Activity in Canada," *Sky and Telescope,* July 1967, 13–14; and S. Wieser, "Calgary's Planetarium and Museum," *Sky and Telescope,* July 1967, 14–15.

48. Quotation from preamble to the constitution of the Planetarium Association of Canada, reproduced in "A Brief History of the Planetarium Association of Canada," PAC folder, IPS Archives.
49. Jerome V. DeGraffe, "The Historic Rochester Convention, October 24–27, 1968," *Great Lakes Planetarium Association (GLPA) Newsletter* 3, no. 4 (1968): 1–2; "The Planetarium Trade," *Sky and Telescope,* December 1968, 359, 375. For an account of the Strasenburgh Planetarium, see Ian McLennan, "A Major Planetarium in Rochester," *Sky and Telescope,* April 1968, 208–210.
50. Unofficial announcement of Chamberlain's intent to host a national meeting appears in "News Notes," *Great Lakes Planetarium Association (GLPA) Newsletter* 3, no. 4 (1968): 2.
51. Norman Sperling, "A Planetarium Educators Conference," *Sky and Telescope,* January 1971, 7–9. CAPE sessions are described in the printed program, "Conference of American Planetarium Educators," Abrams Planetarium (AP) Archives, Michigan State University, East Lansing, Michigan.
52. Statement of purpose, section C, part I., "Proposals of the Pre-Conference Organization Committee, C.A.P.E.[,] 19–20 October 1970." ISPE folder, IPS Archives.
53. Undated draft, Chamberlain to committee members, regarding the two "alternatives." Organization—Pre-Conference Meeting file, AP Archives.
54. Wieser to Chamberlain, 21 September 1970, Organization—Pre-Conference Meeting file, AP Archives.
55. "Decisions Made by the Organization Committee, Monday, 19 October 1970," Organization—Pre-Conference Meeting file, AP Archives.
56. See "Proposals of the ad hoc Publications Committee, CAPE, 19–20 October 1970," ISPE file, IPS Archives.
57. Larry Gilchrist, "The Case for a North American Planetarium Publication," *The Planetarium Journal* 2 (May 1970): 7–9; Gilchrist, "The Case for a North American Planetarium Association," ibid., 6. Further discussion, however, was tabled at the PAC Conference, 9–11 September 1970, pending outcome of the CAPE meeting.
58. "The Case for a North American Planetarium Periodical: Brief," unpublished manuscript, CAPE file, IPS Archives.
59. Jettner's selection likely arose from his performance as proceedings editor of the Conference on Education in Astronomy (CEA) convened at the 130th meeting of the American Astronomical Society, 11–14 August 1969. Before his move to Albany, Jettner had been senior astronomer at the Adler Planetarium (1959–66).
60. "Proposals of the ad hoc Publications Committee," CAPE Publication Com. file, IPS Archives.
61. Audio recording, taped September 1970 by Armand N. Spitz and replayed before the Conference of American Planetarium Educators, 22 October 1970. See "Armand Spitz at CAPE, 1970," *The Planetarian* 1, no. 1 (June 1972): 7.
62. Chamberlain to Abell, 16 September 1970, George Abell file, AP Archives. Abell's lecture, "And Now, May I Wish You All a Very Good Morning," appeared in *The Planetarian* 1, no. 2 (September 1972): 35–40. E. J. Spoehr to Dr. and Mrs. Armand Spitz, 23 October 1970, and Armand N. Spitz to Spoehr, 31 October 1970, George Abell file, AP Archives. Chamberlain to Colleague[s], 27 April 1971, ISPE file, IPS Archives; Von Del Chamberlain, "The Armand N. Spitz Memorial Fund," *The Planetarian* 1, no. 2 (September 1972): 43.

63. Roger L. Geiger, *To Advance Knowledge: The Growth of American Research Universities, 1900–1940* (New York: Oxford University Press, 1986), 22.
64. Prior SWAP editorials decrying member apathy stirred representatives to embrace the establishment of ISPE. SWAP folder, IPS Archives.
65. C.A.P.E. Constitution Committee Correspondence file, IPS Archives.
66. Jettner's criticisms of the provisional bylaws were expressed in Jettner to Howarth, 6 April 1971; a memorandum to the ISPE Constitution Committee, 7 April 1971; and a concurrent memorandum to the ISPE Publications Committee, 7 April 1971. Wieser's rebuttal of Jettner's "emotional epistle" was addressed to the ISPE Council, 16 April 1971. LSU Meeting file, IPS Archives.
67. "Bylaws of the International Society of Planetarium Educators," SWAP file, IPS Archives.
68. Jettner to Donald Zahner (Sky Map Publications), 11 November 1970; Myrl H. Ahrendt to Jettner, 12 November 1970; CAPE Publication Com. file, IPS Archives.
69. See Jettner's questionnaire to "Planetarium Colleague[s]," 20 January 1971, CAPE file; undated manuscript, "Proposed Active Subcommittees," CAPE Publication Com. file. For Jettner's apology, see memorandum, 23 February 1971, CAPE Publication Com. file, IPS Archives.
70. [Sperling], memorandum to CPC members, 7 March 1971, CAPE Publication Com. file, IPS Archives; "[T]houghts from FCJ," *The Planetarian* 1, no. 1 (June 1972): 2.
71. Sperling, *Catalog of North American Planetariums (CATNAP).*
72. Gilchrist to Jettner, 1 February 1972. Jettner's insecurity, however, was still apparent. See his "Editorial Structure" and accompanying chart, 9 September 1971. CAPE Publication Com. file, IPS Archives.

Epilogue

1. Paul Deans, "Magic Under the Dome," *Sky and Telescope,* January 2004, 52–59.
2. Scherer to Griffith, 16 May 1913. Van M. Griffith Papers, Collection 2060, UCLA Library, Los Angeles, California, Box 15, Folder 7.
3. John Mosley, *The Hugo Ballin Murals of the Griffith Observatory* (Los Angeles: Griffith Observatory, n.d.).
4. Knight's triptych was reproduced in *The Sky,* February 1938, 18–19, 32. It was commissioned by A. Cressy Morrison, chairman of the Hayden Planetarium Advisory Committee.
5. Stokley, "America's Fifth Planetarium."
6. Anthony Cook, "Six Decades with Six Astronomers on the Front Lawn of Griffith Observatory," *Griffith Observer,* November 1994, 2–11, 21.
7. Jordan Marché II, "Planetarium," in *Encyclopedia of Literature and Science,* ed. Pamela Gossin (Westport, Conn. and London: Greenwood Press, 2002), 342.
8. *Rebel Without a Cause,* directed by Nicholas Ray. © 1955, Warner Brothers, Inc.; © 1983 Warner Home Video, Inc.
9. Dinsmore Alter, "The Planetarium as a Public School Laboratory," *The Sky,* August 1938, 26.
10. Spencer, "The Stars Are His Playthings," 43, 97.

Appendix

1. Technically speaking, the catalog was inaccurate by its inclusion of the Atwood Celestial Sphere, a non-projection device then housed at the Chicago Academy of Sciences, but since restored and relocated to the Adler Planetarium and Astronomy Museum.
2. Through 1966, 421 permanent American planetaria had been catalogued by NASA's Educational Programs Division. Ahrendt, "Planetarium Installations in the United States."
3. *Space Science Classroom,* 5. A caption accompanying the figure describes a "planetarium chamber . . . suitable for either a 24- or 30-foot diameter dome in a cubical room."
4. The Spitz ISTP (Intermediate Space Transit Planetarium) projector was the first of its kind to feature independent three-axis motions simulating roll, pitch, and yaw (RPY). This technology was later applied to the smaller A3P design and marketed as the Spitz A4 (or System 512 with digital automation).
5. "Planetarium Notes" debuted as the "Planetarium Page" in the April 1938 issue of *The Sky,* but only featured activities at New York's Hayden Planetarium. In January 1940 the column adopted its later title when news of the Buhl Planetarium was added. Notices of programs at the Adler and Fels planetaria, plus the Griffith Observatory, were not printed until the February 1946 issue of *Sky and Telescope.* The column's frequency decreased to quarterly and finally annually. "Planetarium Notes" last appeared in the September 1968 issue.
6. John M. Cavanaugh, director of Franklin and Marshall College's North Museum Planetarium from 1953 to 1970 (and beyond), served in an entirely voluntary capacity.
7. Certain assumptions were followed in the assignment of gender to individuals identified only by first and middle initials. In the absence of further clues, these have been denoted exclusively as men. Because of cultural practices emphasizing marital status, women's first names were often replaced by the names or initials of their husbands, with the abbreviation "Mrs." offering the only hint of their feminine identity.
8. See Marcel C. LaFollette, "Eyes on the Stars: Images of Women Scientists in Popular Magazines," *Science, Technology, and Human Values* 13 (1988): 262–275. Within her survey of literature published between 1910 and 1955, LaFollette found those "few positive aspects to the portrayal of women" strongly outweighed by the "negative aspects—that to be a woman scientist required extraordinary hard work and sacrifices" (on p. 265).
9. National Science Foundation, *American Science Manpower 1960: A Report of the National Register of Scientific and Technical Manpower* (Washington, D.C.: Government Printing Office, 1962), table 2, p. 7. Within all biennial reports issued from 1962–70, astronomy was subsumed under the larger scientific "field" of physics. While the number of astronomers (men and women combined) was reported through 1968, no distribution on the basis of gender within any subfield was presented. "Astronomy" no longer appeared as a subfield in the 1970 *Report.*
10. Rossiter, *Before Affirmative Action,* table 4.4, 81–82.
11. Morton E. Mattson, unpublished "CAPE Opinionnaire," administered 22 October 1970, Lansing Community College Planetarium. CAPE folder, IPS Archives.

12. Cross-correlations of selected scores from Mattson's CAPE survey were performed and published by Frank C. Jettner and John J. Soroka, "The Planetarium in Modern Science Education," in *Education in and History of Modern Astronomy,* ed. Richard Berendzen. *Annals of the New York Academy of Sciences* 198 (25 August 1972): 178–191. Distribution of doctoral degrees (based on 139 responses) occurred most often (21.9%) among university and college planetaria, secondarily (10.5%) among museums and science centers, and least of all (1.4%) among school and district facilities (table 2, p. 184).
13. Nigel O'Connor Wolff directed (1954–58) the Spitz Model B installation at Montevideo, Uruguay, before heading the Maryland Academy of Sciences during the 1960s. See Wolff, "Montevideo Planetarium Inaugurated," *Sky and Telescope,* May 1955, 282–283. A major planetarium was erected at Sao Paulo, Brazil in 1957 and directed by Aristotles Orsini. See O. Dias de Almeida, "Planetarium in Brazil," *Sky and Telescope,* June 1958, 400.
14. The first woman known to oversee a major American planetarium after 1945 was Elizabeth Hill, director of the W. A. Gayle Planetarium, Montgomery, Alabama, in 1970. The "fourth" female director of a major planetarium (*CATNAP* data) was a misnomer. Before heading the U.S. Atomic Energy Commission, Dixie Lee Ray had directed the Pacific Science Center, Seattle, Washington, and nominally its planetarium (then under construction). Ray, however, did not serve as planetarium director when the facility opened in 1972.
15. Degree data are available on only forty-four directors (55%) of major planetaria. These were distributed among sixteen bachelor's, eight master's, and seventeen doctoral degrees. Three individuals received some post-secondary education.
16. Data on years of service are available for seventy-five directors (94%) of major American planetaria. From a total of 343 years accrued, an average of roughly 4.6 years as director is found.
17. Draper's record of service was exceeded by Frank Korkosz, who directed Springfield's Seymour Planetarium (1937–70). That institution, however, was not a major planetarium.
18. Henry C. King directed the London Planetarium (1958–68) before heading the McLaughlin Planetarium, Toronto, Ontario (1968–70). Both facilities are major planetaria. If King's European experience were included, then the number of dual-institution directors of major American planetaria became nine.
19. For data on planetarium authors, I have relied chiefly on Gary Tomlinson and Katherine Becker, eds., *Planetarium Bibliography, 1913 to 1990* (East Lansing, Mich.: Great Lakes Planetarium Association, 1992), supplemented by additional references.

Bibliography

Archives and Manuscript Collections

Abrams Planetarium Archives, Michigan State University, East Lansing, Michigan.

Alumni Biographical Records, Northwestern University Library, Evanston, Illinois.

Board of Managers Minutes, Franklin Institute Archives, Philadelphia, Pennsylvania.

Buhl Foundation Records, Library and Archives Division, Historical Society of Western Pennsylvania, Pittsburgh, Pennsylvania.

Carnegie Corporation of New York Archives, Columbia University, New York, New York.

Central Archives, American Museum of Natural History, New York, New York.

Einstein Papers Project, Boston University, Boston, Massachusetts.

History of Astronomy Research Center, Adler Planetarium and Astronomy Museum, Chicago, Illinois.

International Planetarium Society Archives, Ash Enterprises, Bradenton, Florida.

Julius Rosenwald Correspondence, Museum of Science and Industry Archives, Chicago, Illinois.

Microfilm Edition, The George Ellery Hale Papers, 1882–1937, edited by Daniel J. Kevles. Carnegie Institution of Washington and the California Institute of Technology, 1968.

Morehead Planetarium Archives, University of North Carolina, Chapel Hill, North Carolina.

Morrison Planetarium Archives, California Academy of Sciences, San Francisco, California.

Official Proceedings, South Park Commissioners, Chicago Park District Archives, Chicago, Illinois.

Otterbein College Archives, Westerville, Ohio.

Papers of the Director, Yerkes Observatory Archives, Williams Bay, Wisconsin.

Richard S. Perkin Collection, Department of Astrophysics, American Museum of Natural History, New York, New York.

Rosicrucian Egyptian Museum and Planetarium, San Jose, California.

Samuel Simeon Fels Papers, Historical Society of Pennsylvania, Philadelphia, Pennsylvania.

Smithsonian Institution Libraries, Washington, D.C.

Temple University Library, Philadelphia, Pennsylvania.

University Archives, Northwestern University Library, Evanston, Illinois.

University of Pennsylvania Archives, Philadelphia, Pennsylvania.
Van M. Griffith Papers, University of California at Los Angeles Library, Los Angeles, California.
Walter Sydney Adams Papers, Carnegie Observatory Collections, Huntington Library, San Marino, California.

Audio and Motion Picture Sources

Dedication ceremonies at Griffith Observatory, 14 May 1935. Audio recording of broadcast made over the Columbia-Donnelley Network. Transcribed by Anthony Cook.
Rebel Without a Cause. Directed by Nicholas Ray. © 1955, Warner Brothers, Inc.; © 1983, Warner Home Video, Inc.

Unpublished Theses and Dissertations

Abbatantuono, Brent P. "Armand Neustadter Spitz and His Planetaria, with Historical Notes of the Model A at the University of Florida." M.A. thesis, University of Florida, 1994.
Chamberlain, Joseph Miles. "The Administration of a Planetarium as an Educational Institution." Ed.D. diss., Columbia University, 1962.
Curtin, John Thomas. "An Analysis of Planetarium Program Content and the Classification of Demonstrators' Questions." Ed.D. diss., Wayne State University, 1967.
Emlen, Stephen Thompson. "Experimental Analysis of Celestial Orientation in a Nocturnally Migrating Bird." Ph.D. diss., University of Michigan, 1967.
Emmons, Richard H. "A Report on a School Planetarium: Its Design; Its Development as a Group Project; Its Utility as an Instructional Aid; and Its Program in School Community Relations." M.A. thesis, Kent State University, 1950.
Gronauer, Charles F. "The Planetarium, its History, Functions, and Architecture: With Application for a Proposed Addition to the Florida State Museum." M.A. thesis, University of Florida, 1978.
Kennedy, John Michael. "Philanthropy and Science in New York City: The American Museum of Natural History, 1868–1968." Ph.D. diss., Yale University, 1968.
Korey, Ruth Anne. "Contributions of Planetariums to Elementary Education." Ph.D. diss., Fordham University, 1963.
Marché, Jordan D., II. "Theaters of Time and Space: The American Planetarium Community, 1930–1970." Ph.D. diss., Indiana University, 1999.
Mayer, Frederick Robert. "The Pennsylvania Department of Public Instruction's Earth and Space Science Curriculum Promotional Program and the Status of Earth and Space Science Education in the Public Secondary Schools of Pennsylvania, 1963–1964." Ed.D. diss., Temple University, 1965.
McDonald, Dale E. "The Utilization of Planetaria and Observatories in Secondary Schools." Ed.D. diss., University of Pittsburgh, 1966.
Moore, Maurice Gene. "An Analysis and Evaluation of Planetarium Programming as it Relates to the Science Education of Adults in the Community." Ph.D. diss., Michigan State University, 1965.
Newell, Julie R. "American Geologists and Their Geology: The Formation of the American Geological Community, 1780–1865." Ph.D. diss., University of Wisconsin-Madison, 1993.

Reed, George Francis. "A Comparison of the Effectiveness of the Planetarium and the Classroom Chalkboard and Celestial Globe in the Teaching of Specific Astronomical Concepts." Ed.D. diss., University of Pennsylvania, 1970.

Rosemergy, John Charles. "An Experimental Study of the Effectiveness of a Planetarium in Teaching Selected Astronomical Phenomena to Sixth-Grade Children." Ph.D. diss., University of Michigan, 1967.

Smith, Billy Arthur. "An Experimental Comparison of Two Techniques (Planetarium Lecture-Demonstration and Classroom Lecture-Demonstration) of Teaching Selected Astronomical Concepts to Sixth-Grade Students." Ed.D. diss., Arizona State University, 1966.

Williams, Thomas R. "Getting Organized: A History of Amateur Astronomy in the United States." Ph.D. diss., Rice University, 2000.

Wright, Delivee Loraine Cramer. "Effectiveness of the Planetarium and Different Methods of Its Utilization in Teaching Astronomy." Ed.D. diss., University of Nebraska, 1968.

Published Sources

"AAM Philadelphia Meeting Program Announced." *The Museum News,* 15 May 1951, 1.

Abbatantuono, Brent P. "Armand Spitz—Seller of Stars." *The Planetarian* 24, no. 1 (March 1995): 14–22.

Abell, G. O. "And Now, May I Wish You All a Very Good Morning." *The Planetarian* 1, no. 2 (September 1972): 35–40.

"Agreement, Between South Park Commissioners and Max Adler, Relating to Adler Planetarium and Astronomical Museum, June 20, 1928." *Journal of the Proceedings of the South Park Commissioners* (2 April 1930): 33–38.

Ahrendt, Myrl H. "Services of NASA to Astronomy Education." In *Education in and History of Modern Astronomy,* edited by Richard Berendzen. *Annals of the New York Academy of Sciences* 198 (25 August 1972): 197–201.

Aitken, Robert G. "Philip Fox, 1878–1944." *Publications of the Astronomical Society of the Pacific* 56 (1944): 177–181.

Alexander, Edward P. "Oskar von Miller and the Deutsches Museum: The Museum of Science and Technology." In *Museum Masters: Their Museums and Their Influence,* 341–375. Nashville, Tenn.: American Association for State and Local History, 1983.

———. "Carl Ethan Akeley Perfects the Habitat Group Exhibition." In *The Museum in America: Innovators and Pioneers,* 33–49. Walnut Creek, Calif.: Altamira Press, 1997.

Allen, Charles E., III. "A Golden Celebration." *Sky and Telescope,* May 1997, 102–105.

Alter, Dinsmore. "The Planetarium as a Public School Laboratory." *The Sky,* August 1938, 18–20, 26.

———. "The Planetarium as a University Classroom." *Griffith Observer,* December 1947, 146–148, 154.

———. "A Trip to the Moon." *Griffith Observer,* July 1948, 81–82; August 1948, 86–88, 94.

Ambrose, Stephen E. *Eisenhower: The President.* Vol. 2. New York: Simon and Schuster, 1984.

"Armand Spitz at CAPE, 1970." *The Planetarian* 1, no. 1 (June 1972): 7.

Arthur, Billy. "Please Fasten Seat Belt, Next Stop the Moon." *Tarheel Wheels* 15, no. 2 (February 1958): 1, 3, 5.
"Astronauts End Study Time in NC." *Gastonia Gazette,* 3 March 1960.
"Astronomy de Luxe." *The Outlook,* 3 February 1926, 161.
"Astronomy Professor May Devote Full Time to Delivering Commercials on Television." *Printer's Ink* 234 (2 March 1951): 74.
"Atwater, Navy Man, to Head Planetarium." *New York Times,* 28 August 1945.
Atwood, Wallace W. "A New Way of Studying Astronomy." *Scientific American,* 21 June 1913, 557–558.
———. "Giant Celestial Sphere a Forerunner of Later Planetariums." *The Sky,* April 1938, 12–13.
Baker, R. Robin. *Bird Navigation: The Solution of a Mystery?* New York: Holmes and Meier, 1984.
"Banker to Religion via Stars." *Time,* 15 January 1934, 36–37.
Barbour, Ian C. *Issues in Science and Religion.* New York: Prentice-Hall, 1966.
Barton, D. R. "He Brought the Stars to America." *Natural History,* June 1940, 59–63.
Barton, Wm. H., Jr. "The End of the World." *The Sky,* October 1937, 3–4, 14, 19.
———. "Exploring the Moon." *The Sky,* April 1938, 3–5, 23.
———. Letter to the editor. *The Sky,* July 1938, 20.
Bauersfeld, Walther. "Das Projektions-Planetarium des Deutschen Museums in München." *Zeitschrift des Vereines Deutscher Ingenieure* 68 (1924): 793–797.
———. "The Great Planetarium of the German Museum in Munich." *Proceedings of the United States Naval Institute* 51 (1925): 761–774.
———. "Projection Planetarium and Shell Construction." Parts 1 and 2. *The Engineer,* 17 May 1957, 755–757; 24 May 1957, 796–797.
Belt, Haller. Preface to "The Great Planetarium of the German Museum in Munich," by W. Bauersfeld. *Proceedings of the United States Naval Institute* 51 (1925): 761–762.
Benford, James R. "New Projector for Navigation Stars." *Sky and Telescope,* November 1942, 3–4.
Bennett, Dorothy A. "The Mysterious Moon." *Natural History,* October 1935, 262–264.
Bennot, Maude. "Proper Motions of Forty Stars." *Astronomical Journal* 36 (8 June 1926): 177–181.
Berland, Theodore. "Classroom for the Space Age." *Science Digest,* January 1961, 42–47.
Boehm, Charles H. "Development of the Earth and Space Science Program in Pennsylvania." *The Pennsylvania Geographer* 1, no. 1 (March 1963): 1–6.
Bok, Bart J. "Report on Astronomy." *Popular Astronomy,* August-September 1939, 356–372.
Brackbill, M. T. Letter to the editor. *Sky and Telescope,* December 1947, 35.
Brill, Louis M. "Planetarium Lightshows—Past, Present and Future." *The Planetarian* 13, no. 1 (first quarter 1984): 4–7.
Broman, Lars. "A Spitz-Type Planetarium from 1935." *The Planetarian* 15, no. 4 (fourth quarter 1986): 10–11.
Brooke, John Hedley. *Science and Religion: Some Historical Perspectives.* Cambridge: Cambridge University Press, 1991.
"Buhl Foundation Gives Pittsburgh Planetarium." *Greater Pittsburgh,* May 1937, 34–36.

Bulkeley, Rip. *The Sputniks Crisis and Early United States Space Policy.* Bloomington/Indianapolis: Indiana University Press, 1991.

———. "The Sputniks and the IGY." In *Reconsidering Sputnik: Forty Years Since the Soviet Satellite,* edited by Roger D. Launius, John M. Logsdon, and Robert W. Smith, 125–159. Amsterdam: Harwood Academic Publishers, 2000.

Bunton, George W. "What Price Change?" *Curator* 2, no.1 (1959): 21–26.

Butler, Howard Russell. "An Ideal Astronomic Hall." *Natural History,* July-August 1926: 393–398.

Cantril, Hadley. *The Invasion from Mars: A Study in the Psychology of Panic.* New York: Harper and Row, 1966.

Capshew, James H. *Psychologists on the March: Science, Practice, and Professional Identity in America, 1929–1969.* New York: Cambridge University Press, 1999.

Capshew, James H., and Karen A. Rader. "Big Science: Price to the Present." *Osiris,* 2d ser., 7 (1992): 3–25.

"Central High School Dedicates Planetarium." *Sky and Telescope,* January 1947, 13.

Chamberlain, Joseph M. "A Philosophy for Planetarium Lectures." *Curator* 1, no. 1 (1958): 52–60.

———. "The Development of the Planetarium in the United States." *Annual Report of the Board of Regents of the Smithsonian Institution . . . for the Year Ended June 30, 1957,* 261–277. Washington, D.C.: Government Printing Office, 1958.

———. "Proceedings of the First U.S. Conference on Graduate Education in Astronomy." *Astronomical Journal* 68, no. 3 (April 1963): 215–219.

Chamberlain, Von Del. "Education Dawns in the Planetarium." *Great Lakes Planetarium Association (GLPA) Projector* 1 (1968): 52–53.

———. "The Armand N. Spitz Memorial Fund." *The Planetarian* 1, no. 2 (September 1972): 43.

"Charles F. Lewis, Led Conservancy." *New York Times,* 20 March 1971.

Chartrand, Mark R., III. "A Fifty Year Anniversary of a Two Thousand Year Dream." *The Planetarian* 2, no. 3 (September 1973): 95–101.

"Children's Museums Grow in Number and Quarters." *The Museum News,* 15 December 1947, 1.

"Christmas Star, The." *Museum News,* December 1964, 11–16.

"Cities of Germany Foster Enjoyment of Astronomy by Planetaria." *American City,* June 1928, 102–103.

Clarke, Charles G. "Griffith J. Griffith and His Park." *Griffith Observer,* May 1980, 2–10.

Clotfelter, Jim. "All 14 Astronauts at UNC Scheduled for Space Trips." *Durham Herald,* 28 March 1964.

Clowse, Barbara Barksdale. *Brainpower for the Cold War: The Sputnik Crisis and the National Defense Education Act of 1958.* Westport, Conn.: Greenwood Press, 1981.

Coleman, Laurence Vail. *The Museum in America: A Critical Study.* 3 vols. Washington, D.C.: American Association of Museums, 1939.

"Committee on Science and its Social Relations, The." *Popular Astronomy,* June-July 1938, 351–352.

Conn, Steven. *Museums and American Intellectual Life, 1876–1926.* Chicago: University of Chicago Press, 1998.

Cook, Anthony. "The Secret History of Griffith Observatory: A Belated Acknowledgment to the Patron Saint of the Telescope Makers." *Griffith Observer,* May 1994, 2–18.

———. "Six Decades with Six Astronomers on the Front Lawn of Griffith Observatory." *Griffith Observer,* November 1994, 2–11, 21.

Cook, Robert C., ed. *Who's Who in American Education: An Illustrated Biographical Directory of Eminent Living Educators of the United States and Canada,* 20th ed. Nashville, Tenn.: Who's Who in American Education, 1961–62.

Cooter, Roger, and Stephen Pumfrey. "Separate Spheres and Public Places: Reflections on the History of Science Popularization and Science in Popular Culture." *History of Science* 32 (1994): 237–267.

Corsini, Raymond J., ed. *Encyclopedia of Psychology.* Vol. 2. New York: John Wiley and Sons, 1984.

"Cosmic Rays Start Brilliant Display." *New York Times,* 1 May 1939.

Cremin, Lawrence A. *The Transformation of the School: Progressivism in American Education, 1876–1957.* New York: Random House, 1964.

Crooks, Pamela D. "San Diego's Adventure in Space." *Sky and Telescope,* February 1983, 127–129.

Crull, Harry E. "Astronomy in the Junior High School Curriculum." *School Science and Mathematics* 49 (1949): 371–373.

Curtis, Heber D. "The Importance of the Planetarium to Science, to Education, and to Recreation." *Journal of the Franklin Institute* 216 (July-December 1933): 794–800.

Danilov, Victor J. "IMAX/OMNIMAX: Fad or Trend?" *Museum News,* August 1987, 32–39.

Davis, Donald D. "New Skies for a New City." *Sky and Telescope,* April 1966, 196–198.

"Davison is Named to Head Museum." *New York Times,* 10 January 1933.

Day, Dwayne A. "Cover Stories and Hidden Agendas: Early American Space and National Security Policy." In *Reconsidering Sputnik: Forty Years Since the Soviet Satellite,* edited by Roger D. Launius, John M. Logsdon, and Robert W. Smith, 161–195. Amsterdam: Harwood Academic Publishers, 2000.

de Almeida, O. Dias. "Planetarium in Brazil." *Sky and Telescope,* June 1958, 400.

Deans, Paul. "Magic Under the Dome." *Sky and Telescope,* January 2004, 52–59.

DeGraffe, Jerome V. "The Historic Rochester Convention, October 24–27, 1968." *Great Lakes Planetarium Association (GLPA) Newsletter* 3, no. 4 (1968): 1–2.

DeVorkin, David H. "Organizing for Space Research: The V–2 Rocket Panel." *Historical Studies in the Physical and Biological Sciences* 18, pt. 1 (1987): 1–24.

———. *Science With a Vengeance: How the Military Created the US Space Sciences after World War II.* New York: Springer-Verlag, 1992.

Dickson, Paul. *Sputnik: The Shock of the Century.* New York: Walker Publishing, 2001.

"Directory of Institutions Offering Coursework in Planetarium Education, A." *The Planetarian* 1, no. 1 (June 1972): 24, 32.

Divine, Robert A. *Since 1945.* New York: John Wiley, 1975.

———. *The Sputnik Challenge.* New York: Oxford University Press, 1993.

Douglass, John A. "A Certain Future: Sputnik, American Higher Education, and the Survival of a Nation." In *Reconsidering Sputnik: Forty Years Since the Soviet Satellite,* edited by Roger D. Launius, John M. Logsdon, and Robert W. Smith, 327–363. Amsterdam: Harwood Academic Publishers, 2000.

Downs, Winfield Scott, ed. *Encyclopedia of American Biography,* new ser., vol. 13. New York: American Historical Company, 1941.

Draper, Arthur L. "How Man Controls the Universe." *The Sky,* December 1936, 22–23, 28.

Duschl, Richard A. *Restructuring Science Education: The Importance of Theories and Their Development.* New York: Teachers College Press, 1990.

Early, Roger Neil. *The Use of the Planetarium in the Teaching of Earth and Space Sciences.* Yorklyn, Del.: Spitz Laboratories, 1960.

Earth and Space Guide for Elementary Teachers. Harrisburg, Pa.: Commonwealth of Pennsylvania, Department of Public Instruction, 1961. Reprint, Washington, D.C.: National Aviation Education Council, November 1961.

Earth and Space Science: A Guide for Secondary Teachers. Harrisburg, Pa.: Commonwealth of Pennsylvania, Department of Public Instruction, 1963. Reprint, 1965.

Eberts, Mike. "The Little Known Early History of the Griffith Observatory." *Griffith Observer,* May 1995, 2–18.

———. *Griffith Park: A Centennial History.* Los Angeles: Historical Society of Southern California, 1996.

Edsall, Florence S. "Planetarium." *The Institute News,* May 1944, 4.

Emlen, Stephen T. "Migratory Orientation in the Indigo Bunting, *Passerina cyanea.*" Parts 1 and 2. *The Auk* 84 (1967): 309–342; 463–489.

———. "The Celestial Guidance System of a Migrating Bird." *Sky and Telescope,* July 1969, 4–6.

———. "Celestial Rotation: Its Importance in the Development of Migratory Orientation." *Science* 170 (11 December 1970): 1198–1201.

———. "The Stellar-Orientation System of a Migratory Bird." *Scientific American,* August 1975, 102–111.

"End of the World, The!" *Science Illustrated,* October 1946, 6–7.

Evans, Evan. Preface to *Earth and Space Guide for Elementary Teachers.* Harrisburg, Pa.: Commonwealth of Pennsylvania, Department of Public Instruction, 1961. Reprint, Washington, D.C.: National Aviation Education Council, November 1961.

"Fac-totum." *The Institute News,* December 1943, 3.

Faunce, Wayne M. "Interpreter of the Heavens." *Sky and Telescope,* September 1944, 5–7.

"Federal Money for Education." *American Education,* April 1965, 26–27.

Federer, Charles A., Jr. "The Editor's Note." *Sky and Telescope,* February 1944, 2.

———. "Armand N. Spitz—Planetarium Inventor." *Sky and Telescope,* June 1971, 354–355.

Fels, Samuel S. "Donation of the Planetarium to the Franklin Institute." *Journal of the Franklin Institute* 216 (July-December 1933): 790–792.

———. *This Changing World: As I See Its Trend and Purpose.* Boston: Houghton Mifflin, 1933.

Fernald, Cyrus F. "Thirty-Sixth Annual Meeting of the AAVSO Held at Harvard College Observatory, Cambridge, Mass., October 10–11, 1947." *Variable Comments* 4, no. 18 (n.d.): 83–86.

Field, J. V. "What is Scientific About a Scientific Instrument?" *Nuncius* 3, no. 2 (1988): 3–26.

"Fifth American Planetarium Opened to Public." *School Science and Mathematics* 37 (1937): 1139.

"Fight Woman's Ouster as Planetarium Chief." *Chicago Times,* 27 December 1944.

"'First Work of These Times . . . The': A Description and Analysis of the Elementary and Secondary Act of 1965." *American Education,* April 1965, 13–20.

Fisher, Clyde. "The New Projection Planetarium." *Natural History,* July-August 1926, 402–410.

———. "The Drama of the Skies." *Natural History,* March 1931, 147–154.

———. "The Hayden Planetarium." *Natural History,* May 1934, 247–258.

———. *Meteor Crater, Arizona.* [Guide leaflet ser., no. 92] New York: American Museum of Natural History, 1936.

———. *The Meteor Craters in Estonia. 'Footprints' left by visitors from outer space—Evidence of an astronomical collision that occurred perhaps two thousand years ago.* New York: American Museum of Natural History, 1936.

Fleischer, Robert. "Make Your Own Planetarium." *The Sky,* November 1936, 22–23.

Fowler, James A. "Qualifications for Planetarium Personnel." In *Planetariums and Their Use for Education,* vol. 2, edited by Richard H. Roche, 134–145. Cleveland, Ohio: Cleveland Museum of Natural History, n.d.

Fox, Philip. "The Adler Planetarium and Astronomical Museum of Chicago." Parts 1–4. *Popular Astronomy,* March 1932, 125–155; June-July 1932, 321–351; November 1932, 532–549; December 1932, 613–622.

———. "Auxiliary Apparatus for Planetarium Demonstrations." *Popular Astronomy,* April 1932, 188–194.

———. *Adler Planetarium and Astronomical Museum: An Account of the Optical Planetarium and a Brief Guide to the Museum.* Chicago: Lakeside Press, 1933.

Fraknoi, Andrew, and Donat Wentzel. "Astronomy Education and the American Astronomical Society." In *The American Astronomical Society's First Century,* edited by David DeVorkin, 194–212. Washington, D.C.: American Astronomical Society, 1999.

Fraser, Calvin. "The Zeiss Planetarium: A Fascinating and Instructive Theater of the Stars." *The Mentor,* January 1928, 58–59.

Freitag, Ruth S. *The Star of Bethlehem: A List of References.* Washington, D.C.: Library of Congress, 1979.

Fuchs, Franz. "Die Planetarien des Deutschen Museums." *Berliner Tageblatt,* 7 November 1923.

———. "Der Aufbau der Astronomie im Deutschen Museum (1905–1925)." *Deutsches Museum Abhandlungen und Berichte* 23, no. 1 (1955): 1–68.

Fyfe, George Buchanan. "Time and Space Dramatized at the World's Fair." *The Sky,* May 1939, 13–14, 25.

Geiger, Roger L. *To Advance Knowledge: The Growth of American Research Universities, 1900–1940.* New York: Oxford University Press, 1986.

Gieryn, Thomas F. "Boundaries of Science." In *Handbook of Science and Technology Studies,* edited by Sheila Jasanoff, Gerald E. Markle, James C. Petersen, and Trevor Pinch, 393–443. Thousand Oaks, Calif.: Sage Publications, 1995.

———. *Cultural Boundaries of Science: Credibility on the Line.* Chicago: University of Chicago Press, 1999.

Gilbert, Rose Bennett. "Space Age Education." *Museum News,* November 1965, 25–30.

Gilchrist, Larry. "The Case for a North American Planetarium Association." *The Planetarium Journal* 2 (May 1970): 6.

———. "The Case for a North American Planetarium Publication." *The Planetarium Journal* 2 (May 1970): 7–9.

Goldenson, Robert M. *Encyclopedia of Human Behavior: Psychology, Psychiatry, and Mental Health.* Vol. 1. Garden City, N.Y.: Doubleday, 1970.

Green, Constance McLaughlin, and Milton Lomask. *Vanguard: A History.* Washington, D.C.: Smithsonian Institution Press, 1971.

Griffith, Griffith J. *Parks, Boulevards and Playgrounds.* Los Angeles: Prison Reform League, 1910.

Guide to the Smithsonian Archives. Washington, D.C.: Smithsonian Institution Press, 1996.

Haarstick, Maxine B. "How to Succeed in the Planetarium." *Museum News,* December 1964, 17–23.

Hagar, Charles F. "Through the Eyes of Zeiss." *Griffith Observer,* June 1961, 62–70.

———. "Second International Planetarium Executives Conference." *Sky and Telescope,* October 1966, 205.

———. *Planetarium: Window to the Universe.* Oberkochen: Carl Zeiss, 1980.

Haislip, Bryan. "Astronauts Say Training at Planetarium Valuable." *Greensboro Record,* 30 January 1963.

Hale, George Ellery. "Solar Research for Amateurs." In *Amateur Telescope Making, Book One,* 4th ed., edited by Albert G. Ingalls, 180–214. New York: Scientific American, 1967.

Hall, Donald S. "Astronaut Training." In *What You Should Know About Astronaut Training at Morehead Planetarium.* Chapel Hill, N.C.: Morehead Planetarium, 1966.

———. "What's 'Up' in Space." *Review of Popular Astronomy,* June 1966, 6–8.

Hall, Kay. "Woman Plays Lead in Celestial Show." *New York Times,* Sunday, 16 January 1938, sec. 6.

Hankins, Thomas L., and Robert J. Silverman. *Instruments and the Imagination.* Princeton, N.J.: Princeton University Press, 1995.

Hare, John. "(In-House) Laser Shows . . . A Long Term Proposition." *The Planetarian* 13, no. 1 (first quarter 1984): 8–9.

Harris, Sydney J. "Here is Chicago." *Chicago Daily News,* 24 March 1944.

"Harrisonburg Observatory Expands Facilities." *Sky and Telescope,* July 1956, 396–397.

Hastedt, Glenn P. "Epilogue: Sputnik as Technological Surprise." In *Reconsidering Sputnik: Forty Years Since the Soviet Satellite,* edited by Roger D. Launius, John M. Logsdon, and Robert W. Smith, 401–423. Amsterdam: Harwood Academic Publishers, 2000.

Hatt, Robert T. "Symposium on Planetaria and Their Use for Education." In *Planetaria and Their Use for Education,* edited by Miriam Jagger, 7. Bloomfield Hills, Mich.: Cranbrook Institute of Science, 1959.

"Hayden Planetarium Shows Four Ways in Which the World May End." *Life,* 1 November 1937, 54–58.

Hayes, E. Nelson. *Trackers of the Skies.* Washington, D.C.: Smithsonian Institution, 1967.

Heath, Thomas L. *Greek Astronomy.* New York: Dover Publications, 1991.

Heatherly, Charles. "Spotlight: Morehead Planetarium." In *North Carolina Report.* N.p.: Department of Natural and Economic Resources, n.d.

"Henry Buhl, Jr." In *The Buhl Foundation: A Report by the Director upon its Work to June 30, 1942,* by Charles F. Lewis, 3–7. Pittsburgh: Farmers Bank Building, n.d.

Henson, Pamela M. "'Through Books to Nature': Anna Botsford Comstock and the Nature Study Movement." In *Natural Eloquence: Women Reinscribe Science,* edited by Barbara T. Gates and Ann B. Shteir, 116–143. Madison: University of Wisconsin Press, 1997.

"High School Planetarium, A." *Popular Astronomy,* October 1946, 438–439.

Hilgartner, Stephen. "The Dominant View of Popularization: Conceptual Problems, Political Uses." *Social Studies of Science* 20 (1990): 519–539.

Hill, Joan. "UNC 'Alumni' Reach Moon." *Greensboro Daily News,* 18 July 1976.

Hindle, Brooke. "The Quaker Background and Science in Colonial Philadelphia." In *Early American Science,* 173–180. New York: Science History Publications, 1976.

Hoffman, Paul. "Marsstruck." *Discover,* May 1997, 8.

Hogg, Victor H. "The Abrams Planetarium in Michigan." *Sky and Telescope,* June 1964, 336–338.

"Home-Made Planetarium Reveals Sky-Glories to Springfield, Mass." *New York Times,* Sunday, 7 November 1937, sec. 2.

"How Will the World End?" *The Institute News,* September 1938, 1, 4.

Howard, Henry. "East Carolina Professor Helped Train Astronauts." *High Point Enterprise,* 11 May 1961.

Howe, Ray J. "The Planetarium Program and the School Curriculum." In *Planetaria and Their Use for Education,* edited by Miriam Jagger, 101–104. Bloomfield Hills, Mich.: Cranbrook Institute of Science, 1959.

Howes, Durward, ed. *American Women: The Standard Biographical Dictionary of Notable Women*. Vol. 3. Los Angeles: American Publications, 1939.

Hoyt, William Graves. *Coon Mountain Controversies: Meteor Crater and the Development of Impact Theory.* Tucson: University of Arizona Press, 1987.

Hughes, David. *The Star of Bethlehem: An Astronomer's Confirmation.* New York: Walker, 1979.

Ingalls, Albert G. "Canned Astronomy: What the New Planetariums for Chicago and Philadelphia Will Be Like." *Scientific American,* September 1929, 200–204.

Ingalls, Ruth. "The Hayden Planetarium." *The Sky,* July 1939, 24.

Injejikian, Susan Arminé. "The Universe in Your Mailbox: Observing the *Observer*'s Anniversary." *Griffith Observer,* February 1987, 2–7.

Jagger, Miriam. Preface to *Planetaria and Their Use for Education.* Bloomfield Hills, Mich.: Cranbrook Institute of Science, 1959.

Jettner, Frank C. "[T]houghts from FCJ." *The Planetarian* 1, no. 1 (June 1972): 2.

Jettner, Frank C., and John J. Soroka. "The Planetarium in Modern Science Education." In *Education in and History of Modern Astronomy,* edited by Richard Berendzen. *Annals of the New York Academy of Sciences* 198 (25 August 1972): 178–191.

Johnson, Arthur W., Jr. "Recent Advances in 35-mm Hemispheric Cinematography for the Planetarium." *The Planetarian* 12, no. 2 (second quarter 1983): 5–6.

Johnson, Gaylord. "How Will the World End?" *Popular Science Monthly,* October 1936, 52–53, 123.

Johnston, Don H. "The Hanna Star Dome." *Popular Astronomy,* January 1940, 53–54.

Jones, Bessie Zaban. *Lighthouse of the Skies: The Smithsonian Astrophysical Observatory, Background and History, 1846–1955.* Washington, D.C.: Smithsonian Institution, 1965.

Kaempffert, Waldemar. "Now America Will Have a Planetarium." *New York Times Magazine,* 24 June 1928.

———. "New Planetarium is the First in America." *New York Times,* Sunday, 11 May 1930, sec. 10.

Kevles, Daniel J. *The Physicists: The History of a Scientific Community in Modern America.* Cambridge, Mass.: Harvard University Press, 1987.

Kidger, Mark. *The Star of Bethlehem: An Astronomer's View.* Princeton, N.J.: Princeton University Press, 1999.

Killian, James R., Jr. *Sputnik, Scientists, and Eisenhower: A Memoir of the First Special Assistant to the President for Science and Technology.* Cambridge, Mass.: MIT Press, 1977.

King, Henry C. *Geared to the Stars: The Evolution of Planetariums, Orreries, and Astronomical Clocks.* Toronto: University of Toronto Press, 1978.

Knight, Charles R. [reproduction of triptych] *The Sky,* February 1938, 18–19, 32.

Kogan, Herman. *A Continuing Marvel: The Story of The Museum of Science and Industry.* Garden City, N.Y.: Doubleday, 1973.

Kohler, Robert E. *Partners in Science: Foundations and Natural Scientists, 1900–1945.* Chicago: University of Chicago Press, 1991.

Kohlstedt, Sally Gregory. *The Formation of the American Scientific Community: The American Association for the Advancement of Science, 1848–1860.* Urbana: University of Illinois Press, 1976.

Korey, Ruth Anne. "Making a Planetarium: A Unit on the Sky." *The Grade Teacher* 70 (December 1952): 43, 91–92.

———. "Planetariums in the United States." *Sky and Telescope,* September 1963, 147–148.

"KORKOSZ Planetarium." [advertisement] *Sky and Telescope,* December 1947, 55.

Kosoloski, John E. "Establishing a Space Science Curriculum." In *Planetariums and Their Use for Education,* vol. 2, edited by Richard H. Roche, 8–17. Cleveland, Ohio: Cleveland Museum of Natural History, n.d.

———. "The Pennsylvania Earth and Space Science Program." *Geotimes* 7, no. 3 (October 1962): 15–17.

Kuhn, Thomas S. *The Structure of Scientific Revolutions,* 2d ed. Chicago: University of Chicago Press, 1970.

LaFollette, Marcel C. "Eyes on the Stars: Images of Women Scientists in Popular Magazines." *Science, Technology, and Human Values* 13 (1988): 262–275.

Langfeld, Paul Goerz. "Astronauts' Star Recognition Training Intensifies." *Missiles and Rockets,* 19 November 1962, 28–30.

Lankford, John. *American Astronomy: Community, Careers, and Power, 1859–1940.* Chicago: University of Chicago Press, 1997.

———, ed. *History of Astronomy: An Encyclopedia.* New York: Garland Publishing, 1997.

Lattin, Harriet Pratt. *Star Performance.* Philadelphia: Whitmore Publishing, 1969.

Laurence, William L. "'Tour of Sky' Opens Planetarium; 800 Get a New Vision of Universe." *New York Times,* 3 October 1935.

Lee, Oliver Justin. "Philip Fox, 1878–1944." *Popular Astronomy,* October 1944, 365–370.

Letsch, Heinz. *Captured Stars.* Translated by Harry Spitzbardt. Jena: Gustav Fischer Verlag, 1959.

Lewis, Charles F. "Presentation Address." In *Dedication of the Buhl Planetarium and Institute of Popular Science,* 14–21. N.p., n.d.

———. *The Buhl Foundation: A Report by the Director upon its Work to June 30, 1942.* Pittsburgh: Farmers Bank Building, n.d.

Lindberg, David C., and Ronald L. Numbers, eds. *God and Nature: Historical Essays on the Encounter between Christianity and Science.* Berkeley: University of California Press, 1986.

Lockwood, Marian. "The Hayden Planetarium." *Natural History,* October 1935, 188–190.

———. Review of *The Pinpoint Planetarium* by Armand N. Spitz. *The Sky,* June 1940, 11.

———. "Clyde Fisher: Naturalist and Teacher." *Sky and Telescope,* March 1949, 111–113.

Lora, Ronald. "Education: Schools as Crucible in Cold War America." In *Reshaping America: Society and Institutions, 1945–1960,* edited by Robert H. Bremner and Gary W. Reichard, 223–260. Columbus: Ohio State University Press, 1982.

Lovi, George. "Vienna Planetarium Conference." *Sky and Telescope,* October 1969, 236–239.

Ludhe [sic], Ernest T., et al. "Trail Blazing with Spitz Planetariums." *Sky and Telescope,* January 1949, 66–69.

Lule, Jack. "Roots of the Space Race: Sputnik and the Language of U.S. News in 1957." *Journalism Quarterly* 68, nos. 1/2 (spring/summer 1991): 76–86.

Luyten, W. J. "The New Projection Planetarium." *Natural History,* July-August 1927, 383–390.

Maddry, Lawrence. "Astronauts Praise Course at Chapel Hill Planetarium." *Raleigh Times,* 27 March 1964.

Madison, Harold L., Jr. "The Hanna Star Dome." *The Sky,* June 1938, 8–9.

"Man-Made Moon." *Science Illustrated,* August 1947, 8–9.

Marché, Jordan D., II. "A Bibliography of the Spitz Planetarium Director's Handbook." *The Planetarian* 18, no. 3 (September 1989): 36–37.

———. "Sputnik, Planetaria, and the Rebirth of U.S. Astronomy Education." *The Planetarian* 30, no. 1 (March 2001): 4–9.

———. "Mental Discipline, Curricular Reform, and the Decline of U.S. Astronomy Education, 1893–1920." *Astronomy Education Review* 1 (2002): 58–75.

———. "Planetarium." In *Encyclopedia of Literature and Science,* edited by Pamela Gossin, 342. Westport, Conn. and London: Greenwood Press, 2002.

———. "A Bibliography of Armand N. Spitz (1940–1972)." *The Planetarian* 31, no. 1 (March 2002): 10.

———. "Gender and the American Planetarium Community." *The Planetarian* 31, no. 2 (June 2002): 4–7, 36.

Mark, Kathleen. *Meteorite Craters.* Tucson: University of Arizona Press, 1987.

Marshall, Roy K. "Star of Bethlehem?" *Sky and Telescope,* December 1943, 15.

———. "Franklin Lecturer Perfects Poor Man's Planetarium." *Philadelphia Evening Bulletin,* 25 November 1947.

———. "The Morehead Planetarium." *Sky and Telescope,* August 1949, 243–245.

———. "University of Nevada's Atmospherium-Planetarium." *Sky and Telescope,* December 1963, 318–321.

Martin, Ernest L. *The Birth of Christ Recalculated.* Pasadena, Calif.: Foundation for Biblical Research Publications, 1978.

Matthews, William H., III. *Sources of Earth Science Information.* Englewood Cliffs, N.J.: Prentice-Hall, 1964.

"Maude Bennot Will be Ousted at Planetarium." *Chicago Tribune,* 27 December 1944.

"Max Adler, Donor of Planetarium, 86." *New York Times,* 6 November 1952.

Mayr, Otto. *The Deutsches Museum: German Museum of Masterworks of Science and Technology, Munich.* London: Scala Books, 1990.

McCurdy, Howard E. *Space and the American Imagination.* Washington, D.C.: Smithsonian Institution Press, 1997.

McDougall, Walter A. . . . *The Heavens and the Earth: A Political History of the Space Age.* New York: Basic Books, 1985.

———. "Introduction: Was Sputnik Really a Saltation?" In *Reconsidering Sputnik: Forty Years Since the Soviet Satellite,* edited by Roger D. Launius, John M. Logsdon, and Robert W. Smith, xv–xx. Amsterdam: Harwood Academic Publishers, 2000.

McLennan, Ian C. "Canada's First Public Planetarium." *Sky and Telescope,* August 1962, 86–88.

———. "A Major Planetarium in Rochester." *Sky and Telescope,* April 1968, 208–210.

Menke, David H. "Phillip [sic] Fox and the Adler Planetarium." *The Planetarian* 16, no. 1 (January 1987): 46–48.

———. "Clyde Fisher and the Hayden Planetarium." *The Planetarian* 16, no. 2 (April 1987): 54–58.

———. "James Stokley and the Fels Planetarium." *The Planetarian* 16, no. 3 (July 1987): 39–43.

———. "Dinsmore Alter and the Griffith Observatory." *The Planetarian* 16, no. 4 (October 1987): 68–78.

———. "Arthur L. Draper and the Buhl Planetarium." *The Planetarian* 17, no. 1 (March 1988): 58–61, 73.

———. "George Bunton and the Morrison Planetarium." *The Planetarian* 17, no. 2 (June 1988): 50–55.

———. "Tony Jenzano and the Morehead Planetarium." *The Planetarian* 17, no. 3 (September 1988): 51–55.

Meyer, William F. "Dedication of the Griffith Observatory." *Publications of the Astronomical Society of the Pacific* 47 (1935): 170–171.

Miller, Grace Maddock. "Planetarium." *The Sky,* May 1939, 10.

Miller, Oskar von. "The German Museum in Munich." *Proceedings of the American Society of Civil Engineers* 52 (January 1926): 13–17.

Miller, Robert Cunningham. "Galaxy by the Golden Gate." *Pacific Discovery,* November-December 1952, 11–17.

Molnar, Michael R. *The Star of Bethlehem: The Legacy of the Magi.* New Brunswick, N.J.: Rutgers University Press, 1999.

Montgomery, Scott L. *Minds for the Making: The Role of Science in American Education, 1750–1990.* New York: Guilford, 1994.

"Moon Trip Continued." *The Institute News,* June 1939, 1–2.

"Morehead Director Resigns." *Sky and Telescope,* May 1951, 161.

Morison, G. H. "Die Geheimnisse der Sterne." *Westermanns Monatshefte* 137 (February 1925): 576–580.

———. "Heavens Built of Concrete." *Scientific American,* March 1925, 170–171.

Morrell, Jack, and Arnold Thackray. *Gentlemen of Science: Early Years of the British Association for the Advancement of Science.* Oxford: Clarendon Press, 1981.

Mosley, John. *The Hugo Ballin Murals of the Griffith Observatory.* Los Angeles: Griffith Observatory, n.d.

Mosley, John, and Ernest L. Martin, "The Star of Bethlehem Reconsidered: An Historical Approach." *The Planetarian* 9, no. 2 (summer 1980): 6–9.

"Mr. Spitz Brings Science Down to Earth." *Science Illustrated,* September 1947, 26–29.

Myrover, George. "A Little of This . . . A Little of That." *Asheville Citizen,* 31 January 1963.

National Cyclopedia of American Biography, The. Vol. 56. Clifton, N.J.: James T. White, 1975.

"National MOONWATCH Committee Meets at Smithsonian Observatory Headquarters." *Sky and Telescope,* April 1957, 277.

National Science Foundation. *American Science Manpower 1960: A Report of the National Register of Scientific and Technical Manpower.* Washington, D.C.: Government Printing Office, 1962.

Nerdahl, Rodney M., and Gary Tomlinson. "In Memoriam . . . Maxine B. Haarstick, 1922–1985." *The Planetarian* 14, no. 3 (third quarter 1985): 15–16.

Neufeld, Michael J. "Orbiter, Overflight, and the First Satellite: New Light on the Vanguard Decision." In *Reconsidering Sputnik: Forty Years Since the Soviet Satellite,* edited by Roger D. Launius, John M. Logsdon, and Robert W. Smith, 231–257. Amsterdam: Harwood Academic Publishers, 2000.

"New American Planetarium for Springfield, Massachusetts." *Popular Astronomy,* November 1937, 517–519.

"New Director Named for Planetarium." *Chicago Sun,* 28 December 1944.

"News Notes." *Great Lakes Planetarium Association (GLPA) Newsletter* 3, no. 4 (1968): 2.

Newton, Earl W., and Phillip D. Stern. "New Planetarium Opens at Bridgeport." *Sky and Telescope,* March 1962, 132–134.

Nicholson, Thomas D. "The Planetarium Lecture." *Curator* 2, no. 3 (1959): 269–274.

Noble, Margaret. "A Survey of Planetarium Programs." In *Planetariums and Their Use for Education,* vol. 2, edited by Richard H. Roche, 18–27. Cleveland, Ohio: Cleveland Museum of Natural History, n.d.

"North Canton Planetarium in Ohio, The." *Sky and Telescope,* March 1956, 214.

Norton, O. Richard. "Nevada's Atmospherium-Planetarium . . . The Sky's the Limit!" *Review of Popular Astronomy,* July 1964, 4–7.

———. "New Advances in Planetaria." *Astronomical Society of the Pacific Leaflet* no. 444 (June 1966), 8 pp.

———. *The Planetarium and Atmospherium: An Indoor Universe.* Healdsburg, Calif.: Naturegraph Publishers, 1968.

———. "Techniques of Extreme Wide-Angle Motion-Picture Photography and Projection." *Journal of the Society of Motion Picture and Television Engineers* 78, no. 2 (February 1969): 81–85.

———. "A Major Planetarium for Tucson, Arizona." *Sky and Telescope,* March 1975, 143–146.

"Noted Scientist to Speak at Telescope Dedication." *Pittsburgh Press,* 19 November 1941.

Notes and Working Papers Concerning the Administration of Programs Authorized Under Title III of National Defense Education Act. Prepared for the Subcommittee on Education of the Committee on Labor and Public Welfare. U.S. Senate, 90th Congress, 1st Session. Washington, D.C.: Government Printing Office, 1967.

"Notes on a Convention in Miami." *Sky and Telescope,* September 1956, 485–488.

"NSTA Regions." [map] *The Science Teacher,* October 1964, 5.

Nyhart, Lynn K. *Biology Takes Form: Animal Morphology and the German Universities, 1800–1900.* Chicago: University of Chicago Press, 1995.

Odland, Lisa. "At the Planetarium." *Popular Astronomy,* January 1938, 14.
"$150,000 by Hayden for Planetarium." *New York Times,* 5 January 1934.
Orosz, Joel J. *Curators and Culture: The Museum Movement in America, 1740–1870.* Tuscaloosa: University of Alabama Press, 1990.
Osborn, Henry Fairfield. *The American Museum of Natural History: Its Origin, its History, the Growth of its Departments to December 31, 1909.* 2d ed. New York: Irving Press, 1911.
———. *Forty-Eighth Annual Report of the American Museum of Natural History for the Year 1916.* New York: N.p., 1 May 1917.
Osgood, Kenneth A. "Before Sputnik: National Security and the Formation of U.S. Outer Space Policy." In *Reconsidering Sputnik: Forty Years Since the Soviet Satellite,* edited by Roger D. Launius, John M. Logsdon, and Robert W. Smith, 197–229. Amsterdam: Harwood Academic Publishers, 2000.
Page, Thornton. "Motion Pictures as an Aid in Teaching Astronomy." *Popular Astronomy,* March 1951, 117–128.
"Parks to Oust Maude Bennot at Planetarium." *Chicago Sun,* 27 December 1944.
Patterson, John. "Boston's Planetarium Opens." *Sky and Telescope,* November 1958, 12–15.
Pegram, Howard. "Choosing and Training Planetarium Personnel." In *Planetariums and Their Use for Education,* vol. 2, edited by Richard H. Roche, 165–171. Cleveland, Ohio: Cleveland Museum of Natural History, n.d.
Pendray, G. Edward. "To the Moon via Rocket?" *The Sky,* November 1936, 6–7, 21.
"Personals." *The Museum News,* 1 November 1956, 3.
Phalen, Dale. *Samuel Fels of Philadelphia.* Philadelphia: Samuel S. Fels Fund, 1969.
"Philadelphian Named Planetarium Director." *Chicago Daily News,* 28 December 1944.
"Philadelphian Picked as Head of Planetarium." *Chicago Tribune,* 28 December 1944.
"Planetarian." *Time,* 24 April 1939, 61–62.
"Planetarium Associations." *Sky and Telescope,* May 1969, 299.
"Planetarium Can Enrich Curriculum, A." *Pennsylvania Education,* November-December 1969, 4–5.
"Planetarium Conference." *Sky and Telescope,* March 1965, 140.
"Planetarium Conference Announced for May 30." *The Museum News,* 1 April 1951, 1.
"Planetarium Here Assured by Loan." *New York Times,* 27 June 1933.
"Planetarium, Junior Version." *Business Week,* 20 February 1954, 112–114.
"Planetarium Operators' Symposium at Cranbrook Institute." *Sky and Telescope,* December 1958, 80–81.
"Planetarium Representatives Attend International Conference." *Sky and Telescope,* July 1959, 503.
"Planetarium to be Erected Here by Rosicrucians." *San Jose Mercury Herald,* 19 February 1936.
"Planetarium to Reproduce Ancient Skies." *Science—Supplement* 78 (22 December 1933): 12.
"Planetarium Trade, The." *Sky and Telescope,* December 1968, 359, 375.
"'Playing' the Planetarium." *The Literary Digest,* 12 October 1935, 18.
"Presentation in absentia of Dr. Walther Bauersfeld." *Journal of the Franklin Institute* 216 (July-December 1933): 792–794.
Price, Derek J. de Solla. "An Ancient Greek Computer." *Scientific American,* June 1959, 60–67.

———. *Gears from the Greeks: The Antikythera Mechanism, a Calendrical Computer from ca. 80 B.C.* Philadelphia: American Philosophical Society, 1974.

Prison Reform League. *Crime and Criminals.* Los Angeles: Prison Reform League, 1910.

Rainger, Ronald. *An Agenda for Antiquity: Henry Fairfield Osborn and Vertebrate Paleontology at the American Museum of Natural History, 1890–1935.* Tuscaloosa: University of Alabama Press, 1991.

Ravitch, Diane. *The Troubled Crusade: American Education, 1945–1980.* New York: Basic Books, 1983.

"Received by Franz Fieseler, Esq., on Behalf of Dr. Walther Bauersfeld." *Journal of the Franklin Institute* 216 (July-December 1933): 794.

Recommended Design Criteria for School Facilities. Harrisburg, Pa.: Pennsylvania Department of Education, March 1977.

Reed, George. "Is the Planetarium a More Effective Teaching Device than the Combination of the Classroom Chalkboard and Celestial Globe?" *School Science and Mathematics* 70 (1970): 487–492.

Riesman, David. "The Zeiss Planetarium." *General Magazine and Historical Chronicle,* January 1929, 236–241.

"Rite Dedicates Rosicrucians' Planetarium." *San Jose Mercury Herald,* 14 July 1936.

Roche, Richard H. Preface to *Planetariums and Their Use for Education,* vol. 2. Cleveland, Ohio: Cleveland Museum of Natural History, n.d.

Rodger, David A. "Current Planetarium Activity in Canada." *Sky and Telescope,* July 1967, 13–14.

Rosemergy, John C. "The Use of the Planetarium in Secondary Schools." In *Planetaria and Their Use for Education,* edited by Miriam Jagger, 55–60. Bloomfield Hills, Mich.: Cranbrook Institute of Science, 1959.

———. "The Planetarium—Stars and Space in School." *School Executive* 79 (October 1959): 70–71.

Rossiter, Margaret W. *Women Scientists in America: Struggles and Strategies to 1940.* Baltimore: Johns Hopkins University Press, 1982.

———. *Women Scientists in America: Before Affirmative Action, 1940–1972.* Baltimore: Johns Hopkins University Press, 1995.

Rothschild, Joyce. "The NDEA Decade." *American Education,* September 1968, 4–11.

Rudolph, John L. *Scientists in the Classroom: The Cold War Reconstruction of American Science Education.* New York: Palgrave, 2002.

Rumrill, H. B. "The Revival of Popular Interest in Astronomy." *Popular Astronomy,* June-July 1937, 310–317.

Russell, Henry Norris. "A Catastrophe That Did Not Occur." *Scientific American,* May 1936, 248–249.

Rydell, Robert W. *World of Fairs: The Century of Progress Expositions.* Chicago: University of Chicago Press, 1993.

Salanave, Leon E. "Russell W. Porter [1871–1949]." *Pacific Discovery,* November-December 1952, 35.

———. "San Francisco Planetarium." *Sky and Telescope,* December 1952, 31–34.

Sanderson, Richard. "Frank Korkosz and the First American Planetarium." *Griffith Observer,* May 1986, 14–17.

Sauer, E. G. Franz. "Zugorientierung einer Mönchsgrasmücke (*Sylvia a. atricapilla,* L.) unter künstlichem Sternhimmel." *Die Naturwissenschaften* 43 (1956): 231–232.

———. "Die Sternorientierung nächtlich ziehender Grasmücken (*Sylvia atricapilla, borin* und *curruca*)." *Zeitschrift für Tierpsychologie* 14 (1957): 29–70.

———. "Astronavigatorische Orientierung einer unter künstlichem Sternhimmel verfrachteten Klappergrasmücke, *Sylvia c. curruca* (L.)." *Die Naturwissenschaften* 44 (1957): 71.

———. "Celestial Navigation by Birds." *Scientific American,* August 1958, 42–47.

———. "Nächtliche Zugorientierung europäischer Vögel in Südwestafrika." *Die Vogelwarte* 20 (1959): 4–31.

———. "Further Studies on the Stellar Orientation of Nocturnally Migrating Birds." *Psychologische Forschung* 26 (1961): 224–244.

Sauer, E. G. Franz, and Eleonore M. Sauer. "Zur Frage der nächtlichen Zugorientierung von Grasmücken." *Revue Suisse de Zoologie* 62 (1955): 250–259.

———. "Star Navigation of Nocturnally Migrating Birds: The 1958 Planetarium Experiments." *Cold Spring Harbor Symposia on Quantitative Biology* 25 (1960): 463–473.

Scanlon, Leo J. "Amateur Astronomy in Pittsburgh." *Scientific American,* July 1931, 30–32.

Schilling, Govert. "Eise Eisinga's Novel Planetarium." *Sky and Telescope,* February 1994, 28–30.

Schnurmacher, Emile C. "Explorers of Space." *Popular Mechanics,* November 1939, 680–683, 122A.

Schultz, William, Jr. "A 45-Cent Planetarium Projector." In *Planetaria and Their Use for Education,* edited by Miriam Jagger, 167–169. Bloomfield Hills, Mich.: Cranbrook Institute of Science, 1959.

Science Service. "The 'Planetarium'." *Science—Supplement* 60 (5 September 1924): xii.

———. "The Planetarium." *Science—Supplement* 67 (22 June 1928): x.

Scully, Cornelius D. "Acceptance Address." In *Dedication of the Buhl Planetarium and Institute of Popular Science,* 22–28. N.p., n.d.

Servos, John W. *Physical Chemistry from Ostwald to Pauling: The Making of a Science in America.* Princeton, N.J.: Princeton University Press, 1990.

Shapin, Steven. "Science and the Public." In *Companion to the History of Modern Science,* edited by R. C. Olby, G. N. Cantor, J. R. R. Christie, and M. J. S. Hodge, 990–1007. London/New York: Routledge, 1990.

Shepard, Leslie, ed. *Encyclopedia of Occultism and Parapsychology.* 3d ed., 2 vols. Detroit: Gale Research, 1991.

Shinn, Terry, and Richard Whitley, eds. *Expository Science: Forms and Functions of Popularisation.* Sociology of the Sciences Yearbook, vol. 9. Dordrecht, Netherlands: D. Reidel, 1985.

Shipp, Horace. "In the Planetarium." In *Palimpsest.* London: Sampson Low, Marston, 1930.

Sinclair, Bruce. *Philadelphia's Philosopher Mechanics: A History of the Franklin Institute, 1824–1865.* Baltimore: Johns Hopkins University Press, 1974.

Smith, Robert W. *The Expanding Universe: Astronomy's 'Great Debate' 1900–1931.* Cambridge: Cambridge University Press, 1982.

Smithsonian Astrophysical Observatory. *Bulletin for Visual Observers of Satellites.* Nos. 1–9. July 1956–July 1958.

Socha, Mabel V. "Address at the Formal Opening of the Griffith Observatory and

Planetarium, May 14, 1935." *Publications of the Astronomical Society of the Pacific* 47 (1935): 157–159.

Sorensen, Lorin. *Sears, Roebuck and Co. 100th Anniversary, 1886–1986.* St. Helena, Calif.: Silverado Publishing, 1985.

Sorensen, W. Conner. *Brethren of the Net: American Entomology, 1840–1880.* Tuscaloosa: University of Alabama Press, 1995.

Space Science Classroom, A. Yorklyn, Del.: Spitz Laboratories, n.d.

Spencer, Steven M. "Meet the Donor of Planetarium." *Philadelphia Evening Bulletin,* 17 November 1933.

———. "The Stars are His Playthings." *The Saturday Evening Post,* 24 April 1954, 42–43, 97–98, 100, 102–103.

Sperling, Norman. "A Planetarium Educators Conference." *Sky and Telescope,* January 1971, 7–9.

———, ed. *A Catalog of North American Planetariums (CATNAP).* Special Report #1. N.p.: International Society of Planetarium Educators, June 1971.

Spitz, Armand N. "Who's Who in The Franklin Institute: James Stokley." *The Institute News,* December 1936, 4.

———. "Who's Who in The Franklin Institute: Wagner Schlesinger." *The Institute News,* September 1937, 4.

———. *The Pinpoint Planetarium.* New York: Henry Holt, 1940.

———. "Profile of Prodigy." *The Institute News,* August 1943, 3.

———. "Meteorology in the Franklin Institute." *Journal of the Franklin Institute* 237 (January-June 1944): 271–287, 331–357.

———. "Puerto Rican Visit." *The Institute News,* August 1945, 2.

———. "Kite, Apple, and Tea-Kettle, 1946." *The General Magazine and Historical Chronicle* 48 (spring 1946): 174–184.

———. "Hall of Earth, Air, and Sky of the Polytechnic Institute of Puerto Rico." *Sky and Telescope,* April 1947, 3–5.

———. "MOONWATCH Preparations Swing into High Gear." *Bulletin for Visual Observers of Satellites,* no. 3 (November 1956): 1.

———. "Planetarium: An Analysis of Opportunities and Obligations." *Griffith Observer,* June 1959, 78–82, 87.

———. "Pre-School Children in the Planetarium." In *Planetariums and Their Use for Education,* vol. 2, edited by Richard H. Roche, 180–185. Cleveland, Ohio: Cleveland Museum of Natural History, n.d.

———. "The Philosopher's Point of View." *Museum News,* December 1964, 22.

Spitz, Mrs. Armand. "First Armand Spitz Lecture." *Great Lakes Planetarium Association (GLPA) Projector* 1 (1968): 6–8.

Spitz Planetariums: A Fundamental of Modern Education. Yorklyn, Del.: Spitz Laboratories, n.d.

Spoehr, Jack. "Herbert N. Williams: A Remembrance." *The Planetarian* 14, no. 1 (first quarter 1985): 16.

Spring, Joel. *The Sorting Machine: National Educational Policy Since 1945.* New York: David McKay, 1976.

"Spring Stars a la Disney." *The Institute News,* April 1941, 1.

Stapleton, Sidney. "NASA and NC." *Wachovia,* July-August 1971, 11–16.

"Star Chamber." *Time,* 19 May 1930, 72.

"Star-Level Parley Has No Geneva Woe." *New York Times,* 14 May 1959.
"Starry-eyed." *American Magazine,* January 1946, 138.
Stebbins, Joel. "Philip Fox." *Science* 100 (1 September 1944): 184–186.
Stein, Donna M. *Thomas Wilfred: Lumia, A Retrospective Exhibition.* Washington, D.C.: Corcoran Gallery of Art, 1971.
Stein, Elizabeth A. "The Planetarium." *The Sky,* December 1936, 28.
Steinbaum, Martin J. "The Chosen." *The Science Teacher* 35 (January 1968): 49–50.
Stewart, Charles D. "A Beam from Arcturus." *Atlantic Monthly* 151 (1933): 331–336.
Stokley, James. "Where the Stars Stand Still." *Publications of the Astronomical Society of the Pacific* 40 (1928): 24–30.
———. "Benjamin Franklin Memorial and the Franklin Institute: Opening of the Fels Planetarium." *Popular Astronomy,* December 1933, 535–538.
———. "The Fels Planetarium of the Franklin Institute." *The Telescope,* April 1934, 29–32.
———. "Planetarium Operation." *Scientific Monthly,* October 1937, 307–316.
———. Letter to the editor. *Popular Astronomy,* March 1938, 174–175.
———. "America's Fifth Planetarium." *The Sky,* October 1939, 3–6, 25.
———. "Easter: The Awakening." *The Sky,* March 1940, 5–6.
"Story of Griffith, The." *Griffith Park Quarterly,* April/July 1978, 4–29.
Struve, Otto. "Giant Mystery Star . . . the Great Ghost Companion of Epsilon Aurigae." *The Sky,* March 1938, 3–5, 27.
Suggested Teaching Guide for the Earth and Space Science Course, A. Harrisburg, Pa.: Commonwealth of Pennsylvania, Department of Public Instruction, 1959.
Sullivan, Walter. *Assault on the Unknown: The International Geophysical Year.* New York: McGraw-Hill, 1961.
Taub, Liba. "'Canned Astronomy' vs. Cultural Credibility: The Acquisition of the Mensing Collection by the Adler Planetarium." *Journal of the History of Collections* 7 (1995): 243–250.
Taylor, Rennie. "Special Instruments Will Trace Satellite." *Columbus Dispatch,* 12 June 1956.
Thomas, Shirley. *Satellite Tracking Facilities: Their History and Operation.* New York: Holt, Rinehart, and Winston, 1963.
Thompson, Roy. "Chapel Hill Helped Cooper 'Find Himself' in Space." *Charlotte News,* 18 May 1963.
Todd, David. "A New Optical Projection Planetarium for Visualizing the Motions of the Celestial Bodies, as Seen by the Naked Eye, from the Earth." *Popular Astronomy,* August 1925, 446–456.
Tolischus, O. D. "Seeing Stars: How an Intricate German Machine Reveals the Heavens." *The World's Work,* November 1927, 96–100.
Tolles, Frederick B. *Quakers and the Atlantic Culture.* New York: Macmillan, 1960.
Tomlinson, Gary, and Katherine Becker, eds. *Planetarium Bibliography, 1913 to 1990.* GLPA Tips Booklet #15. East Lansing, Mich.: Great Lakes Planetarium Association, 1992.
"Training Planetarium Operators." *Sky and Telescope,* April 1967, 207.
"Trip to the Moon." *The Institute News,* April 1939, 1–2.
Turner, James. *Without God, Without Creed: The Origins of Unbelief in America.* Baltimore: Johns Hopkins University Press, 1985.

"Two Eastern Planetariums Re-Equipped." *Sky and Telescope,* March 1960, 271–273.

Tyson, Neil de Grasse. "Onward to the Edge." *Natural History,* July 1996, 60–63.

United States Code, 1958 Edition. Containing the General and Permanent Laws of the United States, in Force on January 6, 1959. Vol. 4. Title 16—Conservation to Title 21—Food and Drugs. Washington, D.C.: Government Printing Office, 1959.

United States Code, 1958 Edition. Cumulative Supplement V. Containing the General and Permanent Laws of the United States Enacted During the 86th and 87th Congresses and 88th Congress, First Session. Vol. 2. Title 20—Education to Title 39—The Postal Service. Washington, D.C.: Government Printing Office, 1964.

"Universe in Your Classroom, The." [advertisement] *Sky and Telescope,* January 1947, 27.

Van Dyke, Gretchen J. "Sputnik: A Political Symbol and Tool in 1960 Campaign Politics." In *Reconsidering Sputnik: Forty Years Since the Soviet Satellite,* edited by Roger D. Launius, John M. Logsdon, and Robert W. Smith, 365–400. Amsterdam: Harwood Academic Publishers, 2000.

Van Helden, Albert. "The Birth of the Modern Scientific Instrument, 1550–1750." In *The Uses of Science in the Age of Newton,* edited by John G. Burke, 49–84. Berkeley and Los Angeles: University of California Press, 1983.

Van Helden, Albert, and Thomas L. Hankins. "Introduction: Instruments in the History of Science." *Osiris,* 2d ser., 9 (1994): 1–6.

Vanek, Polly H. "The Role of Astronomers in Elementary and Secondary Education. I." In Proceedings of the Conference on Education in Astronomy, held at the 130th Meeting of the American Astronomical Society, 11–14 August 1969, at the State University of New York at Albany, edited by Frank C. Jettner. *Bulletin of the American Astronomical Society* 2 (1970): 272–275.

Villiger, W. *The Zeiss Planetarium.* London: W. and G. Foyle, 1926.

Wagner, Bartlett A., ed. *The Planetarium: An Elementary-School Teaching Resource.* Bridgeport, Conn.: University of Bridgeport, 1966.

Wallace, Mike. "Mickey Mouse History: Portraying the Past at Disney World." *Radical History Review* 32 (1985): 33–57.

Ward, F. A. B., ed. *The Planetarium of Giovanni de Dondi, Citizen of Padua.* London: Antiquarian Horological Society, 1974. A manuscript of 1397 translated by G. H. Baillie with additional material from another Dondi manuscript translated by H. Alan Lloyd.

Warneking, Glenn E. "Planetarium Education in the 1970's—Time for Assessment." *The Science Teacher* 37 (October 1970): 14–15.

Warner, Deborah Jean. "What is a Scientific Instrument, When Did It Become One, and Why?" *British Journal for the History of Science* 23 (1990): 83–93.

Watson, Fletcher G. "A Tin Can Planetarium." *The Science Teacher,* November 1950, 180–183.

Werner, Helmut. *From the Aratus Globe to the Zeiss Planetarium.* Translated by A. H. Degenhardt. Stuttgart: Gustav Fischer Verlag, 1957.

"What is Astrology?" *The Institute News,* April 1941, 1.

Who's Who of American Women: A Biographical Dictionary of Notable Living American Women. Chicago: A. N. Marquis, 1958–71.

Widney, Edwin [sic]. "Bank, as Trustee, Builds Greek Theatre and Astronomical Observatory." *Trust Companies* 53 (1931): 334–336.

Wieser, S. "Calgary's Planetarium and Museum." *Sky and Telescope,* July 1967, 14–15.
Wilford, John Noble. *We Reach the Moon.* New York: Bantam Books, 1969.
Will of Henry Buhl, Jr., Deceased. Pittsburgh: Buhl Foundation, 1929.
Willard, Berton C. *Russell W. Porter: Arctic Explorer, Artist, Telescope Maker.* Freeport, Maine: Bond Wheelwright, 1976.
Williams, Herbert N. "Recent Developments in Planetarium Projectors and Accessories." In *Planetariums and Their Use for Education,* vol. 2, edited by Richard H. Roche, 86–88. Cleveland, Ohio: Cleveland Museum of Natural History, n.d.
———. "The Planetarium in Modern Education." *American School Board Journal* 141 (September 1960): 40.
Williams, Thomas R. "Albert Ingalls and the ATM Movement." *Sky and Telescope,* February 1991, 140–143.
Wilson, J. Tuzo. *I.G.Y., The Year of the New Moons.* New York: Alfred A. Knopf, 1961.
Wolff, Nigel O'Connor. "Montevideo Planetarium Inaugurated." *Sky and Telescope,* May 1955, 282–283.
"Woman Likely to Succeed Fox as Adler Planetarium Head." *Chicago Herald and Examiner,* 27 April 1937.
"Wrap-Around Film Projection System Helps Skygazers to Learn Wonders of Atmosphere." *Business Screen* 28 (October 1967): 54.
Wright, Helen. *Explorer of the Universe: A Biography of George Ellery Hale.* New York: E. P. Dutton, 1966.
Wyatt, Stanley P., Jr. "Astronomy at the Lower School Levels." In *Education in and History of Modern Astronomy,* edited by Richard Berendzen. *Annals of the New York Academy of Sciences* 198 (25 August 1972): 178–191.
Zilversmit, Arthur. *Changing Schools: Progressive Educational Theory and Practice, 1930–1960.* Chicago: University of Chicago Press, 1993.

INDEX

About the Author

Jordan D. Marché II is a lecturer in the department of astronomy at the University of Wisconsin-Madison. He received his Ph.D. in the History and Philosophy of Science from Indiana University in 1999. He was professionally employed for over ten years in the American planetarium community. He lives with his wife, Teri, in Oregon, Wisconsin.